Artificial Intelligent Approaches in Petroleum Geosciences

Constantin Cranganu

Editor

Artificial Intelligent Approaches in Petroleum Geosciences

Second Edition

 Springer

Editor
Constantin Cranganu
Earth and Environmental Sciences
CUNY City University of New York,
Brooklyn College
Brooklyn, New York, NY, USA

ISBN 978-3-031-52717-3   ISBN 978-3-031-52715-9 (eBook)
https://doi.org/10.1007/978-3-031-52715-9

This Springer imprint is published by the registered company Springer Nature Switzerland AG
The registered company address is: Gewerbestrasse 11, 6330 Cham, Switzerland

Paper in this product is recyclable.

Preface to the Second Edition

The first edition of "Artificial Intelligent Approaches in Petroleum Geosciences," published in 2015, can be considered a success because "it is among the 20% most downloaded Springer books and interest in it remains high." Consequently, Springer has approached me asking if I would "consider publishing a new edition, with some updated/new material that could keep the book as relevant to researchers as it has been up to now."[1]

The petroleum industry has been at the forefront of technological innovation for decades, driven by the constant quest to optimize exploration, extraction, and production processes. Long before the public opinion learned about ChatGPT or Bard, Artificial Intelligence (AI) and Machine Learning (ML) were revolutionizing the field of petroleum geosciences, offering new avenues for improving exploration, reservoir management, and operational efficiency.

The petroleum geosciences industry is complex and data intensive. Geoscientists rely on a wide range of data to understand the subsurface, including seismic surveys, well logs, and core samples. AI and ML can help geoscientists make better sense of this data by identifying patterns and trends that would be difficult or impossible to spot manually.

In the first chapter of the current edition, Ashkan Jahanbani Ghahfarokhi presented the decision analysis in sequential decision problems such as waterflooding optimization under geological uncertainties (Part One), followed by a predictive modeling for well production forecast based on real field data from Volve field (North Sea) (Part Two). Proxy models are used in a variety of optimization problems, including Water-Alternating Gas (WAG) injection, where they are used to assess design parameters to maximize oil recovery and/or CO_2 storage. Proxy models have also been developed for simulating CO_2 injection and storage at the Svelvik CO_2 Field Lab in Norway and the Smeaheia CO_2 storage site in the North Sea. Accurate estimation of rock and fluid properties is essential for the design and monitoring of CO_2 storage. The author

---

[1] I quoted some lines from the invitation sent to me by Mr. Anthony Doyle, Executive Editor, to whom I express my gratitude and thanks for making available this new edition.

also discusses the utilization of machine learning approaches in modeling interfacial tension of hydrogen–brine system, predicting wax deposition under extensive production conditions, and estimating the interwell connectivities and production rates during gas injection.

Total Organic Carbon (TOC) is a key parameter for predicting the gas and oil production potential of shale formations. However, TOC measurements are often not available or are only available at discrete locations. Machine learning (ML) can be used to generate continuous TOC depth distributions from limited data.

In the second chapter, Danijela Dimitrijevic and Constantin Cranganu present a novel comparative study of the performance of three different ML methodologies in generating TOC content of the Marcellus shale in New York State: Multilayer Perceptron Neural Networks (MLPNN), Genetic Algorithms (GA), and Support Vector Machines (SVM). Their results can help further exploration of areas of interest for their natural gas content.

Subsurface lithofacies are an important parameter that can be used to characterize the compartmentalization of hydrocarbon reservoirs. In the third chapter, authored by Runhai Feng, Gated Recurrent Units (GRU neural networks are used to classify sequential lithofacies in the subsurface, taking into account the geological dependency between data samples along the vertical direction. GRUs can learn how to update or reset hidden states (in this case, lithofacies) to regulate the information flow through the system. The *softmax* function is used at the output layer to map the probability values over various possible lithofacies, and the associated uncertainty can be analyzed. Hidden Markov Models (HMM) are used as a benchmark for the performance of GRUs, with the embedded transition matrix enforcing the conditional probability between different lithofacies.

GRU and HMM are applied to a synthetic model of Book Cliffs and a real dataset from the Vienna Basin. Instead of using well logs, elastic rock properties from a non-linear inversion scheme are proposed as inputs for classification purpose. This could help to overcome the location limitations of cored wells, and 2D sections of reservoir lithofacies can then be obtained.

The remaining chapters are briefly discussed in the Preface to the 1st edition.

Overall, artificial intelligence and machine learning are transforming the petroleum geosciences industry, offering unparalleled advantages in exploration, reservoir management, operational efficiency, environmental sustainability, and data-driven decision-making. These technologies are essential tools for ensuring the long-term viability and sustainability of the industry. As the field of AI and ML continues to evolve, its application in petroleum geosciences will only become more central to the industry's success, efficiency, and environmental responsibility. Embracing these technological advancements is not only a choice but also a necessity for petroleum companies looking to thrive in an ever-changing world.

The advent of robust computing systems, sophisticated AI and ML algorithms, and the substantial data output from various industry tools heralds a promising era for devising solutions to intricate challenges in the oil and gas sector. These were problems that had previously eluded analytical solutions and numerical simulations.

While AI and ML tools do have their constraints, they are not bound by the restrictive assumptions of analytical solutions or the specific data and computational demands inherent to numerical simulators. They have the capability to encompass every intricacy present in log data and all the pertinent information linked to the target data.

In conclusion, AI and ML offer great potential in solving problems in almost all areas of the oil and gas industry involving prediction, classification, and clustering. Given the vast volume of data generated in routine operations within the oil and gas industry, the adoption of machine learning and advanced big data management methods has become imperative for achieving greater efficiency in the sector. AI and ML will continue to provide remarkable achievements, including enhanced exploration and reservoir characterization, predictive maintenance and operational efficiency, enhanced reservoir management, data-driven decision-making, environmental sustainability, reduced costs, improved safety, and accelerated innovation.

Brooklyn, New York, USA   Constantin Cranganu
October 2023

Preface to the First Edition

Integration, handling data of immense size and uncertainty, and dealing with risk management are among crucial issues in petroleum geosciences. The problems one has to solve in this domain are becoming too complex to rely on a single discipline for effective solutions, and the costs associated with poor predictions (e.g., dry holes) increase. Therefore, there is a need to establish new approaches aimed at proper integration of disciplines (such as petroleum engineering, geology, geophysics, and geochemistry), data fusion, risk reduction, and uncertainty management.

This book presents several artificial intelligent approaches[2] for tackling and solving challenging practical problems from the petroleum geosciences and petroleum industry. Written by experienced academics, this book offers state-of-the-art working examples and provides the reader with exposure to the latest developments in the field of artificial intelligence methods applied to oil and gas research, exploration, and production. It also analyzes the strengths and weaknesses of each method presented using benchmarking while also emphasizing essential parameters such as robustness, accuracy, speed of convergence, computer time, overlearning, or the role of normalization.

The reader of this book will benefit from exposure to the latest developments in the field of modern heuristics applied to oil and gas research, exploration, and production. These approaches can be used for uncertainty analysis, risk assessment, data fusion and mining, data analysis and interpretation, and knowledge discovery, from diverse data such as 3D seismic, geological data, well logging, and production data. Thus, the book is intended for petroleum scientists, data miners, data scientists and professionals, and postgraduate students involved in the petroleum industry.

Petroleum Geosciences are—like many other fields—a paradigmatic realm of difficult optimization and decision-making real-world problems. As the number, difficulty, and scale of such specific problems increase steadily, the need for diverse, adjustable problem-solving tools can hardly be satisfied by the necessarily limited number of approaches typically included in a curriculum/syllabus from academic

---

[2] Artificial Intelligence methods, some of which are grouped together in various ways, under names such as *Computational Intelligence*, *Soft Computing*, *Meta-heuristics*, or *Modern heuristics.*

fields other than Computer Science (such as Petroleum Geology). Therefore, the first three chapters of this volume aim at providing working information about modern problem-solving tools, in particular in machine learning and in data mining, and also at inciting the reader to look further into this thriving topic.

Traditionally, solving a given problem in mathematics and in sciences at large implies the construction of an abstract model, the process of proving theoretical results valid in that model, and eventually, based on those theoretical results, the design of a method for solving the problem. This problem-solving paradigm has been and will continue to be immensely successful. Nevertheless, an abstract model is an approximation of the real-world problem; there have been failures triggered by a tiny mismatch between the original problem and the proposed model for it. Furthermore, a problem-solving method developed in this manner is likely to be useful only for the problem at hand. While, ultimately, any problem-solving technique may be—in various degrees—subject to these two observations, some relatively new approaches illustrate alternative lines of attack; it is the editors' hope that the first three chapters of the book illustrate this idea in a way that will prove to be useful to the readers.

In the first chapter, Simovici presents some of the main paradigms of intelligent data analysis provided by machine learning and data mining. After discussing several types of learning (supervised, unsupervised, semi-supervised, active, and reinforcement learning), he examines several classes of learning algorithms (naïve Bayes classifiers, decision trees, support vector machines, and neural networks) and the modalities to evaluate their performance. Examples of specific applications of algorithms are given using System R.

The second and third chapters, by Luchian, Breaban, and Bautu, are dedicated to *meta-heuristics*. After a rather simple introduction to the topic, the second chapter presents, based on working examples, evolutionary computing in general and, in particular, genetic algorithms and differential evolution; particle swarm optimization is also extensively discussed. Topics of particular importance, such as multimodal and multi-objective problems, hybridization, and also applications in petroleum geosciences are discussed based on concrete examples. The third chapter gives a compact presentation of genetic programming, gene expression programming, and also discusses an R package for genetic programming and applications of GP for solving specific problems from the oil and gas industry.

Ashena and Thonhauser discuss Artificial Neural Networks (ANNs), which have the potential to increase the ability of problem-solving in geosciences and in the petroleum industry, particularly in case of limited availability or lack of input data. ANN applications have become widespread because they proved to be able to produce reasonable outputs for inputs they have not learned how to deal with. The following subjects are presented: artificial neural networks basics (neurons, activation function, ANN structure), feed-forward ANN, back-propagation and learning, perceptrons and back-propagation, multilayer ANNs and back-propagation algorithm, data processing by ANN (training, overfitting, testing, validation), ANN, and statistical parameters. An applied example of ANN, followed by applications of ANN in geosciences and petroleum industry, completes the chapter.

Al-Anazi and Gates present the use of support vector regression to accurately estimate two important geomechanical rock properties, Poisson's ratio and Young's modulus. Accurate prediction of rock elastic properties is essential for wellbore stability analysis, hydraulic fracturing design, sand production prediction and management, and other geomechanical applications. The two most common required material properties are Poisson's ratio and Young's modulus. These elastic properties are often reliably determined from laboratory tests by using cores extracted from wells under simulated reservoir conditions. Unfortunately, most wells have limited core data. On the other hand, wells typically have log data. By using suitable regression models, the log data can be used to extend knowledge of core-based elastic properties to the entire field. Artificial neural networks (ANNs) have proven to be successful in many reservoir characterization problems. Although nonlinear problems can be well resolved by ANN-based models, extensive numerical experiments (training) must be done to optimize the network structure. In addition, generated regression models from ANNs may not perfectly generalize to unseen input data. Recently, Support Vector Machines (SVMs) have proven successful in several real-world applications for their potential to generalize and converge to a global optimal solution. SVM models are based on the structural risk minimization principle that minimizes the generalization error by striking a balance between empirical training errors and learning machine capacity. This has proven superior in several applications to the empirical risk minimization principle adopted by ANNs that aims to reduce the training error only. Here, support vector regression (SVR) to predict Poisson's ratio and Young's modulus is described. The method uses a fuzzy-based ranking algorithm to select the most significant input variables and filter out dependency. The learning and predictive capabilities of the SVR method are compared to that of a back-propagation neural network (BPNN). The results demonstrate that SVR has similar or superior learning and prediction capabilities to that of the BPNN. Parameter sensitivity analysis was performed to investigate the effect of the SVM regularization parameter, the regression tube radius, and the type of kernel function used. The result shows that the capability of the SVM approximation depends strongly on these parameters.

The next three chapters introduce the active learning method (ALM) and present various applications of it in petroleum geosciences.

First, Cranganu, and Bahrpeyma use ALM to predict a missing log (DT or sonic log) when only two other logs (GR and REID) are present. In their approach, applying ALM involves three steps: (1) supervised training of the model, using available GR, REID, and DT logs; (2) confirmation and validation of the model by blind-testing the results in a well containing both the predictors (GR, REID) and the target (DT) values; and (3) applying the predicted model to wells containing the predictor data and obtaining the synthetic (simulated) DT values. Their results indicate that the performance of the algorithm is satisfactory, while the performance time is significantly low. The quality of the simulation procedure was assessed by three parameters, namely mean square error (MSE), mean relative error (MRE), and Pearson product momentum correlation coefficient (R). The authors employed both the measured and simulated sonic log DT to predict the presence and estimate the depth intervals where overpressured fluid zone may develop in the Anadarko Basin, Oklahoma. Based on interpretation of the sonic log trends, they inferred that overpressure regions are developing between ~1,250 and 2,500 m depth and the overpressured intervals have thicknesses varying between ~700 and 1,000 m. These results match very well previous published results reported in the Anadarko Basin, using the same wells, but different artificial intelligence approaches.

Second, Bahrpeyma et al. employed ALM to estimate another missing log in hydrocarbon reservoirs, namely the density log. The regression and normalized mean squared error (MSE) for estimating density log using ALM were equal to 0.9 and 0.042, respectively. The results, including errors and regression coefficients, proved that ALM was successful in processing the density estimation. In their chapter, the authors illustrated ALM by an example of a petroleum field in the NW Persian Gulf.

Third, Bahrpeyma et al. tackled the common issue when reservoir engineers should analyze the reservoirs with small sets of measurements (this problem is known as the small sample size problem). Because of small sample size problem, modeling techniques commonly fail to accurately extract the true relationships between the inputs and the outputs used for reservoir properties prediction or modeling. In this chapter, small sample size problem is addressed for modeling carbonate reservoirs by using the active learning method (ALM). Noise injection technique, which is a popular solution to small sample size problem, is employed to recover the impact of separating the validation and test sets from the entire sample set in the process of ALM. The proposed method is used to model hydraulic flow units (HFUs). HFUs are defined as correlatable and mappable zones within a reservoir controlling the fluid flow. This research presents quantitative formulation between flow units and well-log data in one of the heterogeneous carbonate reservoirs in the Persian Gulf. The results for R and $nMSE$ are 85 % and 0.0042, respectively, which reflect the ability of the proposed method to improve generalization ability of the ALM when facing with sample size problem.

Dobróka and Szabó carried out a well-log analysis by global optimization-based interval inversion method. Global optimization procedures, such as genetic algorithms and simulated annealing methods, offer robust and highly accurate solution to several problems in petroleum geosciences. The authors argue that these methods can be used effectively in the solution of well-logging inverse problems. Traditional inversion methods are used to process the borehole geophysical data collected at a given depth point. As having barely more types of probes than unknowns in a given depth, a set of marginally overdetermined inverse problems has to be solved along a borehole. This single inversion scheme represents a relatively noise-sensitive interpretation procedure. To reduce the noise, the degree of overdetermination of the inverse problem must be increased. This condition can be achieved by using a so-called interval inversion method, which inverts all data from a greater depth interval jointly to estimate petrophysical parameters of hydrocarbon reservoirs to the same interval. The chapter gives a detailed description of the interval inversion problem, which is then solved by a series expansion-based discretization technique. The high degree of overdetermination significantly increases the accuracy of parameter estimation. The quality improvement in the accuracy of estimated model parameters often leads to a more reliable calculation of hydrocarbon reserves. The knowledge of formation boundaries is also required for reserve calculation. Well logs contain information about layer thicknesses, which cannot be extracted by the traditional local inversion approach. The interval inversion method is applicable to derive the layer boundary coordinates and certain zone parameters involved in the interpretation problem automatically. In this chapter, the authors analyzed how to apply a fully automated procedure for the determination of rock interfaces and petrophysical parameters of hydrocarbon formations. Cluster analysis of well-logging data is performed as a preliminary data-processing step before inversion. The analysis of cluster number log allows the separation of formations and gives an initial estimate for layer thicknesses. In the global inversion phase, the model including petrophysical parameters and layer boundary coordinates is progressively refined to achieve an optimal solution. The very fast simulated reannealing method ensures the best fit between the measured data and theoretical data calculated on the model. The inversion methodology is demonstrated by a hydrocarbon field example, with an application for shaly sand reservoirs.

Finally, Mohebbi and Kaydani undertake a detailed review of meta-heuristics dealing with permeability estimation in petroleum reservoirs. They argue that proper permeability distribution in reservoir models is very important for the determination of oil and gas reservoir quality. In fact, it is not possible to have accurate solutions in many petroleum engineering problems without having accurate values for this key parameter of hydrocarbon reservoir. Permeability estimation by individual techniques within the various porous media can vary with the state of in situ environment, fluid distribution, and the scale of the medium under investigation. Recently, attempts have been made to utilize meta-heuristics for the identification of the relationship that may exist between the well-log data and core permeability. This chapter overviews the different meta-heuristics in permeability prediction, indicating the advantages of each method. In the end, some suggestions and comments about how to choose the best method are presented.

December 2014
Constantin Cranganu
Henri Luchian
Mihaela Elena Breaban

About this Book

Overall, Artificial Intelligence and Machine Learning are transforming the petroleum geosciences industry, offering unparalleled advantages in exploration, reservoir management, operational efficiency, environmental sustainability, and data-driven decision-making. These technologies are essential tools for ensuring the long-term viability and sustainability of the industry. As the field of AI and ML continues to evolve, its application in petroleum geosciences will only become more central to the industry's success, efficiency, and environmental responsibility. Embracing these technological advancements is not just a choice but a necessity for petroleum companies looking to thrive in an ever-changing world.

In conclusion, AI and ML offer great potential in solving problems in almost all areas of the oil and gas industry involving prediction, classification, and clustering. Given the vast volume of data generated in routine operations within the oil and gas industry, the adoption of machine learning and advanced big data management methods has become imperative for achieving greater efficiency in the petroleum sector, whose main features include enhanced exploration and reservoir characterization, predictive maintenance and operational efficiency, enhanced reservoir management, data-driven decision making, environmental sustainability, reduced costs, improved safety, and accelerated innovation.

Contents

Applications of Data-Driven Techniques in Reservoir Simulation and Management

Ashkan Jahanbani Ghahfarokhi

Abstract Advanced algorithms and supercomputers enhance the speed and accuracy of traditional modeling techniques (numerical and analytical); however, they are incapable of transforming the models to new levels of computational footprint. Application of artificial intelligence and machine learning (AI&ML) in reservoir engineering has shown significant potential for transforming traditional modeling approaches. By leveraging data-driven techniques, AI&ML can develop powerful and fast tools that can model complex and uncertain physics. Such tools perceive the relationship among relevant data and develop models based on the available measurements or simulated data. Promising results have been obtained from the application of data-driven techniques for resolving a wide variety of reservoir management problems such as history matching, well control and placement, injection strategies optimization, production forecasting, CO_2 storage, and many more. In this chapter (part 1), we aim to discuss some recent applications of ML-based data-driven models in reservoir simulation and management, highlighting the benefits of such approaches in capturing high nonlinearity. We often refer to these data-driven models as proxy models since they act on behalf of the "actual" or physics-based models. The successfully trained proxy models are coupled with, for example, metaheuristic algorithms in solving optimization problems in reservoir management. We show examples of fast well control optimization in waterflooding using proxy models including two-stage (local and global) proxy modeling to handle problems with high dimensions. Some relevant topics include integration of sampling techniques, adaptive sampling and retraining to address the geological uncertainties, and use of more complex reservoir models. To further illustrate the applications of machine learning, we discuss:

- decision analysis in sequential decision problems such as waterflooding optimization under geological uncertainties;
- predictive modeling for well production forecast based on real field data from the Volve field.

A. Jahanbani Ghahfarokhi (✉)
Department of Geoscience and Petroleum (IGP), Norwegian University of Science and Technology (NTNU), Trondheim, Norway
e-mail: ashkan.jahanbani@ntnu.no

C. Cranganu (ed.), *Artificial Intelligent Approaches in Petroleum Geosciences*,
https://doi.org/10.1007/978-3-031-52715-9_1

There is a need to develop advanced modeling techniques within the area of carbon capture, utilization, and storage (CCUS) for sustainable utilization of the subsurface. Such tools can quickly model and investigate the impact of uncertainties and decision variables on processes. In addition to the economic value of CCUS related to producing the remaining resources, improved modeling and optimization techniques will be useful when the subsurface is to be repurposed for storage. In this chapter (part 2), we aim to discuss further applications of data-driven models. We present examples where proxy models are employed in mono- and multi-objective optimization problems, for example in Water Alternating Gas (WAG) injection, in which design parameters are assessed to maximize the oil recovery and/or the CO_2 stored. Development of proxy models are also discussed for simulation of CO_2 injection and storage in Svelvik CO_2 Field Lab in Norway, and Smeaheia CO_2 storage in the North Sea. Accurate estimation of rock and fluid properties has a significant effect on the design and monitoring of CO_2 storage. We will discuss data-driven techniques used to develop robust paradigms to accurately estimate the CO_2 thermal conductivity and diffusivity in brine under various operating conditions using the representative experimental database. We also briefly discuss the use of machine learning techniques in:

- modeling interfacial tension of hydrogen–brine system;
- predicting wax deposition under extensive production conditions;
- estimating the interwell connectivities and production rates in gas injection.

Overall, the application of data-driven proxy modeling offers significant advancements in reservoir modeling, addressing computational limitations and enabling efficient optimization and decision-making. Proxy modeling is however still subject to limitations including data sampling efficiency, learning techniques, dimensionality, uncertainty, etc., that need further investigations.

Keywords Artificial intelligence · Machine learning · Data-driven techniques · Proxy models · Reservoir simulation · Reservoir management · Waterflooding · Decision analysis · Optimization · Water alternating gas · Gas injection · Carbon capture utilization and storage · CO_2 storage · H_2 storage

1 Introduction

Advancements in computing tools have enabled the successful handling of challenges in reservoir modeling and management. The conventional process of reservoir simulation involves constructing a geological model of the reservoir using static data, incorporating fluid flow principles to create a dynamic model, calibrating it using production history and other measurements, and employing it, for instance, in field development planning. With real-time data coming in, computational requirements of such numerical simulations restrict their use for in-depth analysis, uncertainty assessment, and optimization. This becomes more of an issue for large reservoir models when swift decision-making is imperative.

A new approach has emerged which involves the development of fast and accurate data-driven proxy models through a combination of optimization, statistics-based methods (e.g., the surface response method), and Machine Learning (ML). Data-driven modeling may alternatively be known as proxy modeling, while the latter englobes other approaches such as reduced-order modeling which are not purely data-driven. The term "proxy" implies acting on behalf of another; in numerical simulation, the primary role of the proxy is to reproduce numerical simulator's behavior and results. Data is the primary building block of such models, which allows for rapid proxy construction compared to numerical models. Moreover, proxies maintain reasonable CPU times and run in the order of a few seconds (Mohaghegh 2017).

Injection of CO_2 into deep saline aquifers is the major option for storage in CCS (Carbon Capture and Storage) to mitigate the carbon footprints (Krevor et al. 2023). Availability of CO_2 for storage can provide opportunities to practice Carbon Capture, Utilization, and Storage (CCUS), such as injecting the captured CO_2 for both Enhanced Oil Recovery (EOR) and storage in depleted/depleting fields with huge volumes of remaining hydrocarbons. Optimized CO_2 injection for storage and enhanced recovery from the existing fields will provide time for a smooth energy transition. Subsurface modeling is vital for safe CO_2 storage in geological formations and optimized development strategies. ML in conjunction with numerical simulators can be used to develop fast models and serve as smart decision support for CO_2 injection projects.

Noticeable progress has been observed in the application of data-driven techniques to address a broad range of modeling challenges using proxy models. A survey was conducted on the use of ML proxy models and their coupling with metaheuristic algorithms for fast reservoir simulation and optimization studies (Ng et al. 2023a). The study presented successful applications and discussed the benefits and limitations of methods in intelligent proxy modeling. The applications include well control and placement, monitoring production parameters, forecasting, gas storage, optimization, history matching, waterflooding, miscible gas injection, water-alternating gas (WAG), and other enhanced oil recovery (EOR) techniques and uncertainty studies. The fundamentals of building ML proxy models are summarized in Fig. 1. These steps are also described below.

Simulation studies: Numerical simulation is used to generate a database for model training. This data is used later as benchmark to validate the developed proxy models. Different modeling techniques have been developed to describe the physics involved: blackoil, compositional, streamline, and thermal. The blackoil model is generally used for the cases in which the characteristics of the fluid and reservoir are not influenced (like primary and secondary recovery). The streamline approach is recognized as a visualization paradigm for better analysis of improved recovery techniques such as waterflooding and infill drilling. The compositional and thermal models are mainly used for cases with change of the fluid and reservoir rock (e.g., tertiary recovery

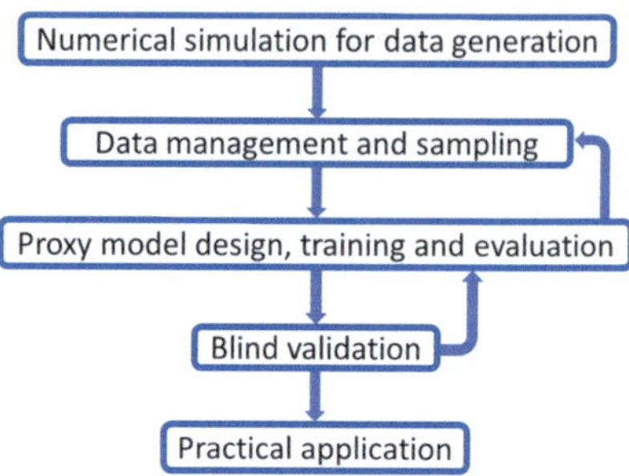

Fig. 1 Workflow of proxy model development

with chemical injection). Simulation results give parameters like pressures, phase saturations, and compositions (molar fractions) as a function of time and position.

Database generation: Design of experiments (DoE) is used to execute numerical simulations using different assumptions according to the objectives of the study such as number of wells, rate of injection, number of data points, etc. Sampling technique is employed to retrieve information from the simulations to form the spatiotemporal database to be used in the model-building phase. To have good coverage of the information with respect to the design variables, well-formulated techniques such as Latin hypercube sampling (LHS), Hammersley, and Sobol, as well as adaptive sampling (which iteratively adds data to the training set with restructuring and repartitioning) may be used. The most influential parameters on the outcomes are ranked and extracted so that the dimensionality of problem is reduced in the selection of input/output for model training. The data may be partitioned into training, calibration, and validation sets (plus blind set, i.e., data not seen in training) to learn from the numerical model.

Sufficient sample size is needed to "mimic" the response of the actual system, but there is a trade-off between size of training database (accuracy) and computational efficiency. Also, a very non-linear system might produce inaccurate results despite a sufficiently large sample size in one-shot approach. To mitigate this, two-stage or adaptive approach can be applied, which involves training and updating the database.

Proxy model building: Using the database, multiple proxy models are built by data-driven techniques coupled with training algorithms to mimic the response of the simulator. Data-driven techniques can recognize highly complicated patterns in an efficient way, and hence, an accurate and simple-to-use relationship is identified between the decisive variables of the system and the required outcomes. Based on the scale and objective of the study, field-based, well-based, and grid-based proxies may be built.

Different terminologies are used in the field of proxy modeling. An illustration is displayed in Fig. 2 to outline the relationship between proxy modeling and other terminologies. In supervised ML, the identification is made according to a pre-acquisition of information from the system by means of known data, whereas, in unsupervised learning, the identification is done by self-organization of the inputs (i.e., the targets are unknown). Artificial Neural Networks (ANN), Support Vector Regression (SVR), Least Square Support Vector Machine (LSSVM), and Genetic Programming (GP) are among robust supervised learning methods. A Deep Neural Network (DNN) is a reliable supervised technique, more appropriate for modeling highly complicated systems. In addition, Kriging and Response Surface Models (RSM) are among the widely applied statistics-based techniques for building proxy models.

A review by Bahrami et al. (2022) argued that the existing classifications in the literature could not cover all proxy models. They proposed a new classification for proxy models: Multi-Fidelity Models (MFM), Reduced-Order Models (ROM), Traditional Proxy Models (TPM), and Smart Proxy Models (SPM). MFMs are constructed based on simplifying physics assumptions, and ROMs are based on dimensional reduction. SPM differs from TPM in employing an additional "feature generation/engineering" step in which new parameters are extracted and added to the database and are used in model development. Figure 3 summarizes this classification.

The training algorithms are used to minimize the error function (also known as loss or cost function). Examples of the loss function for model performance monitoring are Coefficient of Determination (R^2), Mean Squared Error (MSE), Root Mean Square Error (RMSE), Mean Absolute Error (MAE), Average Percent Relative Error

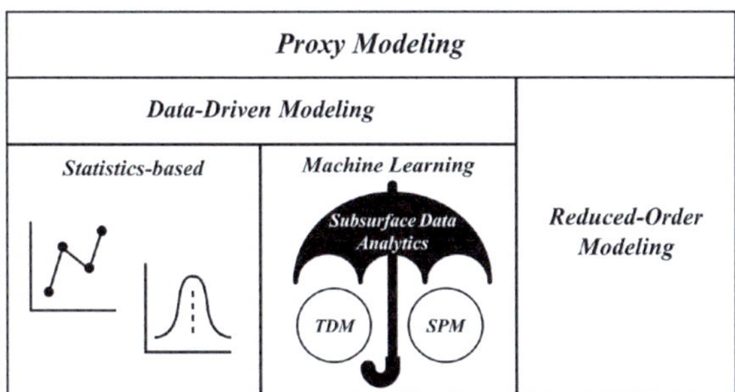

Fig. 2 Different types of proxy models (Ng et al. 2023a)

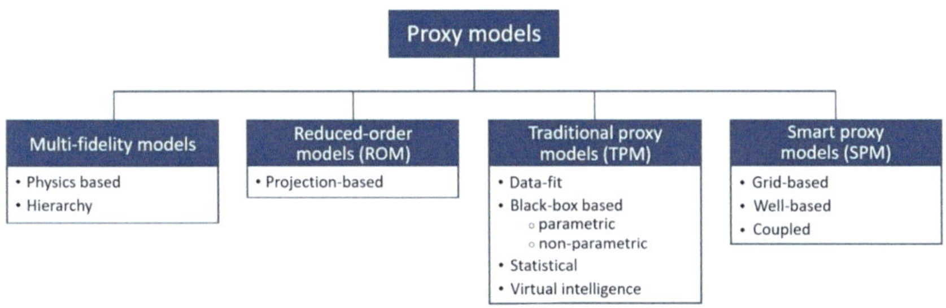

Fig. 3 Different proxy model categories (Bahrami et al. 2022)

(APRE), Absolute Relative Prediction Error (ARPE), Average Absolute Relative Deviation (AARD%), and Average Absolute Percent Relative Error (AAPRE). A committee of machines may be generated by combining different models to select the most performant developed models to improve the prediction accuracy.

Optimization study: Maximization/minimization under various constraints is vital in recovery/storage processes. Multi-objective optimization is employed for performance predictions in, for example, targeting maximum oil recovery and minimum CO_2 production (maximum storage). Classical optimization methods may fail in reaching the global optima in multimodal functions. Examples of derivative-based algorithms include stochastic gradient descent and scaled conjugate gradient. Nature-inspired algorithms (global optimizer with capabilities to include the constraints in the optimization process) offer flexibility to avoid the convergence to local optima by creating a balance between exploration and exploitation over the search space. In addition, employing these algorithms will improve the learning process of the data-driven techniques. Examples of these derivative-free algorithms include Genetic Algorithm (GA), Differential Evolution (DE), Particle Swarm Optimization (PSO), Ant Colony Optimization (ACO), Artificial Bee Colony (ABC), Firefly Algorithm (FA), Imperialist Competitive Algorithm (ICA), Simulated Annealing (SA), Gray Wolf Optimization (GWO), and Cuckoo Optimization Algorithm (COA). Proxy-based optimization is used in handling optimization problems with high dimensionality (number of decision variables) where models are trained along with an update of the data during the optimization (adaptive training and optimization, as mentioned earlier).

This chapter aims at showing some recent studies for developing and implementing proxy modeling mainly using data-driven techniques. Part 1 presents some applications in the domain of reservoir simulation for waterflooding optimization and decision analysis. Part 2 will focus on the applications mainly in water alternating gas injection, CO_2 storage and predictive tools for fast and accurate property estimations. Most of the methods and concepts used in this chapter are introduced in the other chapters of the book; however, the reader is referred to the cited references for more details about the topics and the aspects of the studies.

2 Part 1: Waterflooding

This part mainly focuses on the recent applications of machine learning for proxy model development in waterflooding optimization and decision analysis studies, and also developing predictive models for production forecasting based on field data.

The first study demonstrated the procedure of building proxy models in two synthetic waterflooded fractured Dual Porosity Dual Permeability (DPDP) reservoir models (Ng et al. 2021a). Feedforward neural network was the selected ML technique. Backpropagation algorithm and PSO were implemented to train the proxy

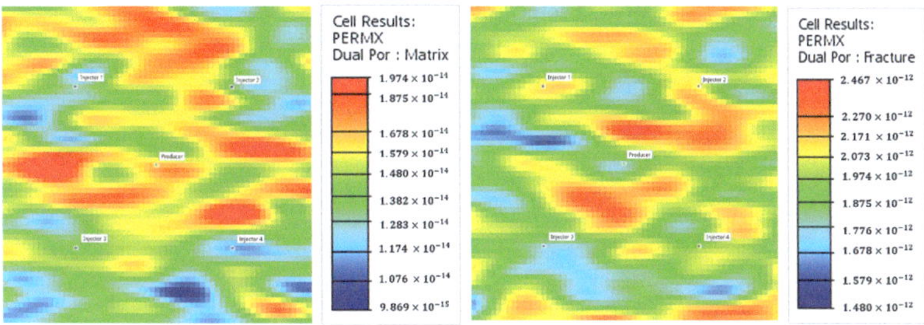

Fig. 4 The x-permeability distribution in the heterogenous model layer 1 in both matrix and fracture (Ng et al. 2021a)

models for comparative studies. Stochastic Gradient Descent (SGD) and Adaptive Moment Estimation (Adam) algorithms were both used to conduct the backpropagation algorithm. The study then presented the application of the developed proxies in production optimization. An additional procedure was also proposed in proxy modeling by integrating the probabilistic application to examine the overall performance of the proxies. In this work, metaheuristic algorithm (PSO) exhibited better training and predictive performance than the backpropagation algorithms. Both homogeneous and heterogenous, 3D two-phase (black oil and water) reservoir models were used as the "true" model (1525 m × 1525 m × 45.7 m with 61 × 61 × 3 blocks) in a five-spot pattern. We only show the results for the heterogeneous model (Fig. 4).

In this work, no proxy was developed for the injection rates as they remain constant throughout the production period. Only two proxy models were discussed to predict oil and water production rates (on monthly basis). The optimization study determines the field injection rate that can maximize the net present value (NPV):

$$NPV(\mathbf{u}) = \sum_{i=1}^{n_{total}} \frac{\left(Q_o^i(\mathbf{u})P_o - Q_w^i(\mathbf{u})P_w - Q_{wi}^i(\mathbf{u})P_{wi}\right) \times \Delta t_i}{(1 + \text{interest rate})^{t_i/D}} \tag{1}$$

where $\mathbf{u}$ is the control vector, Q^i is the field production/injection rate at timestep i and P represents price/cost. The subscripts o, w, and wi, indicate oil, water, and water injected, respectively. t_i is the cumulative time until timestep i, and the reference period for discounting cashflow is 365 days. Workflow of the study is shown in Fig. 5.

The database for the neural network training was formed using sampling to obtain scenarios with different control rates (five injection rates between 636 m³/day and 795 m³/day). After running the simulations, dynamic inputs were extracted and combined with the static inputs to create the database (Table 1). The database was normalized (before being used in neural network training) and partitioned into training, validation, testing, and blind validation sets.

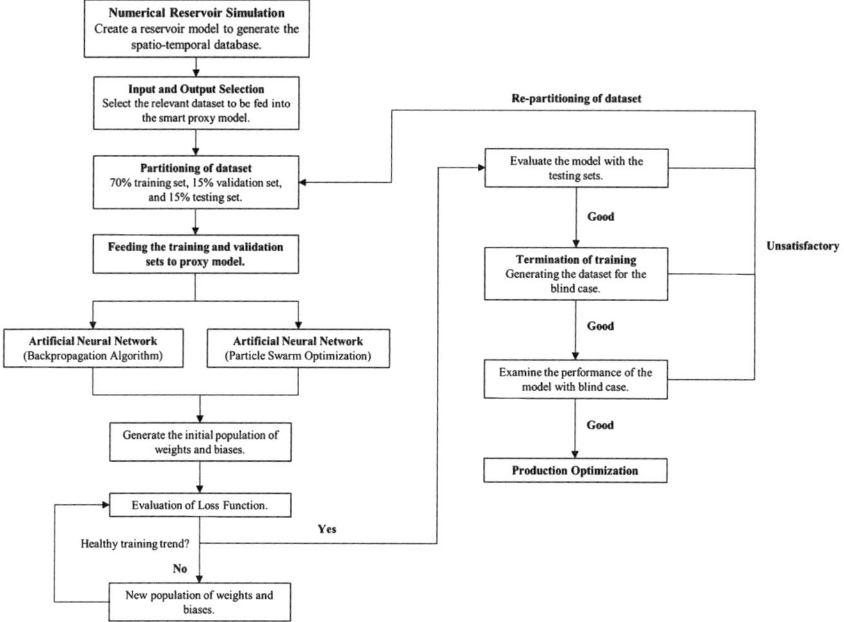

Fig. 5 Workflow of smart proxy model building and optimization (Ng et al. 2021a)

Regarding the topology of the proxy models, there were 62, 10, and 1 node in input, hidden, and output layers, respectively (all with one layer). SGD and Adam used 2000 epochs and a step size of 0.01. Adam used 0.9, 0.99, and 10^{-7} for exponential decay rates of the estimates of the first and second moments, and constant of numerical stability, respectively. PSO used 2000, 100, 0.8, 1.005, and 1.05 for epochs, number of particle swarms, inertial weight, cognitive weight, and social weight, respectively.

The results of the evaluation of the ANNs are presented in Table 2 and Table 3 for oil production rate prediction during training and blind validation, respectively. The models trained by all the algorithms exhibited excellent training results for both oil and water production rates. Adam generally showed the best results. It was observed that PSO deterministically can outperform the other methods in blind validation for oil production rates (also testing results). However, for the estimation of water production rate, Adam produced the predictive model with the best results. Resulting oil production profiles from the proxies trained by Adam are presented and compared against simulator in Fig. 6.

After obtaining the flow rates by the proxy models, production optimization was performed. The optimal NPV obtained by the proxy models are given in Table 4. The proxy models built by Adam produced the optimal NPV with the least percentage error for two injection scenarios (0.117% for a rate of 676 m^3/day and 0.329% for a rate of 755 m^3/day). Furthermore, all the proxy models reached the same decision that the injection rate of 755 m^3/day is economically preferable.

Table 1 Selected input and output data (Ng et al. 2021a)

Inputs			Output
Indexes	Simulation scenario	Scenarios 1, 3, and 5	Field oil production rate at time t
Static inputs	Grid block ith position	Well group (grid block kth denotes the perforated kth grid block)	
	Grid block jth position		
	Grid block kth position		
	Porosity	Average values of layers with well perforation, layers of matrix media, layer of fracture media	
	Permeability		
	Matrix block height	Matrix media (parameters in DPDP modeling)	
	Shape factor		
Dynamic inputs	Time	Monthly basis (timesteps 0–360)	
	Bottom hole pressure (BHP)	For 4 injectors and 1 producer at time t	
	Field water injection rate	At time t	
	Field oil production rate	At time t−1	

Table 2 Results of the performance of the smart proxy for oil rate prediction for the training, validation, and testing sets (Ng et al. 2021a)

		RMSE	R^2
Stochastic gradient descent	Training (758 data)	7.855	0.9977
	Validation (163 data)	4.700	0.9992
	Testing (162 data)	8.202	0.9978
Particle swarm optimization	Training (758 data)	3.846	0.9995
	Validation (163 data)	3.918	0.9995
	Testing (162 data)	2.739	0.9997
Adam	Training (758 data)	3.154	0.9997
	Validation (163 data)	2.410	0.9998
	Testing (162 data)	3.391	0.9996

Probabilistic application was implemented to further evaluate the performance of the models. To this end, we performed the proxy modeling iteratively 200 times (hence, 200 samples). Then, the normalized cumulative frequency distribution (NCFD) for R^2 (that ranges between 0 and 1) was computed over the 200 samples. NCFD refers to the cumulative number of times for a sample to be within a range

Table 3 Results of the performance of the smart proxy for the two blind cases (oil rate prediction) (Ng et al. 2021a)

	Injection rate (m³/day)	RMSE	R^2
Stochastic gradient descent	676	12.45	0.9939
	755	13.04	0.9944
Particle swarm optimization	676	2.097	0.9998
	755	3.827	0.9995
Adam	676	4.489	0.9992
	755	5.468	0.9990

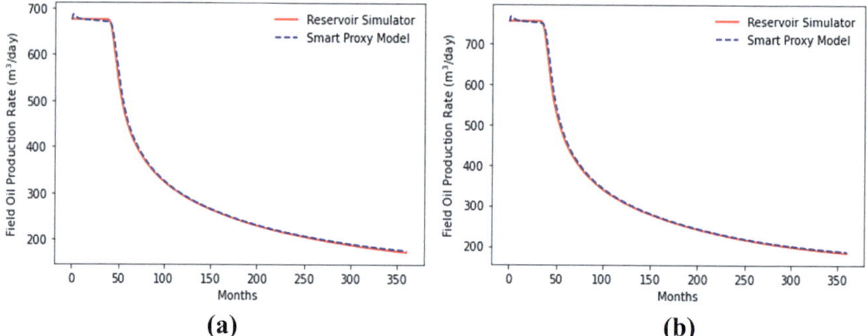

Fig. 6 Comparison of the rates predicted by the proxy and simulator for blind cases: **a** injection rate of 676 m³/day and **b** injection rate of 755 m³/day (oil rate prediction—Adam) (Ng et al. 2021a)

Table 4 The optimal NPV generated by all the models (Ng et al. 2021a)

Injection rates	676 m³/day				755 m³/day			
Models	Simulator	SGD	PSO	Adam	Simulator	SGD	PSO	Adam
$NPV_{optimal}$ (million USD)	428.92	421.77	424.31	429.43	447.22	440.97	444.04	448.69

of values of R^2 over 200 times. The results are demonstrated in Fig. 7 for the rate of 755 m³/day. PSO exhibited better performance than SGD and Adam to produce accurate predictive models (R^2 exceeding 0.99).

Another study (Ng et al. 2021b) applied the procedure described in the previous work for proxy modeling in another setting. Feedforward neural network (using Adam for training) was applied to develop data-driven proxies for 2D and 3D models, in waterflooding well control optimization using PSO and GWO. To further confirm the applicability of the proxy models, an optimization task was also carried out by coupling these algorithms with the simulator. The training and blind validation results revealed the coefficient of determination, R^2 of about 0.99. For both cases, the optimization results obtained using the proxy models were within 5% accuracy

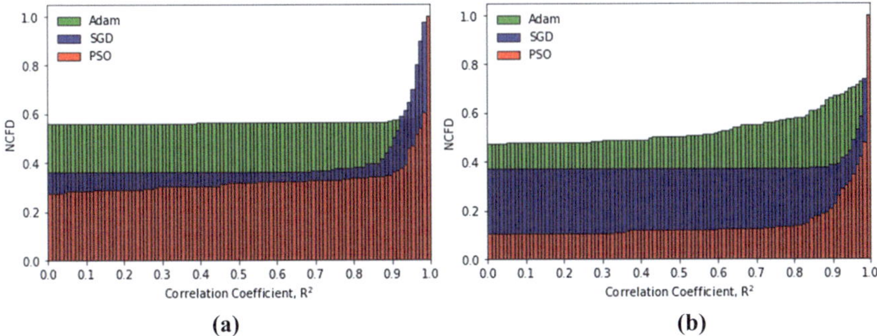

Fig. 7 NCFD of R^2 for rate prediction by the smart proxy models: **a** oil rate and **b** water rate (Ng et al. 2021a)

compared with simulator results at much lower computational costs. Also, the proxy-optimized well controls were fed into the simulator to further illustrate the robustness of the developed models. The workflow of the study is summarized in Fig. 8 which includes model training and optimization study to maximize the NPV by adjusting the injection rate of each well periodically (every 150 days) over 3000 days (within 40 and 100 m^3/day).

A heterogeneous 2D, two-phase (water and black oil) model was first developed and studied ($40 \times 40 \times 1$ grid blocks). Excellent training and blind validation results were obtained, and proxies were applied in optimization study. Here, we only present the results of the 3D Egg Model (Jansen et al. 2014), with $60 \times 60 \times 7$ grid blocks and 18,533 active grid blocks, Fig. 9. The model has eight vertical injectors and four vertical producers, controlled by BHP with the lower limit of 395 bar.

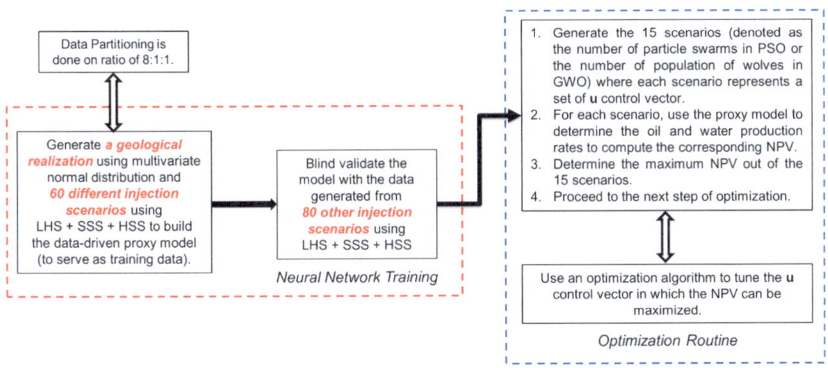

Fig. 8 Workflow of proxy modeling (Ng et al. 2021b)

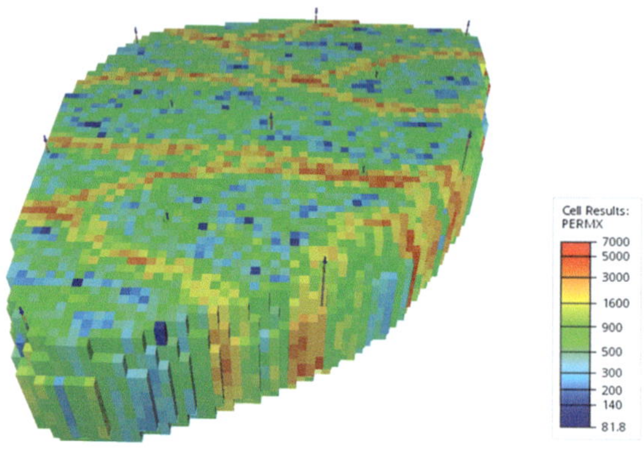

Fig. 9 Egg model permeability distribution (mD) (Ng et al. 2021b)

Two proxy models were developed to predict the field liquid production rates and field water cut at specific timesteps (oil and water production rates for the optimization can then be obtained). The input variables included the number of days at each timestep, t_j; the harmonic mean of grid absolute permeability for each layer, $\bar{k}$; the standard deviation of grid absolute permeability for each layer, k_{SD}; the permeabilities of perforated grid blocks (injectors/producers), $k_{injector}/k_{producer}$; control vector, $\mathbf{u}$; the output at the previous timestep, y_{j-1}. The proxy can be formulated as

$$y_j = f\left(t_j, \bar{k}, k_{SD}, k_{injector}, k_{producer}, \mathbf{u}, y_{j-1}\right) \tag{2}$$

Some parameters used in the training of the neural network and architecture of proxies are tabulated in Tables 5 and 6.

Table 5 Parameters for neural network training (Ng et al. 2021b)

Adam parameters	Values
Number of iterations (epochs)	2000
Learning rate	0.001
Exponential decay rates for the 1st moment estimates, β_1	0.9
Exponential decay rates for the 2nd moment estimates, β_2	0.999
Numerical stability constant, ε	10^{-7}

Table 6 The architecture of ANN (Ng et al. 2021b)

Layers	Field liquid production rate		Field water cut	
	Number of layers	Number of nodes	Number of layers	Number of nodes
Input	1	29	1	29
Hidden	1	100	2	50
Output	1	1	1	1

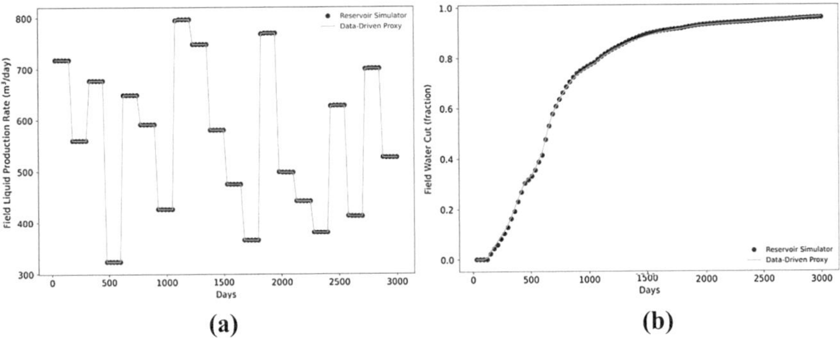

Fig. 10 Results of blind validations of LHS sample set 11 (out of 80). **a** FLPR, **b** FWCT (Ng et al. 2021b)

Regarding prediction performance, R^2 of training, validation, and testing of all the proxies was 0.9999. Comparisons between the actual and the predicted field liquid production rates (FLPR) and field water cut (FWCT) are shown in Fig. 10 (blind case).

Table 7 illustrates the optimization results for three scenarios: scenario 1 represents optimization by the simulator (NPV_{sim}), scenario 2 represents optimization by feeding the proxy-optimized control into the simulator ($NPV_{sim\text{-}proxy}$), and scenario 3 represents optimization by the proxies (NPV_{proxy}). NPV overestimation by the proxies for both algorithms is noticed, though not significant. GWO generally outperformed PSO for optimization studies (except for scenario 2), but PSO gave slightly more accurate rates.

Plots of PSO-optimized field oil production rates for scenario 2 and scenario 3 are shown in Fig. 11.

Comparison of the computational time for the optimization showed both PSO and GWO performed at similar level. While the numerical model took about 13 h to perform the task, the proxies only used 2 h (6 times faster). To further demonstrate the robustness of the approach, field oil production rate for scenario 2 is compared with an unoptimized case (with a constant field injection rate of 560 m³/day over the production period) in Fig. 12. The NPV of the base case was 152.57 million USD.

A continuation of the previous work is presented in Ng et al. 2023b. Long Short-Term Memory (LSTM), an advanced recurrent neural network (RNN), was applied to develop proxies of the 3D reservoir model which were coupled with PSO to conduct

Table 7 Optimization results with PSO and GWO (Ng et al. 2021b)

Optimization algorithm	Scenario 1		Scenario 2		Scenario 3	
	GWO	PSO	GWO	PSO	GWO	PSO
$NPV_{optimal}$ (million USD)	157.14	155.78	154.29	154.73	161.42	159.57

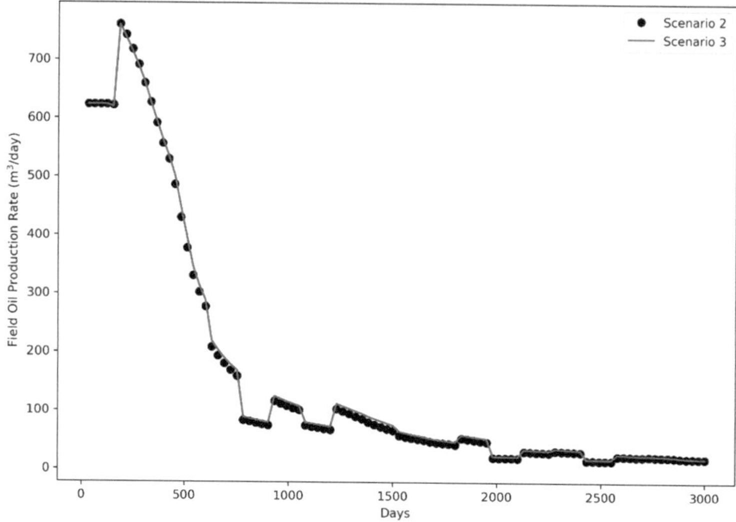

Fig. 11 PSO-optimized field oil production rates (Ng et al. 2021b)

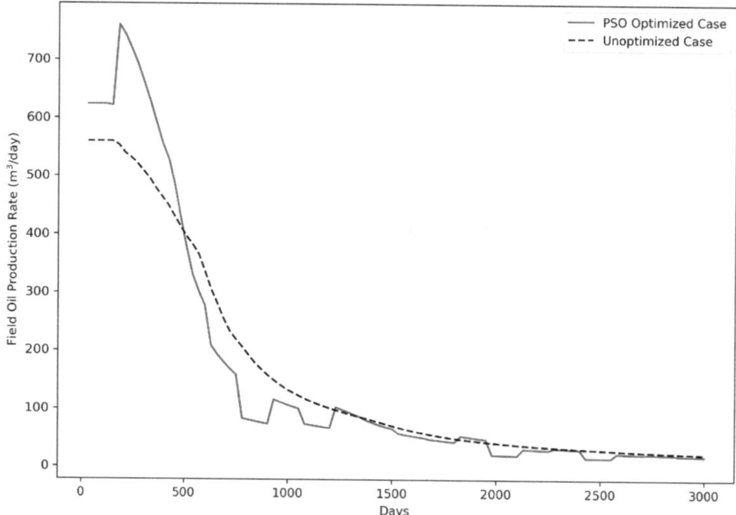

Fig. 12 Field oil production rates for scenario 2 (Ng et al. 2021b)

the production optimization. Both training and blind validation results showed excellent results. The optimization results were also slightly more accurate compared with the previous study as shown.

Regarding the best topology of proxies, FLPR proxy used one input, hidden, and output layer, with 50 nodes in the hidden layer. FWCT proxy used a similar architecture with an additional hidden layer, each with 50 nodes. The backpropagation algorithm Adam was used for training with the same configuration as in the previous study.

Figure 13 shows good agreement between the actual and predicted values for both proxies. Overfitting was prevented as the validation performances of proxies were as good as the training.

Both proxies were successfully blind-validated. The results of a sample retrieved using Latin Hypercube sampling are displayed in Fig. 14.

The proxies were then coupled with PSO to periodically adjust the field water injection rate (FWIR) to maximize the NPV. For PSO, the inertial weight was 0.8, both the social and cognitive learning factors were 1.05. In Table 8, "Dynamic Proxies"

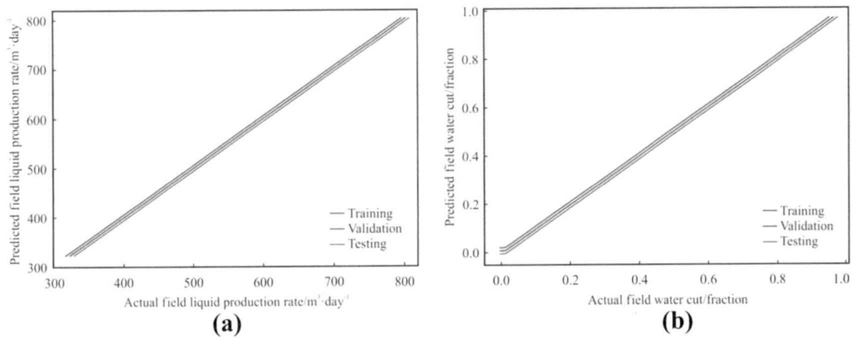

Fig. 13 Cross-plot of actual and predicted values for training, validation, and testing: **a** FLPR, **b** FWCT (Ng et al. 2023b)

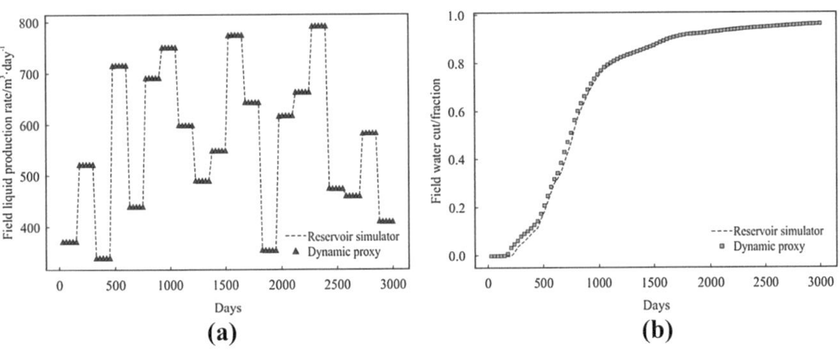

Fig. 14 Blind validation for sample set 32 (out of 80): **a** FLPR, **b** FWCT (Ng et al. 2023b)

refs to the optimization by the proxies. The optimized FWIR from "Dynamic Proxies" were fed into the simulator to compute NPV. This is referred to as "Simulator-Dynamic Proxies". The simulator was also coupled with PSO to conduct the optimization, referred to as "Simulator". The optimal NPV obtained from these cases is shown in Table 8.

Comparison with the optimization results by the simulator showed that the proxies achieved accuracy within 3% while they were computationally 3 times faster, reducing the computational time from about 12–4 h. For illustration, plot of the optimized field oil production rates of "Simulator-Dynamic Proxies" and "Dynamic Proxies" are compared in Fig. 15.

Proxy-based optimization with the aid of ML-based proxy models and global optimizer (derivative-free algorithms) was demonstrated in another study (Ng and Jahanbani Ghahfarokhi 2022a). The methodology implemented in the previous works was refined to consider the geological uncertainty, studying 10 geological realizations (considering permeability) of the Egg model in the preparation of the database for proxy modeling using multilayer perceptron (MLP). Adaptive sampling was integrated into the framework for better model training by adding extra samples (using the optimal control obtained from iterative optimization with the proxy models) to the database for retraining. Proxy models were then coupled with PSO and GWO

Table 8 Optimal NPVs (Ng et al. 2023b)

Models	Simulator	Simulator-dynamic proxies	Dynamic proxies
$\text{NPV}_{\text{optimal}}$ (million USD)	155.89	154.39	158.34

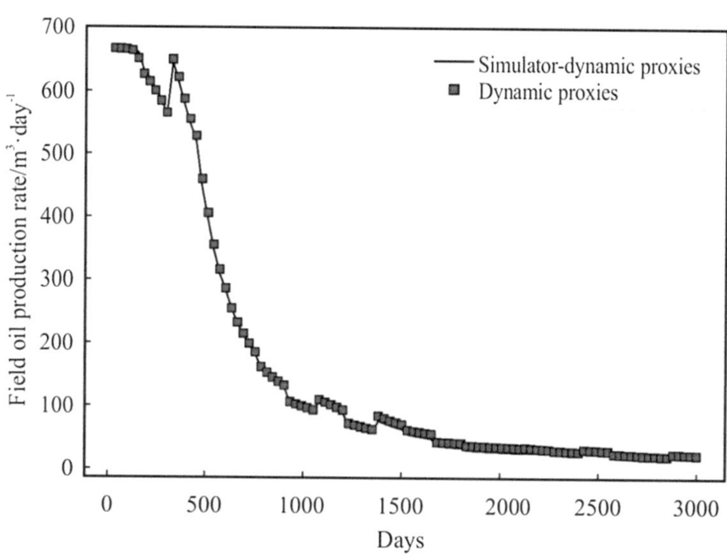

Fig. 15 Optimized FOPR (Ng et al. 2023b)

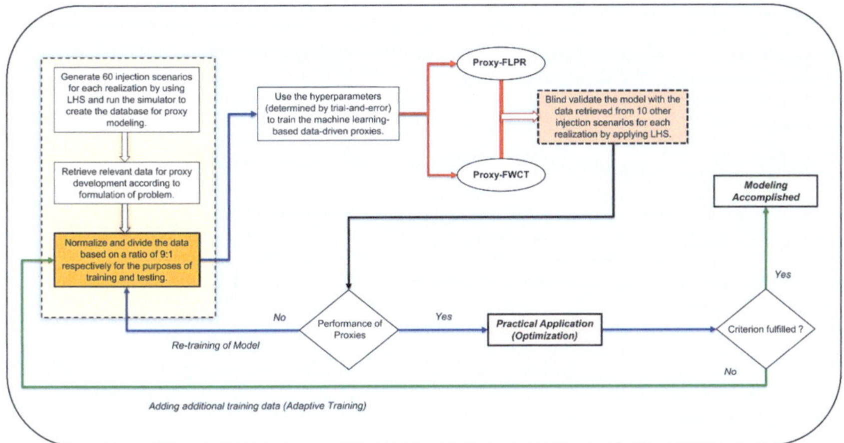

Fig. 16 Details of the main framework of AP-Ropt (Ng and Jahanbani Ghahfarokhi 2022a)

for waterflooding optimization study. The framework of the Adaptive Proxy-based Robust production Optimization (AP-ROpt) is illustrated in Fig. 16.

Two proxy models were developed for field liquid production rates (FLPR) and field water cut prediction (FWCT) using the following parameters in Table 9 (output at the previous timestep, y_{i-1}, is used as an input). 60,000 refers to 60 injection scenarios $\times$ 100 timesteps $\times$ 10 realizations.

The following values were used for FLPR, learning rate: 0.001, number of hidden layers: 4 (each layer with 50 nodes), and tolerance: 10^{-6}. For FWCT, learning rate: 0.005, number of hidden layers: 4 (each layer with 15 nodes), and tolerance: 10^{-6}. Rectified Linear Unit (ReLU) was the activation function. The maximum number of iterations was 1000 in which the early stopping mechanism was activated. After training and testing, 10 extra injection scenarios were generated using LHS for each realization (total 100 cases) to perform the blind validation. Then, comparison of simulator and proxy results decided if the proxies should undergo re-training or proceed to optimization. Results of training, testing, and blind validation phases are shown in Table 10 where MLP-FLPR generally shows a better performance than MLP-FWCT.

The objective function for the optimization study was the expected net present value (ENPV) considering the field water injection rate (between 320 and 400 Sm^3/day) as the control parameter:

$$\text{ENPV}(\mathbf{u}) = \frac{\sum_{r=1}^{n_r}\left(\sum_{i=1}^{n_{\text{total}}} \frac{\Delta t_i \times \left(Q_{i,\text{oil}}(\mathbf{u})P_{\text{oil}} - Q_{i,\text{watprod}}(\mathbf{u})P_{\text{wat prod}} - Q_{i,\text{watinj}}(\mathbf{u})P_{\text{wat inj}}\right)}{(1+\text{interest rate})^{t_i/365}}\right)_r}{n_r} \qquad (3)$$

n_r is the total number of realizations which is 10, here.

Table 9 Summary of the database for proxy modeling (Ng and Jahanbani Ghahfarokhi 2022a)

Types of data	Number of data points	Maximum value	Minimum value	Mean value	Standard deviation
Static data					
Cumulative days until timestep i	$1 \times 60{,}000$	3000	30	1515	865.98
Harmonic mean of grid absolute permeability	$7 \times 60{,}000$	749.41	577.57	641.71	37.88
Standard deviation of grid absolute permeability	$7 \times 60{,}000$	1701.24	654.44	1149.07	252.72
Arithmetic mean of permeabilities of perforated grid blocks (injectors)	$8 \times 60{,}000$	3994.57	132.99	1109.78	963.85
Arithmetic mean of permeabilities of perforated grid blocks (producers)	$4 \times 60{,}000$	5000	200	1581.12	1372.47
Dynamic data					
Field water injection rate (FWIR)	$1 \times 60{,}000$	800	320	559.96	138.34
Previous output and current output (FLPR)	$2 \times 60{,}000$	798.67	0	557.24	143.43
Previous output and current output (FWCT)	$2 \times 60{,}000$	1	0	0.7067	0.3401

Table 10 Results of training, testing, and blind validation (Ng and Jahanbani Ghahfarokhi 2022a)

		ROpt-MLP-FLPR	ROpt-MLP-FWCT
Training	R^2	0.9999	0.9995
	RMSE	0.9288	0.0073
Testing	R^2	0.9999	0.9995
	RMSE	0.9485	0.0074
Blind validation	R^2	0.9999	0.9872
	RMSE	0.9459	0.0328

As explained, the proxy-optimized control is seen as a new injection scenario for which the response of the simulator is compared with the proxy. If the criterion is not met, this scenario is added to the training database. The workflow is iterated until the criteria are fulfilled. PSO and GWO used 100 iterations and 20 swarm particles (or populations).

The optimization was performed using simulator and proxy models. The proxy-optimized control rates were also fed back to the reservoir simulator to calculate

ENPV. The ENPVs for three cases of simulator or "ground truth", simulator-proxies (using proxy-optimized control in the simulator), and proxies are computed as in Table 11. ENPV of the base case (with maximum constant field injection rate) was 155.76 million USD. The optimized NPV of each realization using GWO for the cases of simulator and simulator-proxies are shown in Fig. 17.

The updated methodology presented in this study generated reliable models indicated by excellent proxy performance (coefficient of determination, R^2 exceeding 0.98 in training, testing, and blind validation) and accurate optimization results (less than 1% difference between simulator and proxy NPVs). Though both optimizers worked very well, proxy models coupled with GWO gave slightly more accurate results than PSO in this work. Computational efficiency was also ensured considering optimization under geological uncertainty. MLP-GWO using 54 and MLP-PSO using 66 additional simulations for adaptive training, spent respectively about 13 and 16 h to perform training and optimization. The optimization study using the simulator took 159 h with PSO and 238 h with GWO. Plots of the optimized field oil rates with GWO considering 10 realizations are presented in Fig. 18. Table 12 evaluates the optimized FWPR and FOPR (using 10 different geological realizations).

Table 11 Optimized ENPV (in million USD) (Ng and Jahanbani Ghahfarokhi 2022a)

Optimization algorithm	Reservoir simulator	Simulator-proxies	Dynamic proxies
PSO	160.06	158.93	160.37
GWO	161.60	159.80	159.48

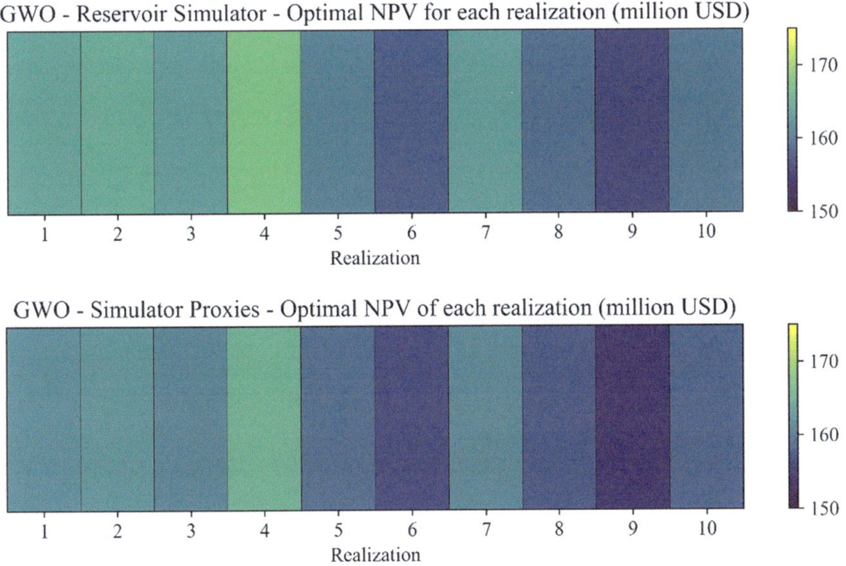

Fig. 17 Optimized NPV for each realization (Ng and Jahanbani Ghahfarokhi 2022a)

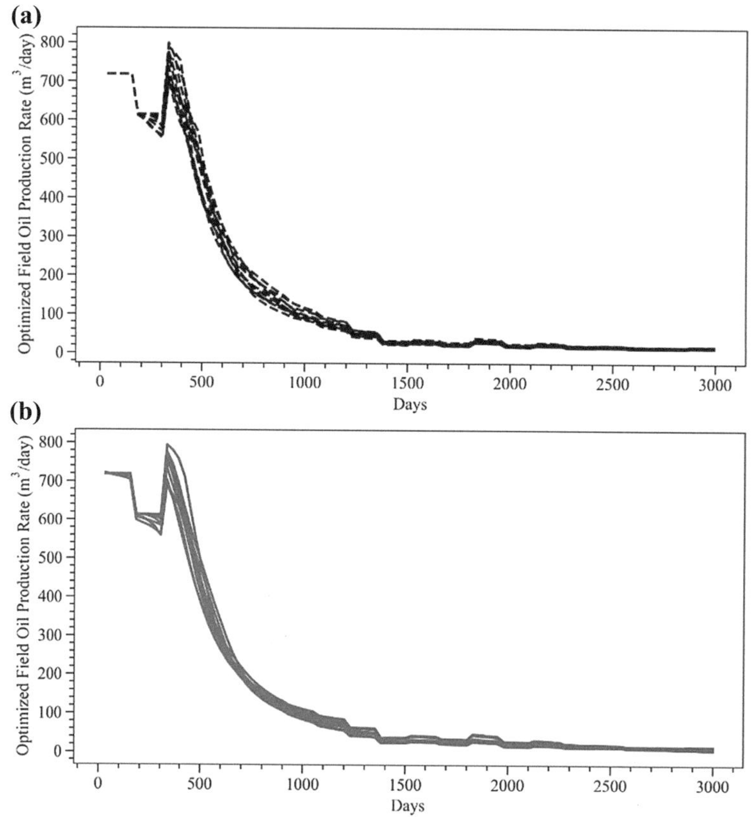

Fig. 18 Optimized FOPR. **a** simulator-proxies, **b** proxy models (Ng and Jahanbani Ghahfarokhi 2022a)

Table 12 Performance evaluation of optimized FWPR and FOPR models (by comparing the results of simulator-proxies and proxy models) (Ng and Jahanbani Ghahfarokhi 2022a)

		ROpt-FWPR	ROpt-FOPR
MLP-PSO	Mean R^2	0.9932	0.9954
	Mean RMSE	13.59	13.11
MLP-GWO	Mean R^2	0.9956	0.9972
	Mean RMSE	11.21	11.31

A workflow of using Automated Machine Learning (AutoML) also known as Tree-based Pipeline Optimization tool (Olson et al. 2016) was presented in Ng and Jahanbani Ghahfarokhi (2022b), Ng et al. (2023c) to build proxy models for fast well control optimization in waterflooding (Fig. 19). The methodology in Ng and Jahanbani Ghahfarokhi (2022b), leveraging the local search in the solution space, used the genetic programming to optimize the ML pipelines in the OLYMPUS model (Fonseca et al. 2020). OLYMPUS was developed for benchmark study on field development optimization under geological uncertainty with 50 model realizations (Fig. 20). The model consists of an oil–water system with an approximate dimension of 9×3 km with properties typical of the North Sea fields with Brent-type oil. The model has a thickness of 50 m with two zones separated by an impermeable shale layer. It has 341,728 grid blocks with an average block size of $50 \times 50 \times 3$ m (192,750 active grid blocks).

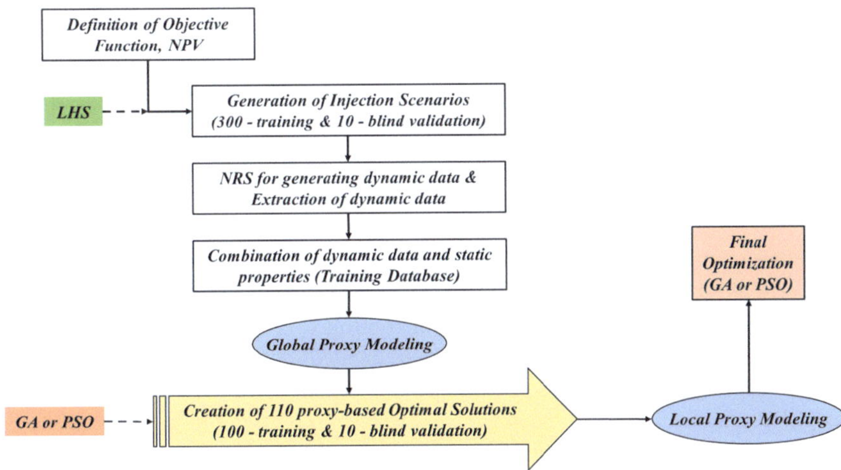

Fig. 19 Workflow of the proposed methodology (Ng et al. 2023c)

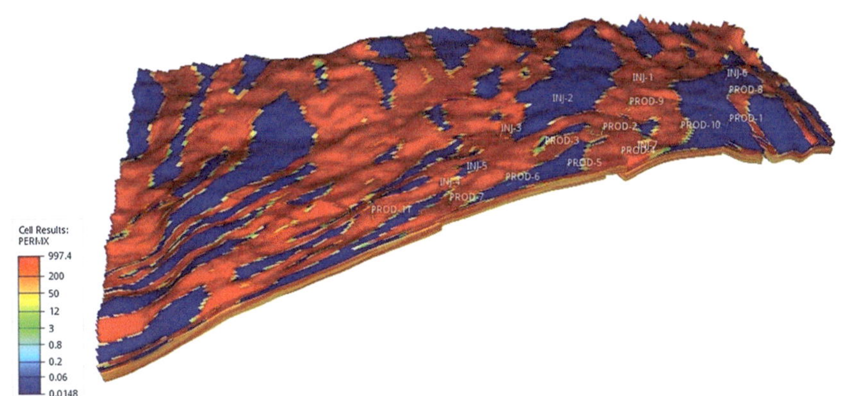

Fig. 20 OLYMPUS model permeability in x-direction (mD) (Ng and Jahanbani Ghahfarokhi 2022b)

Table 13 Prediction performance of proxy models in blind validation (Ng and Jahanbani Ghahfarokhi 2022b)

Global proxy models			
Metrics	**FLPR**	**FWCT**	**FWIR**
R^2	0.9908	0.9953	0.9870
RMSE	84.36	0.0136	98.84
Local proxy models			
R^2	0.9889	0.9997	0.9808
RMSE	70.69	0.0039	87.18

During production of about 20 years, the injection rates of injectors were adjusted every 365 days within 0–2000 Sm^3/day with the maximum Bottom Hole Pressure (BHP) of 250 bars. For the producers, a minimum BHP of 150 bars was used. The intended proxy models gave Field Oil Production Rates (FOPR), Field Water Production Rates (FWPR), and Field Water Injection Rates (FWIR). The input variables for database comprised time, static properties, control parameters, and output at the previous timestep. The static properties were the arithmetic mean of grid block permeability in x-, y-, and z-directions and porosity (for each layer and for the perforated grid blocks). Latin Hypercube Sampling (LHS) was used to generate and run 300 injection scenarios with different sets of controls to create the database for global proxy training. Ten additional scenarios were created for blind validation.

The resulting global proxy models were then coupled with PSO for an optimization study. Results were used to create a new database consisting of the optimal cases (100 iterations and 20 populations) for which the calculated NPVs (by feeding the optimized settings to the simulator) were higher than the initial maximum NPV. Then, this new database was used to train and validate the local proxy models before using them in the final optimization task without running the repetitive process of optimization. Predictability of the developed models is verified in Table 13.

The optimized FOPR for proxy models and reservoir simulator are compared in Fig. 21 (R^2 and RMSE of 0.9983 and 36.62, respectively).

The local proxy models were coupled with PSO to find the optimal NPV, 2198.13 million USD. The resulting optimal controls were fed back to the simulator to compute the respective NPV, 2167.20 million USD; so, the percentage error is 1.4%. The computational time of the whole workflow was about 5 days, while the optimization using the reservoir simulator took about 17 days. This comparison further highlights the advantage of the application of proxy models.

We further studied the application of two-stage proxy modeling in optimization studies (Ng et al. 2023c). The global and local models were developed using MLP, GA, and PSO algorithms, for the UNISIM-I-D model which was developed for the Namorado Field in the Campos Basin in Brazil (Model-based decision analysis applied to petroleum field development and management. 2019). The upscaled model (used in this study) has $81 \times 58 \times 20$ cells (36,739 active cells) with a resolution of $100 \times 100 \times 8$ m (Fig. 22).

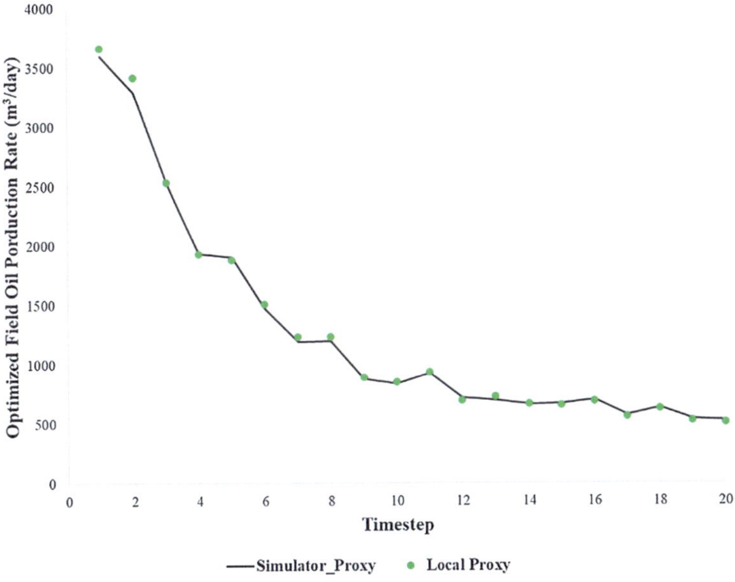

Fig. 21 The optimized FOPR profile (Ng and Jahanbani Ghahfarokhi 2022b)

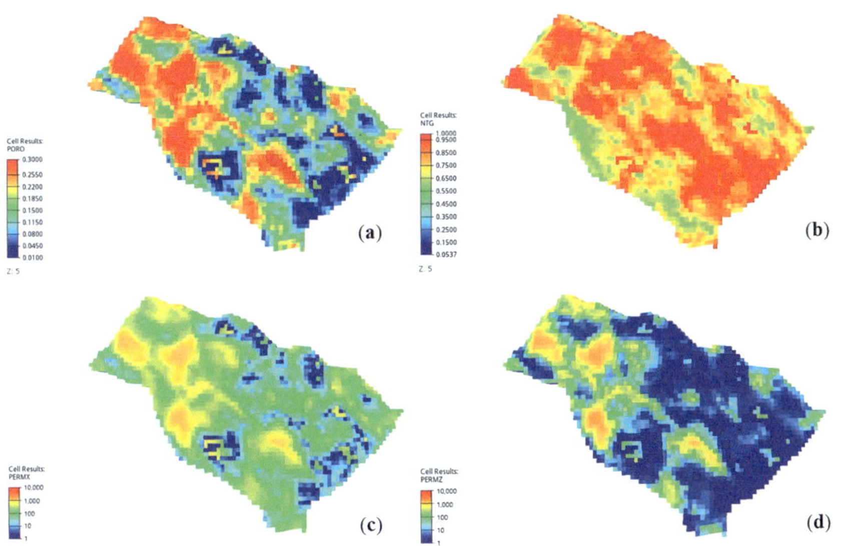

Fig. 22 The upscaled UNISIM-I-D model properties **a** porosity, **b** net-to-gross, **c** permeability I-direction, and **d** permeability K-direction (Ng et al. 2023c)

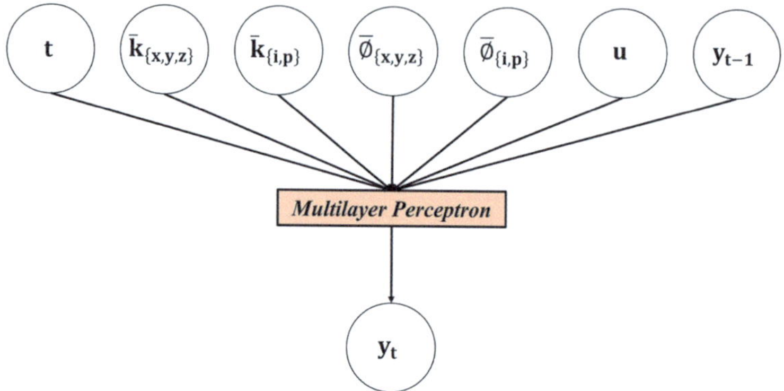

Fig. 23 Relationship between inputs and output (Ng et al. 2023c)

This work differs from the previous studies in terms of the number of control variables for the optimization study. NPV was maximized every 365 days by adjusting injection rates (within 0 and 2500 Sm^3/day) and bottom hole pressure (BHP) of producers (within 175 and 200 bar) during a production time of 9125 days. With four injectors and four producers, there were 200 control parameters (8 variables/ timestep × 25 timesteps).

Three proxy models predicted the FOPR, FWPR, FWIR, and Field Gas Production Rate (FGPR) at each timestep using gas–oil ratio of 113.45 Sm^3/Sm^3. The proxy models were formulated as shown in Fig. 23 according to static and dynamic parameters.

The maximum NPV from all the simulation scenarios generated was 5456.70 million USD. The topologies of global and local proxy models are explained in Table 14. The activation function in the output layer was linear. Adam was used for training with 2000 iterations, learning rate of 0.001, and tolerance of 10^{-6}. For the final optimization with local proxies using both GA and PSO, the number of iterations was 200 and the population size was 30.

The training, testing, and blind validation results of global and local proxy modeling are demonstrated in Tables 15 and 16.

The cumulative density frequency (CDF) of absolute percentage error between the actual NPV (calculated by feeding the optimized control variables obtained by proxy models to the simulator) and the NPV predicted by global and local proxy models are plotted in Fig. 24. The local proxy models demonstrated more accurate results within 3% error compared to the global proxy models with 0–5% error range. Also, proxy models coupled with PSO were more accurate than coupling with GA.

The actual NPVs obtained (using optimized rates of the local proxy models) are shown in Fig. 25. The highest NPVs achieved were 5976.20 and 6105.79 million USD using PSO and GA, respectively. The proxy-optimized NPVs using PSO and GA were 5854.37 and 6131.79 million USD, respectively. The optimized NPVs obtained using the simulator coupled with PSO and GA were also 5832.55 and 6054.61 million USD, respectively.

Table 14 Topology of the MLP. "Local Proxy-GA or PSO" refers to the local proxy models built from the database generated using the global proxy models coupled with GA or PSO. Data extracted from Ng et al. (2023c)

Type of proxy models	Number of hidden layers	Number of hidden nodes	Activation functions (hidden layers)
FLPR			
Global proxy model	3	250	ReLU
Local proxy-GA	3	250	ReLU
Local proxy-PSO	3	200	ReLU
FWCT			
Global proxy model	3	150	ReLU
Local proxy-GA	3	150	ReLU
Local proxy-PSO	3	150	ReLU
FWIR			
Global proxy model	3	200	ReLU
Local proxy-GA	3	200	ReLU
Local proxy-PSO	3	200	ReLU

Table 15 The results of global proxy modeling. Data extracted from Ng et al. (2023c)

Models (training)	R^2	RMSE	AAPE
FLPR	0.9510	150.06	3.357
FWCT	0.9933	0.0074	3.196
FWIR	0.9982	54.93	0.842
Models (testing)	R^2	RMSE	AAPE
FLPR	0.9516	153.01	3.440
FWCT	0.9920	0.0081	3.435
FWIR	0.9980	59.37	0.874
Models (blind validation)	Mean R^2	Mean RMSE	Mean AAPE
FLPR	0.9267	183.18	4.274
FWCT	0.9892	0.0092	4.075
FWIR	0.9974	64.03	1.169

The optimized rates showed good accuracy. Plots of PSO-optimized FOPR, FWPR, FWIR, and FGPR are shown in Fig. 26. Again, "Simulator-Proxies" refers to the case where the optimal controls produced by local proxy models are used in the reservoir simulator. The results of the local proxy models matched well with the simulator-proxies results.

In conclusion, we achieved a faster optimization compared with the optimization performed by the numerical simulator. Global and local proxy modeling, as well as optimization with either PSO or GA took about 2 days, versus 12 days needed

Table 16 The results of local proxy modeling. Data extracted from Ng et al. (2023c)

GA

Models (training)	R^2	RMSE	AAPE
FLPR	0.9660	108.69	1.959
FWCT	0.9961	0.0086	2.403
FWIR	0.9975	43.08	0.534
Models (testing)	R^2	RMSE	AAPE
FLPR	0.9659	119.79	2.123
FWCT	0.9956	0.0094	2.774
FWIR	0.9978	46.51	0.588
Models (blind validation)	Mean R^2	Mean RMSE	Mean AAPE
FLPR	0.9418	152.38	3.012
FWCT	0.9905	0.0112	3.155
FWIR	0.9971	51.76	0.681

PSO

Models (training)	R^2	RMSE	AAPE
FLPR	0.9632	124.66	2.383
FWCT	0.9962	0.0076	2.276
FWIR	0.9974	52.09	0.620
Models (testing)	R^2	RMSE	AAPE
FLPR	0.9630	128.53	2.442
FWCT	0.9953	0.0086	2.396
FWIR	0.9962	66.73	0.666
Models (blind validation)	Mean R^2	Mean RMSE	Mean AAPE
FLPR	0.9578	118.80	2.262
FWCT	0.9935	0.0105	3.037
FWIR	0.9975	42.46	0.581

to finish the optimization with the simulator. The methodology showed very good capability to perform optimization tasks with high dimensions (200 optimization variables).

Another study Jahanbani Ghahfarokhi and Chaturvedi (2023) presented the results of the workflow devised in Chaturvedi (2021) to develop adaptive proxy models for well control optimization using a synthetic model (OLYMPUS). Model realization number 49 was studied (base case scenario with 10 producers and 6 injectors). ANN was used to build proxy models which were coupled with GA to optimize the well control settings to yield the highest NPV. The field development optimization workflow, illustrated in Fig. 27, allows for building and multiple quality checks and retraining of the three proxy models for FOPR, FWPR, and FWIR. BHP and

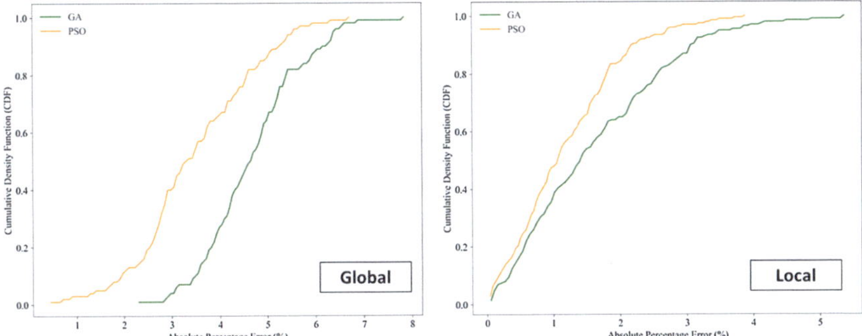

Fig. 24 Cumulative density frequency plot of absolute percentage error between actual and predicted NPV by global and local proxy models (Ng et al. 2023c)

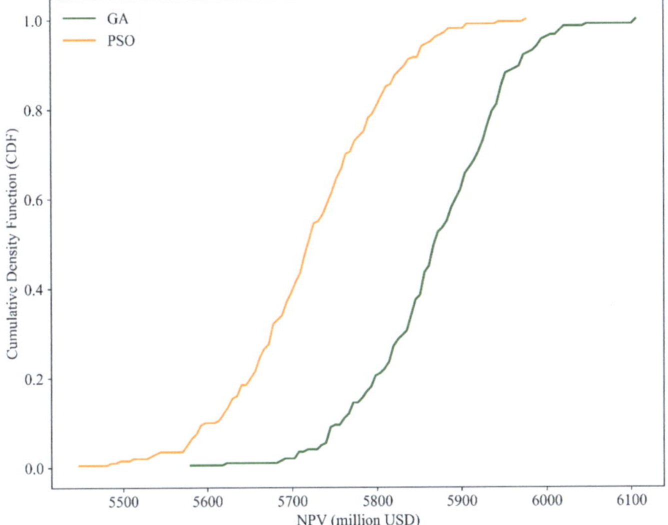

Fig. 25 Cumulative density frequency plot of NPVs (Ng et al. 2023c)

timesteps were the input parameters to develop and train the models. Eighty-five simulations were designed to train the proxy models, together with 5 blind runs.

After the quality checking, the developed models were applied in optimization study maximizing NPV, using BHP of producers and injectors as control variables. The 20 years of production time was divided into 20 time intervals to control the BHP every year; so, in total 320 optimization variables. Sensitivity analysis of GA parameters was performed; 13 combinations were tested, with 5 optimization runs for each combination. BHP configurations from all these cases were run using the

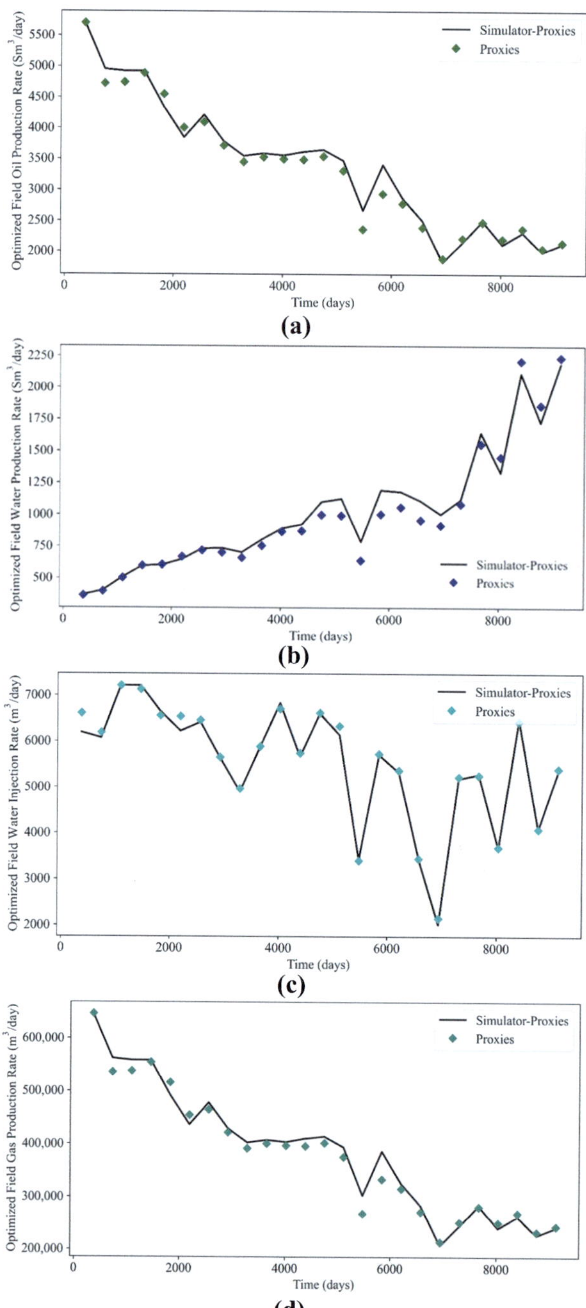

Fig. 26 PSO-optimized rates: **a** FOPR, **b** FWPR, **c** FWIR, and **d** FGPR (Ng et al. 2023c)

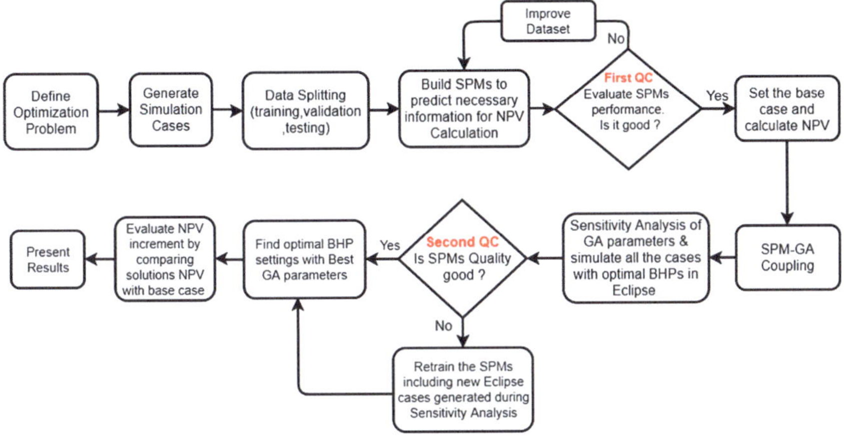

Fig. 27 The main workflow of the study (Chaturvedi 2021)

simulator and the resulting NPVs were compared against NPV calculated from the proxy models. Proxy models were then retrained with 85 (previously designed) plus 65 (new) samples followed by blind validation before being applied in the optimization study. Figure 28 shows the FOPR proxy performance during blind validation. Table 17 shows the optimal hyperparameters for retraining that minimized the loss function after a series of Bayesian optimizations. The table also analyzes the performance of the retrained proxy models.

After retraining and selecting the best GA configuration (population size of 20, mutation probability of 75%, crossover probability of 25%), optimization study was performed until 100 generations to find the optimum controls for all wells for 20

Fig. 28 Blind test results for FOPR proxy (Chaturvedi 2021)

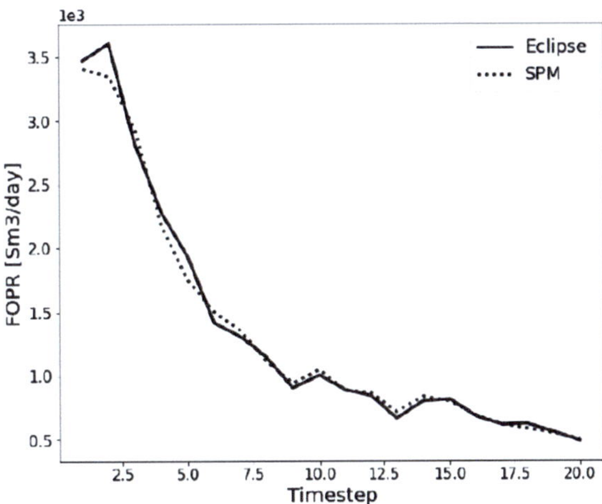

Table 17 The optimum hyperparameter configuration and performance for the retrained SPMs (Chaturvedi 2021)

SPM	Optimum values of hyperparameters			
	Learning rate	**Number of layers**	**Neuron count**	**Dropout rate**
FOPR	3.9e−02	3	35	1.17e−02
FWIR	1.0e−02	5	43	6.89e−02
FWPR	3.9e−02	3	35	1.16e−02

SPM	R^2			MAE [Sm³/day]			NRMSE [%]		
	Training	**Validation**	**Test**	**Training**	**Validation**	**Test**	**Training**	**Validation**	**Test**
FOPR	0.99	0.99	0.98	7.9	8.5	8.7	6.6	7.8	8.9
FWIR	0.99	0.93	0.95	7.4	8.9	8.7	2.0	2.8	2.7
FWPR	0.99	0.99	0.99	10.5	9.5	10	6.1	5.0	5.6

SPM	Blind test		
	R^2	MAE [Sm³/day]	NRMSE [%]
FOPR	0.99	7.6	6.46
FWIR	0.93	7.5	1.83
FWPR	0.97	13.6	7.3

timesteps. Figure 29 shows the NPV_{SPM} development over 100 generations, with the maximum NPV achieved, 2.188 billion USD.

The obtained BHP configuration was also run on the simulator to obtain $\text{NPV}_{\text{Eclipse}}$, which is 2.17 billion USD, or 17.5 million USD lower than the NPV_{SPM}. Relative difference is only 0.81%, which represents a small error. In comparison with the initially trained proxy models, an increase in R^2 of the NPV cross-plot (proxy versus simulator) was observed (from 0.87 to 0.91) using the retrained proxy models. This further shows the success of the retraining workflow (which took 15 min for

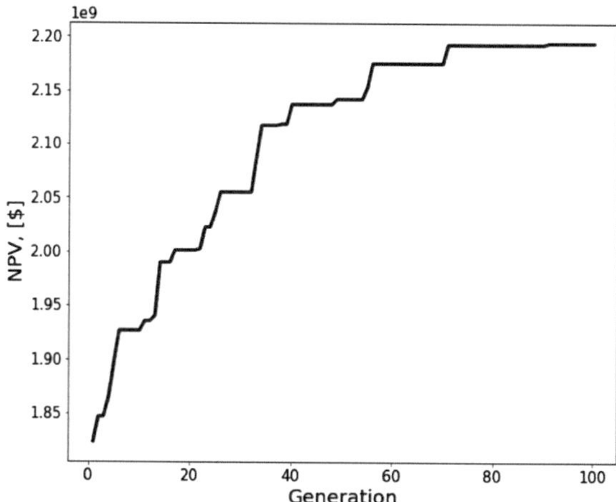

Fig. 29 NPV_{SPM} improvement over 100 generations (Chaturvedi 2021)

each proxy). A maximum reduction of 95% in optimization computational time was observed.

Reservoir management (RM) is an example of sequential decision-making under uncertain conditions. Decision analysis (DA) tools are applied to such problems. The value of information (VOI) is a tool used to evaluate if information acquisition and uncertainty quantification are valuable and can improve the decisions. A comprehensive review of VOI applications in the petroleum industry was presented in Bratvold et al. (2009). VOI can be represented as

$$\text{Value of Information} = \max\{0, \gamma\}$$

$$\gamma = \begin{bmatrix} \text{Expected Value with} \\ \text{Information} \end{bmatrix} - \begin{bmatrix} \text{Expected Value without} \\ \text{Information} \end{bmatrix} \tag{4}$$

We demonstrated in Ng and Jahanbani Ghahfarokhi (2023), the coupling of VOI analysis and ML methods such as Gaussian Process Regression (GPR) and Support Vector Regression (SVR) to optimize the waterflooding initiation time in the OLYMPUS model with 50 different geological realizations with uncertain facies, porosity, permeability, net-to-gross ratio, initial water saturation, and transmissibility across the faults.

The initiation time was decided based on the information acquired from oil and water production data. VOI was estimated using the modified Least-Squares Monte Carlo (LSM) algorithm (Hong et al. 2019), a simulation-regression approach considering the effects of previous and current information. The decisions to be made for each realization were then analyzed. This approximate method is important in sequential decision-making with high dimensionality in the presence of many potential outcomes of the uncertainties.

The production period was assumed to be 10 years. A decision was needed each year to start (or not) the waterflooding by applying the VOI analysis. The termination time of production was also optimized. There were 7 injectors and 11 producers (injectors controlled by the maximum rate of 2000 sm^3/day with bottom hole pressure target of 250 bars, and the producers by the maximum bottom hole pressure of 150 bars). The economic model was represented by net present value (NPV).

DWOI consisted of 2 years of primary recovery and then 8 years of waterflooding and corresponded to the alternative with the highest ENPV over all realizations. The respective EVWOI was 1479.06 million USD, without acquiring any production data. DWPI corresponded to the alternative with the highest NPV for each realization. Averaging the NPVs resulted in EVWPI of 1625.54 million USD. Then, VOPI was 146.47 million USD.

Normalized cumulative frequency distributions (NCFD) of DWPI for the lifetime of primary recovery, and for waterflooding, and a total lifetime of production revealed that majority of the realizations recommend a short lifetime of primary recovery (less than 2 years) to achieve EVWPI, in addition to 10 years of total lifetime. For DWII of waterflooding for each realization, Fig. 30 shows the observed and approximated

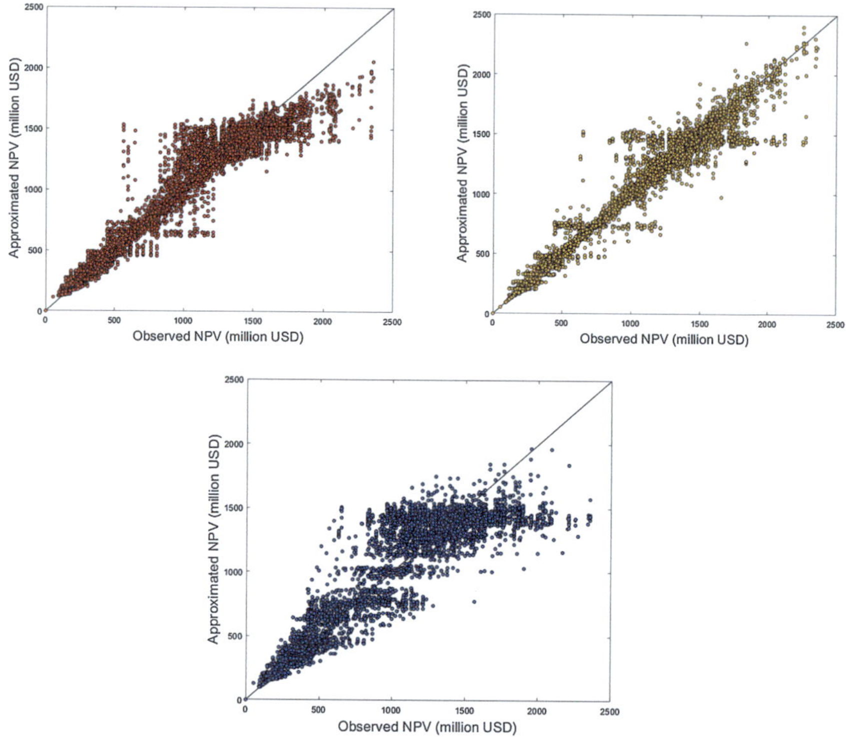

Fig. 30 Plots of observed against approximated NPV for each alternative with **a** LR, **b** GPR, and **c** SVR (Ng and Jahanbani Ghahfarokhi 2023)

NPV for each alternative during regression analysis at each decision point in time with LR, GPR, and SVR, where GPR outperformed the other methods in terms of NPV approximation.

LR in the modified LSM algorithm gave an EVWII of 1490.58 million USD. GPR and SVR resulted in the corresponding EVWIIs of 1490.23 million USD and 1491.52 million USD. The corresponding VOIs estimated by LR, GPR, and SVR were 11.52 million USD, 11.17 million USD, and 12.46 million USD. SVR displayed an improvement of 8.18% compared to the VOI by LR. This shows the high potential of integrating ML techniques into the SRDM paradigm.

EVWIIs were higher than EVWOIs and this implies that it is worthwhile to include the effect of future information in decision-making. EVWIIs estimated by LR, GPR, and SVR were 1490.58 million USD, 1490.23 million USD, and 1491.52 million USD, respectively. SVR improved ENPV the most, by 0.84% despite displaying the lowest accuracy during regression analysis. VOI approximated by GPR (with the highest accuracy of regression analysis) showed improvement of the ENPV by 0.76%. This emphasizes the benefit of acquiring additional data. Based on the NCFD plots of DWII, there was 80% chance that 2 years of primary recovery could produce

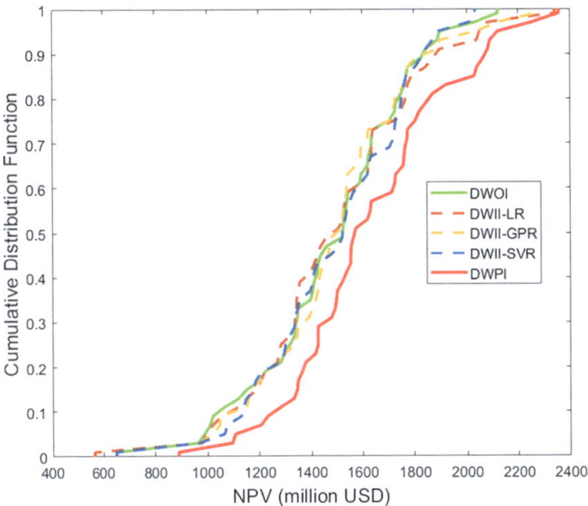

Fig. 31 NPVs corresponding to DWOI, DWII, and DWPI (Ng and Jahanbani Ghahfarokhi 2023)

the optimal results. For the lifetime of waterflooding, 90% of all the realizations proposed a water injection duration of at most 8 years. This resulted in 66% chance of 10 years of total lifetime. Analysis of NCFD plots showed that the three techniques (LR, GPR, and SVR) mostly resulted in the optimal decision of 2 years of primary recovery and 8 years of waterflooding that contributed to a total of 10 years of production.

Figure 31 compares the cumulative distribution function (CDF) of the NPVs corresponding to DWOI, DWII (considering all three techniques), and DWPI. The CDF of NPV_{DWOI} and the three CDFs of NPV_{DWII} are close to each other.

Figure 32 shows the plots of the mean oil and water production rates corresponding to DWOI and DWII. For DWOI, the mean oil production rate starts increasing after Year 2 because of the start of waterflooding (also reflected by the increase in the mean water production rate after Year 2). For DWIIs, the initiation of waterflooding is generally different for different realizations based on the acquisition of information under the framework of SRDM. A sharp increase in oil rate after Year 2 was observed. This could be explained by the fact that, previously more than 50% of the NFD revealed a lifetime of 2 years for primary recovery.

We used real production data from a well in Volve field (Equinor. 2018) to suggest predictive ML-based models as an alternative to decline curve analysis (DCA). Volve field is a 2 km by 3 km oil-bearing sandstone reservoir at a depth between 2750 and 3210 m below sea level with average permeability of about 1000 mD, porosity of 0.21, and net-to-gross ratio of 0.93, a reservoir pressure of 340 bar and temperature of 110 °C. The data from the well NO159-F-14H was used in Ng et al. (2022b), only between July 2013 and July 2016 (10,930 data points). Selected data used for modeling is presented in Table 18 with the mean and standard deviation of each parameter in Table 19. Oil production profile of the well is plotted in Fig. 33.

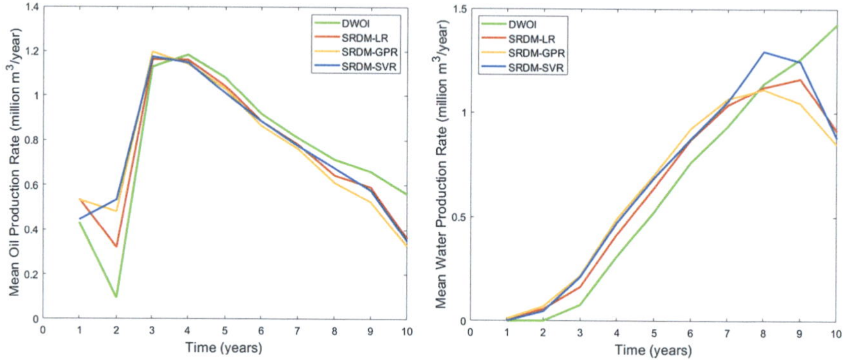

Fig. 32 Mean oil and water production rates for DWOI and DWII (Ng and Jahanbani Ghahfarokhi 2023)

Table 18 Selected input and output for modeling (Ng et al. 2022b)

Parameters	
Input data	*Units*
Time	Days
On-stream hours	hours
Average downhole pressure	bar
Average downhole temperature	°C (degree Celsius)
Average choke size percentage	%
Average wellhead pressure	bar
Average wellhead temperature	°C (degree Celsius)
Gas volume from well	m^3 (daily)
Water volume from well	
Output data	*Units*
Oil volume from well	m^3 (daily)

The (normalized) data were divided into two sets for modeling and prediction, based on a ratio of 7.5:2.5 (as shown in the figure). 70% of the modeling set was used for the training and 30% was equally divided between validation and testing sets.

The data-driven models used to predict the oil production included FNN with backpropagation algorithm (FNN-BP), FNN trained with PSO (FNN-PSO), SVR tuned with trial-and-error approach (SVR-TE), hybrid model of SVR and PSO (SVR-PSO), simple RNN, Long Short-Term Memory (LSTM), and Gated Recurrent Units (GRU). The developed models were trained, validated, tested, and blind validated.

During the training phase, the weights and biases were iteratively adjusted to minimize the cost function, MSE. The FNNs used one input layer with 9 nodes, one hidden layer with 30 nodes, and one output layer with only one node. The RNNs

Table 19 Mean and standard deviation of input and output parameters considering all the data points (Ng et al. 2022b)

Baseline information		
Input and output	Mean	Standard deviation
Time	547	315.67
On-stream hours	23.02	3.89
Average downhole pressure	261.01	15.54
Average downhole temperature	99.38	5.14
Average choke size percentage	90.44	21.88
Average wellhead pressure	30.73	4.21
Average wellhead temperature	86.25	8.47
Gas volume from well	49,263.63	30,342.37
Water volume from well	3171.60	674.34
Oil volume from well	326.88	204.97

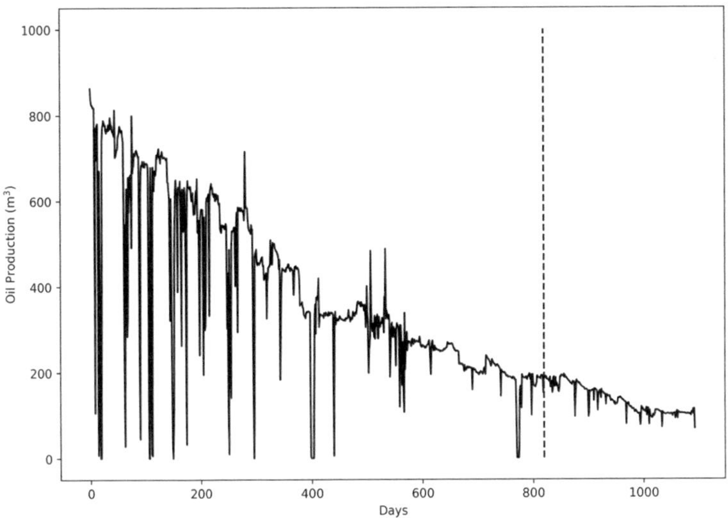

Fig. 33 Oil production of the well NO159-F-14H (Ng et al. 2022b)

used one hidden layer and one output layer with 30 hidden nodes and 1 output node. Adam was only used to train the RNNs and FNN-BP. Some parameters used for the training are presented in Table 20.

The performance evaluation of the models is presented in Table 21, demonstrating excellent results. Also, model evaluations for all the 1093 data points are tabulated in Table 22.

Table 20 Parameters used in model training (Ng et al. 2022b)

Adam parameters	Values
Number of iterations	2000
Learning rate	0.01
Exponential decay rates for the 1st moment estimates, β_1	0.9
Exponential decay rates for the 2nd moment estimates, β_2	0.999
Numerical stability constant, ε	10^{-7}
PSO parameters	**Values**
Number of iterations	2000
Number of particle swarms	100
Inertial weight, ω	0.8
Cognitive learning factor, c_1	1.05
Social learning factor, c_2	1.05

The actual and predicted oil production (for all the data points) are presented in Fig. 34 showing the robustness of ML techniques in capturing the fluctuating trend, while the conventional DCA approach is only able to perform a "curve fitting".

RNN-based models in general outperformed the SVR-based and FNN-based models in training and prediction. LSTM outperformed the other models in training. GRU was the best model in validation and simple RNN was the best model in testing. Nevertheless, the training performance and predictability of SVR-based and FNN-based models were very good. PSO contributed to the accuracy of SVR in training, but not for FNN due to outliers. LSTM produced the best results in prediction. Considering all the data points, GRU showed the best results. The study also concluded that the error distribution produced by predictive models showed a normal distribution with center close to zero which confirmed the reliability of the models.

3 Part 2: Water Alternating Gas Injection, CO_2 Storage, and Property Estimations

We summarized some use of artificial intelligence/machine learning (AI&ML) in carbon capture, utilization, and storage (CCUS) in Jahanbani Ghahfarokhi et al. (2022). Improved modeling and optimization techniques are of value when the subsurface is repurposed for storage purposes in addition to the benefits of extracting the remaining hydrocarbons. The study discussed development of proxy models for optimizing water alternating gas (WAG) in real fields, WAG optimization using CO_2 gas injection and immiscible gas injection.

Experimental methods for property estimations can sometimes be challenging and time-consuming. Empirical models are either simple or cannot ensure accurate

Table 21 Performance metrics of the results estimated for the training, validation, testing, and blind validation. Data extracted from Ng et al. (2022b)

Datasets	Models	R^2	RMSE
Training	SVR-TE	0.9951	13.88
	SVR-PSO	0.9944	14.68
	FNN-BP	0.9948	14.00
	FNN-PSO	0.9945	14.92
	Simple RNN	0.9945	14.46
	LSTM	0.9962	12.03
	GRU	0.9962	12.17
Validation	SVR-TE	0.9880	21.37
	SVR-PSO	0.9889	20.79
	FNN-BP	0.9911	19.13
	FNN-PSO	0.9923	15.75
	Simple RNN	0.9921	18.27
	LSTM	0.9910	19.51
	GRU	0.9940	15.75
Testing	SVR-TE	0.9764	30.83
	SVR-PSO	0.9936	16.61
	FNN-BP	0.9936	16.44
	FNN-PSO	0.9898	19.91
	Simple RNN	0.9941	15.37
	LSTM	0.9922	17.64
	GRU	0.9915	18.24
Blind validation	SVR-TE	0.9476	7.34
	SVR-PSO	0.9644	6.04
	FNN-BP	0.9538	6.89
	FNN-PSO	0.9574	6.61
	Simple RNN	0.9665	5.87
	LSTM	0.9712	5.45
	GRU	0.9700	5.56

Table 22 Performance metrics of all the models considering all data points (Ng et al. 2022b)

Datasets	Models	R^2	RMSE
All	SVR-TE	0.9935	16.52
	SVR-PSO	0.9952	14.21
	FNN-BP	0.9956	13.65
	FNN-PSO	0.9952	14.15
	Simple RNN	0.9957	13.51
	LSTM	0.9961	12.69
	GRU	0.9964	12.28

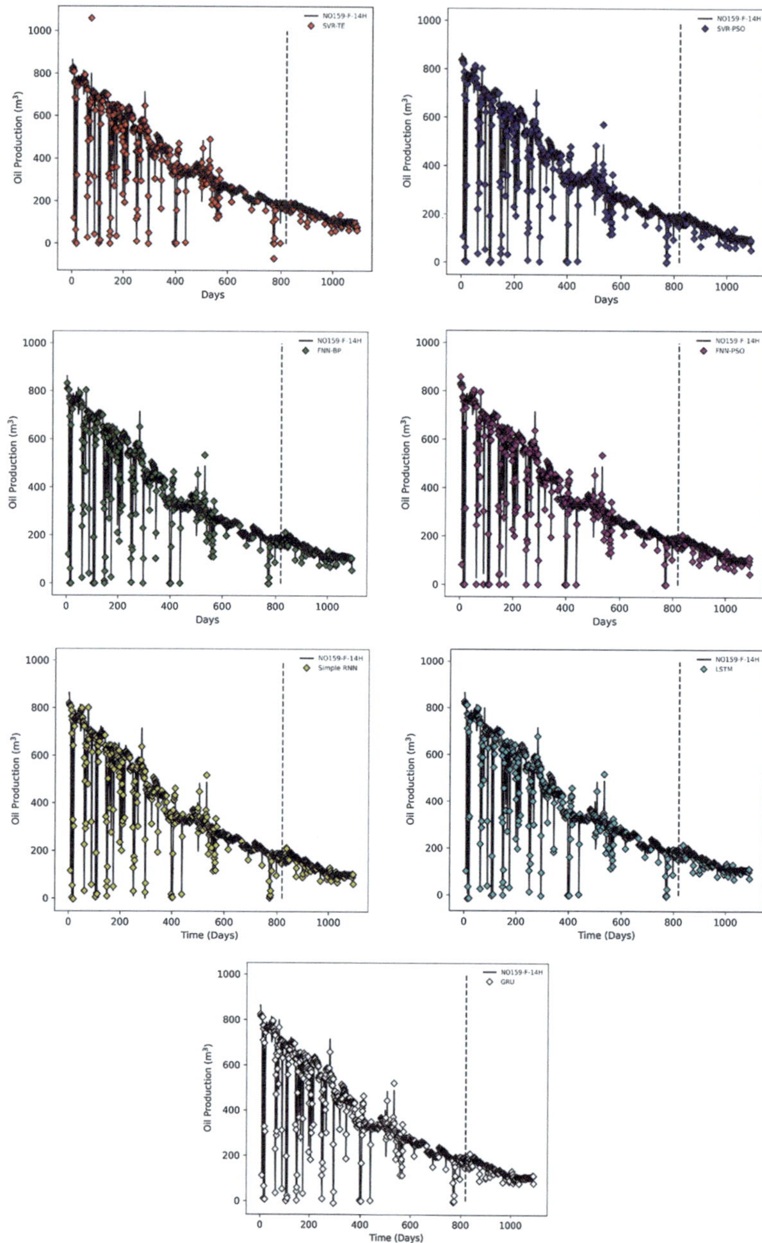

Fig. 34 Oil production profile using the developed models (Ng et al. 2022b)

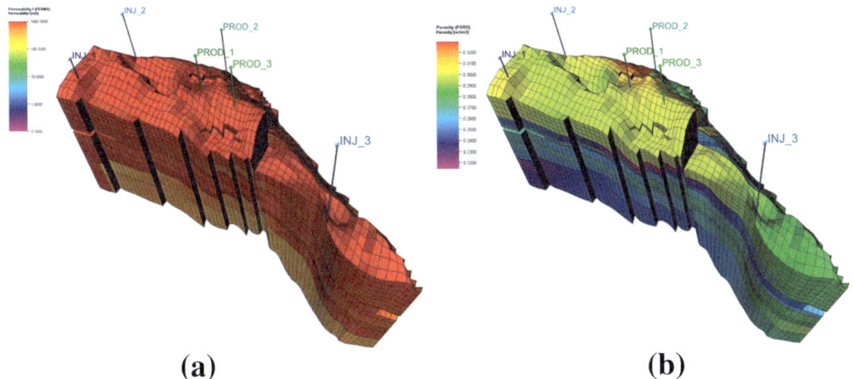

Fig. 35 **a** Permeability distribution; **b** porosity distribution of the segment K1/K2 Gullfaks. (Nait Amar et al. 2021)

predictions. Application of data-driven techniques was also discussed to develop accurate predictive correlations to estimate some CO_2 properties such as diffusivity in brine and thermal conductivity. This part also discusses proxy models developed for CO_2 storage as well as models for predicting wax deposition, and interfacial tension (IFT) of hydrogen–brine system for hydrogen storage.

Water alternating gas (WAG) is an important method for enhancing oil recovery, improving both microscopic and macroscopic displacement efficiencies. We developed accurate proxy models to optimize the design of a miscible WAG process in the K1/K2 segment of Gullfaks in the North Sea (Nait Amar et al. 2021). The study maximized the total oil production at a reduced simulation time compared with the numerical simulator. The K1/K2 is a part of the Gullfaks reservoir with an average thickness of 200 m with its top at 1870 m. Distribution of the properties and location of the wells are illustrated in Fig. 35. Details of the model and compositions of the reservoir fluid and injected hydrocarbon gas are stated in the paper (Nait Amar et al. 2021).

Field oil production total (FOPT) was the objective function during the optimization study with a 50% limit for Field Water Cut (FWCT). The control parameters were water and gas injection rates, half-cycle time (the time during which the gas/ water is injected), and the downtime parameter (corresponding to the years in which the WAG is active).

Two ANN models, multilayer perceptron (MLP) and radial basis function neural network (RBFNN), were implemented to estimate the parameters required for the optimization: the field oil and water production rates (FOPR and FWPR), and the field pressure (FPR), while the other parameters (FOPT, FWCT) were calculated based on the rates and time. Seventy-two runs were generated and used in the development of the proxy models. Latin Hypercube Sampling (LHS) attributed half-cycle time, gas and water injection rates to these runs. Ten supplementary runs were selected for blind validation.

MLP model was optimized using the Levenberg–Marquardt Algorithm (LMA) for learning and trial and error technique was applied for investigating the proper topology (activation function, number of hidden layers, and associated neurons). The control parameters of RBFNN were optimized using Ant colony optimization (ACO) and grey wolf optimization (GWO). The best ML proxy model was then coupled with ACO and GWO to optimize the design parameters of the WAG process.

Figure 36 compares the performance of the developed proxy models for training and blind runs and shows that these models demonstrate descent predictions. MLP-LMA was the best-performing proxy model for predictions, selected for further evaluation and optimization study.

Figure 37 compares the results (FOPR and FOPT as a function of time) obtained by the MLP-LMA model and the numerical simulator for a blind run. Besides the blind validation phase shown here, MLP-LMA also exhibited good predictions in the learning phase. The relative importance of the main input parameters on the outputs of the proxy models was investigated and it was found that water injection rate had the biggest impact.

After confirming the accuracy of the MLP-LMA proxy model, it was coupled with ACO and GWO in order to find the optimal parameters for WAG, including water and gas injection rates, injection half-cycle, downtime, WAG ratio, and gas slug size. The results obtained from the two hybridizations are presented in Table 23 and in Fig. 38.

The optimization results of the two hybridizations were similar with a slight difference in the convergence speed (11 iterations for GWO and 14 for ACO). The results showed that a field water injection rate of 5652 sm^3/d and a field gas injection rate of 10^6 sm^3/d with an injection half-cycle of 6 months and a downtime of 3 years were the optimal parameters for the WAG process studied.

The proxy's resulting FOPT is compared with the result of the numerical simulator optimization (using the best control parameters found) in Table 24 which shows a very good match between the results.

Regarding the computational time, the suggested proxy took only 1.2 s to generate the parameters needed for one realization, while the numerical simulator needed more than 15 min. This demonstrates the robustness and efficiency of the proposed proxy.

In another study, we discussed WAG using CO_2 gas injection (Nait Amar et al. 2020a) as a promising enhanced recovery and subsurface CO_2 storage technique (Fig. 39). We developed fast proxy models using support vector regression (SVR) and genetic algorithm (GA) for optimizing a WAG CO_2 injection using dynamic and PVT data from a real field, and a static model. The permeability distribution of the model is presented in Fig. 40. The model was based on a corner-point grid with 25 × 28 × 10 grid blocks and 54 × 54 × (3–27) m grids, and had 12 wells (4 injectors and 8 producers). For the compositions and details of the equation of state (EOS) modeling, refer to the paper (Nait Amar et al. 2020a).

Multiple SVRs were trained to develop fast dynamic proxy models to mimic numerical simulator outputs, mainly FWPR and FOPR. GA was used to first optimize the SVR hyper parameters and then the WAG CO_2 design parameters, subjected to time-dependent constraints. These parameters included the initialization time (IT),

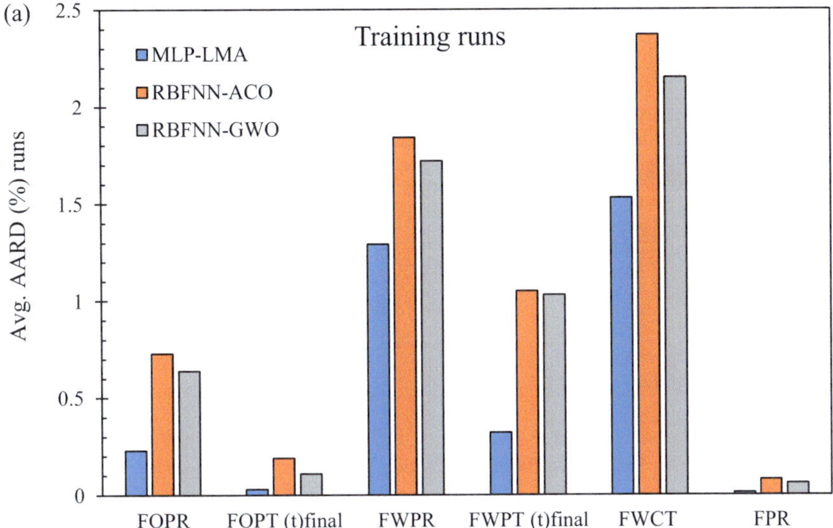

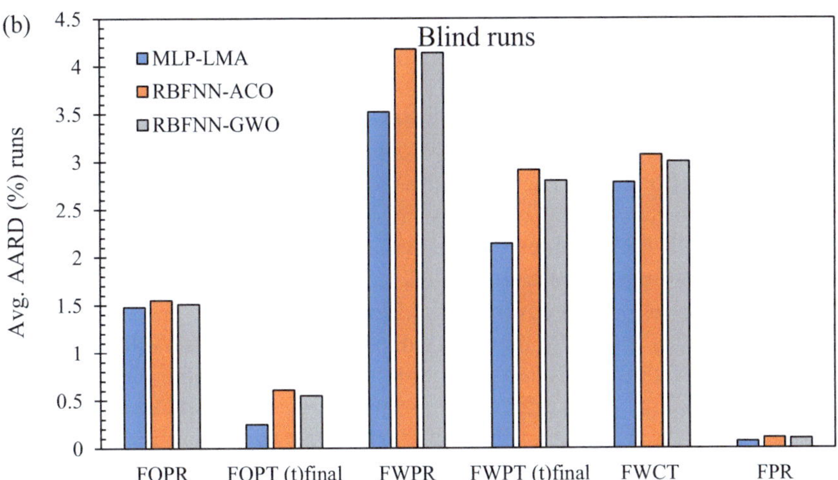

Fig. 36 Evaluation of the prediction performance of the established proxy models: **a** training runs and **b** blind runs (Nait Amar et al. 2021)

Field water injection rate (FWIR), Field gas injection rate (FGIR), the half-cycle time (HCT) of water and gas injection, WAG ratio and slug size, which were adjusted to maximize the field oil production total (FOPT) given the field water-cut (FWCT) constraint of 92% and domain constraints. Figure 41 shows how LHS was applied to design 75 total runs for the training database with three supplementary runs to design the interactions between the limits of the variables. Ten additional runs were chosen randomly for blind validation.

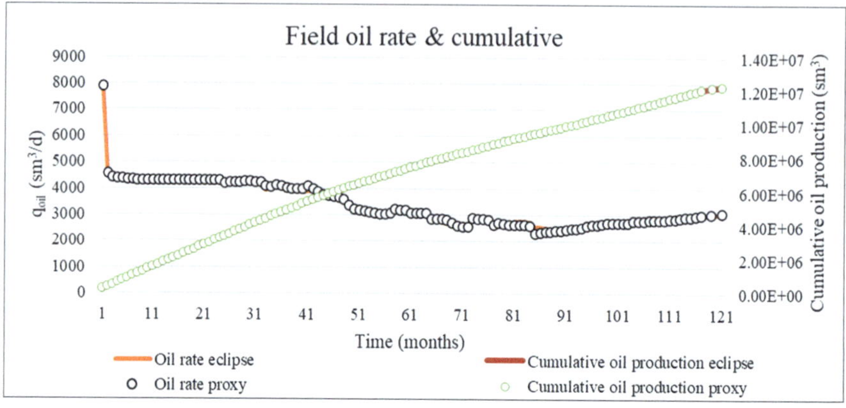

Fig. 37 Demonstration of the MLP-LMA proxy model's ability to emulate the outcomes of one blind run, FOPR, and FOPT (Nait Amar et al. 2021)

The generated samples for the database were divided into two sets for the time-steps with gas and water as the injected fluid. For each outcome, two SVRs were built for the periods with gas and water injection. The global SVR of a parameter then comprised the two SVRs. The inputs and outputs for development of the multiple SVRs are illustrated in Table 25.

GA was used to optimize the hyperparameters of SVR-FOPR and SVR-FWPR during gas and water injections. Figure 42 explains the SVR-GA model coupling.

The proxies showed satisfactory distributions of the predicted results achieved for all the parameters during training with very small error (compared with the simulator results). ARPE values were 1.13% and 0.07% for FOPR and $FOPT_{t\_final}$ (at the final time-step), respectively; 1.46% and 0.14% for FWPR and $FWPT_{t\_final}$ (at the final time-step), respectively; and 0.27% for FWCT. The proxy models also provided promising results for the blind validation with very low APRE. Figure 43 compares the FOPT obtained from proxy and simulator for the training and validation phases at the final time-step, which demonstrates a good agreement between the two approaches.

Figure 44 compares the results of SVR proxy model with the simulator in a blind run. The comparisons show very good matches for all the parameters.

After confirming the excellent performance of the SVR proxy model, it was coupled with GA to optimize the WAG injection. Checking the maximum number of iterations as a stopping criterion, the best-fit individual was chosen as the optimal WAG CO_2 scenario. Figure 45 shows the optimization results, where the optimum FOPT is 11791290.316 sm^3. The optimal control parameters are listed in Table 26 which shows that the optimal WAG is initiated after 6 years (classified as middle-time IT), with gas and water injection half-cycle time of 6 months, high water rate, and low gas rate with high WAG ratio and small slug size.

Table 27 compares the simulator and proxy values for FOPT. All the outputs produced by the proxy model were in good agreement with simulator results.

Table 23 Summary of the best results achieved (Nait Amar et al. 2021)

Algorithm	Number of iterations to reach best FOPT	Best FOPT (10^6 sm^3)	q_{injw} (sm^3/d) (best FOPT)	q_{injg} (sm^3/d) (best FOPT)	half$_{cycg\text{-}w}$ (month) (best FOPT)	WAG ratio	Slug size (PV)	Down time (years)	Max WCT (%)	Min and Max FPR (bar)
ACO	14	13.68	5652	10^6	6	1.98	0.04	3	35.42	305–310
GWO	11	13.68	5652	10^6	6	1.98	0.04	3	35.42	305–310

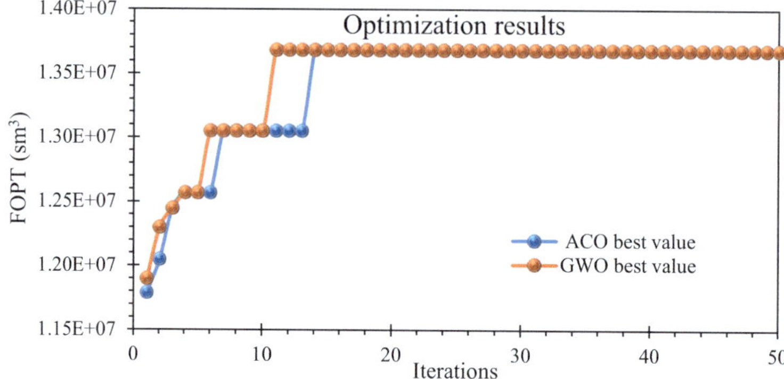

Fig. 38 Optimization results with the proposed hybridizations: proxy-ACO and proxy-GWO (Nait Amar et al. 2021)

Table 24 Comparison of the results of the proxy model and the simulator for the best scenario. Data extracted from Nait Amar et al. (2021)

Parameter	Proxy run (for best GF[*] parameters)	Eclipse 300 run (for best GF parameters)	AARD (%)
$FOPT_{t\_final}$ (sm^3)	13,685,903	13,654,497	0.23

* Global Function

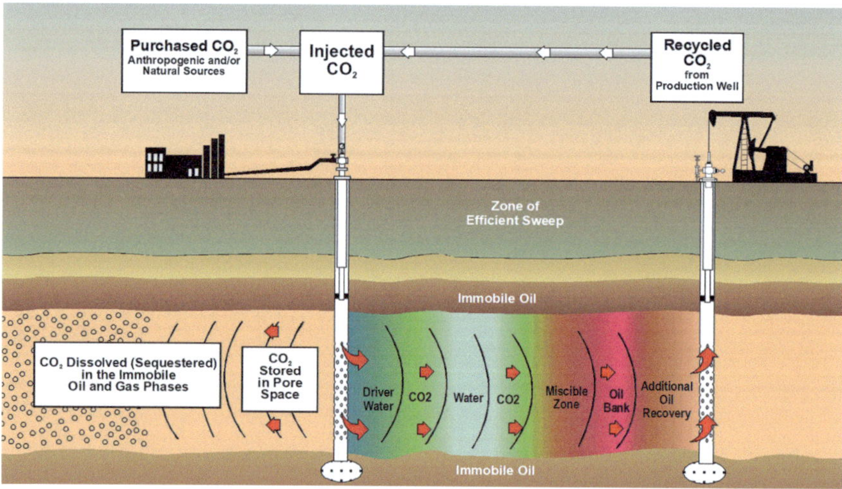

Fig. 39 CO$_2$ WAG illustration (Lake et al. 2019)

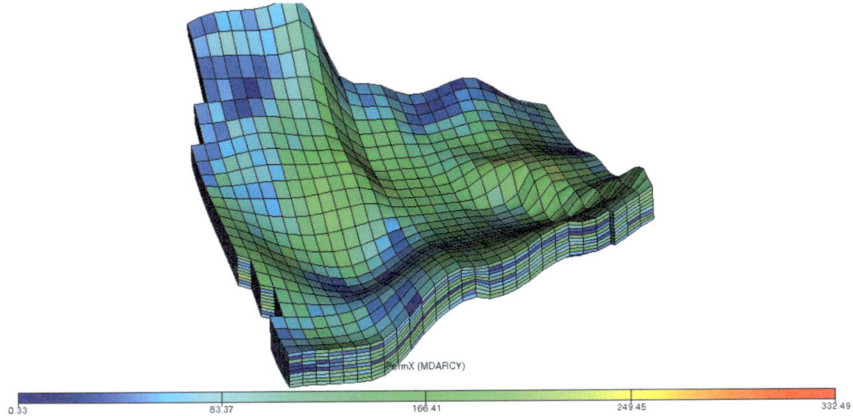

Fig. 40 Permeability distribution in the studied reservoir (Nait Amar et al. 2020a)

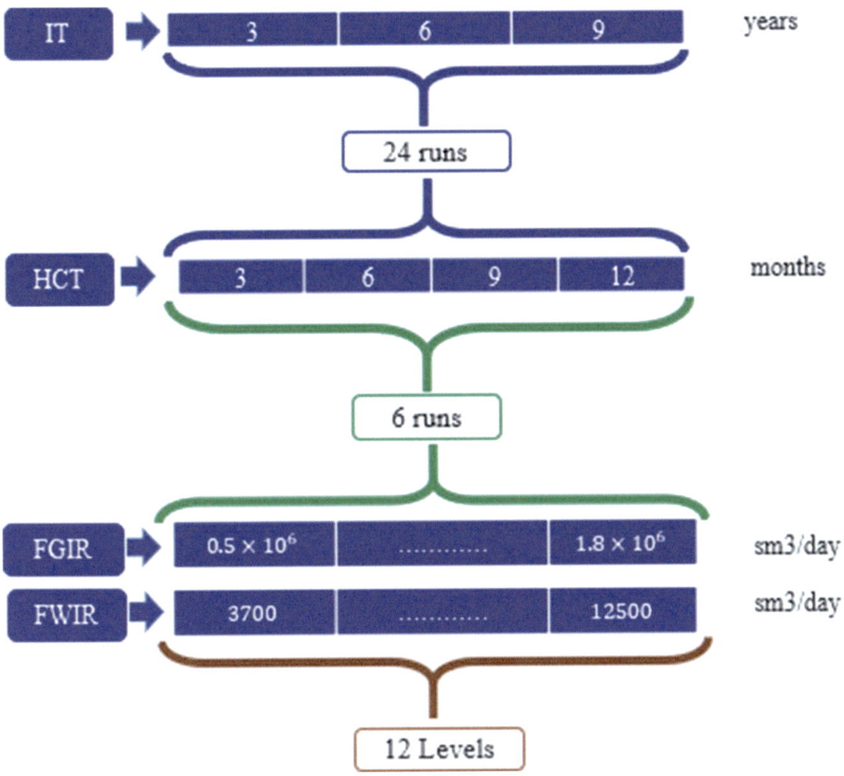

Fig. 41 Summary of the design of the runs (Nait Amar et al. 2020a)

Table 25 Inputs and outputs of the multiple SVR (Nait Amar et al. 2020a)

Multiple SVRs	Inputs	Time
		FWIR
		FGIR
		The value of the needed parameter at the previous time-step, i.e. (t-1)
	Outputs	FOPR and FWPR (FOPT and FWCT are deduced from the flow rates, i.e., FOPR and FWPR and the time)

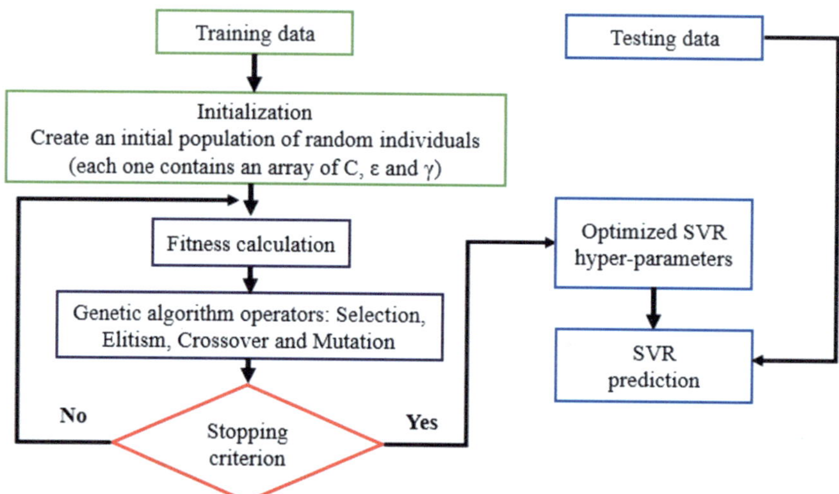

Fig. 42 Flowchart of the implementation of GA in the optimization of SVR hyper-parameters (Nait Amar et al. 2020a)

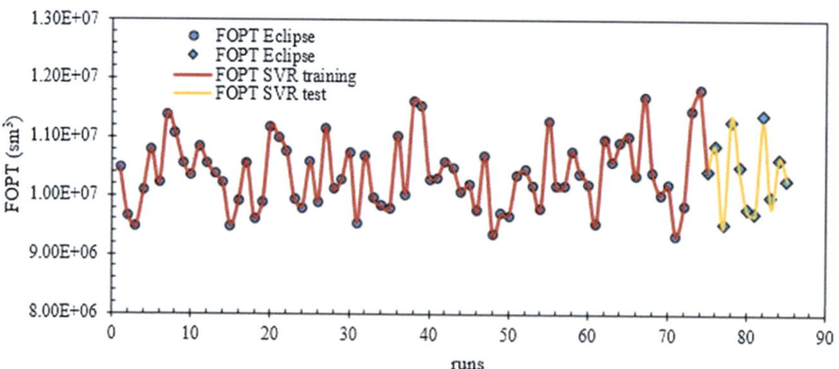

Fig. 43 Results of the training and blind validation: comparison of FOPT_eclipse vs. FOPT_proxy (after 1.2 PV gas injection) (Nait Amar et al. 2020a)

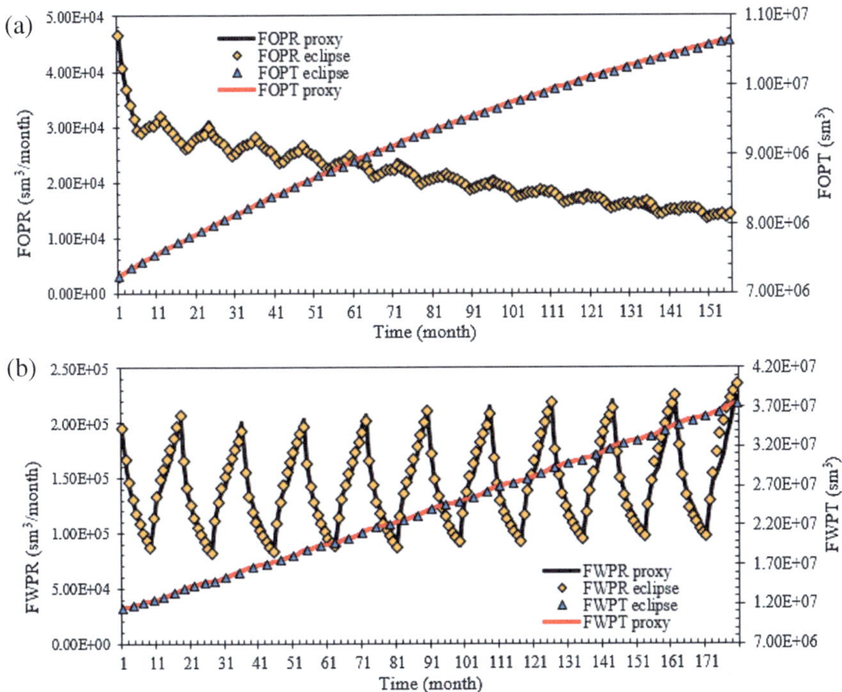

Fig. 44 Results of a random blind test (after 1.2 PV gas injection): **a** FOPR and FOPT, **b** FWPR and FWPT as functions of time (Nait Amar et al. 2020a)

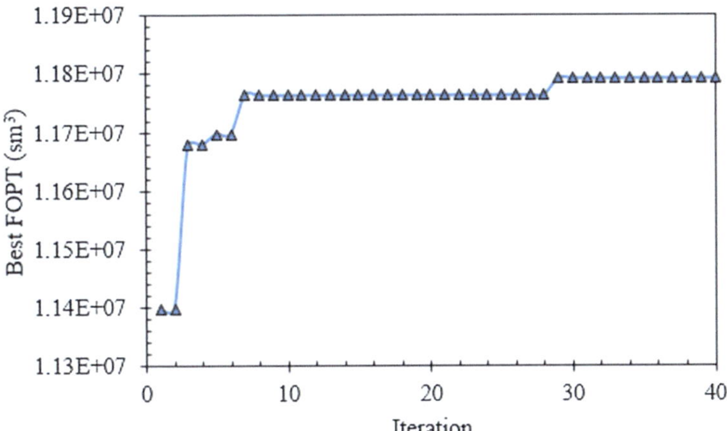

Fig. 45 Optimization results using hybridization SVR-GA (Nait Amar et al. 2020a)

Table 26 WAG CO_2 optimum design parameters (Nait Amar et al. 2020a)

Parameter	IT (years)	FWIR ($sm^3\ day^{-1}$) (of the best FOPT)	FGIR ($sm^3\ day^{-1}$) (of the best FOPT)	HCT_{w-g} (month) (of the best FOPT)	WAG ratio	Slug size (PV)	Problem constraint (Max. FWCT, %) (of the best FOPT)
Value	6	12,261	10^6	6	2.5	0.09	91.45

Table 27 Comparison of the dynamic proxy optimum parameters and numerical simulation results. Data extracted from Nait Amar et al. (2020a)

Parameter	Multiple SVR proxy (for the best parameters)	Eclipse 300 (for the best parameters)	APRE (%)
$FOPT_{t\text{-final}}$ (sm^3)	11,791,290.316	11,735,840.160	0.47

In terms of reduction of the computational time, one run using the simulator took 15–23 min, while this was only 1.086 s for the developed proxy.

In other studies (Matthew 2021; Matthew et al. 2023), we developed proxy models for a multi-objective optimization study concerning WAG (Water Alternating Gas) CO_2 injection. In this study, gas and water injection rates and half-cycle lengths were optimized to maximize both the field total oil production (FOPT) and the CO_2 storage using NSGA-II (Non-dominated Sorting Genetic Algorithm II). Two reservoir models were studied. One represented a simple geological model (the Egg model), and the other represented a more complex model (the Gullfaks K1/K2 model). An overview of the models is given in Table 28 and Fig. 46.

The fluid model (developed with the Peng–Robinson equation of state) has a low viscosity (based on EOS tuning results) and fulfills the criteria of CO_2-EOR. Both models have similar initial oil in place (about 17 million sm^3 after modification). Both models were depleted for eight years with water injection for pressure support. Equal injection/production rate limits were applied in the two cases. An overview of the study is shown in Fig. 47.

LHS generated 68 samples for the Egg Model and 97 samples for the Gullfaks Model. WAG has distinct behavior during the water and gas injection phases. The database was then divided based on the injection phase (the same for the segmentation) to develop proxies for production rates, FCO_2PR, and FOPR. Other strategies

Table 28 Overview of the two reservoir models in this study (Matthew et al. 2023)

Parameter	Egg model	Gullfaks model
Permeability	Channel distribution	Heterogeneous
Porosity	Homogeneous	Heterogeneous
Faults	0	14
Fluid	From Negahban et al. (2010)	From Negahban et al. (2010)
Transmissibility	No multiplier	Heterogeneous multiplier
Relative permeability	Sand preset	Sand preset
Grid system	Cartesian	Cornerpoint
Grid size	Homogeneous	Heterogeneous
Initial condition	320 bar, 120.85 °C, 1850 m	320 bar, 120.85 °C, 1850 m
Wells	Three injectors, three producers	Three injectors, three producers
Perforations	Throughout all layers	Different for each well

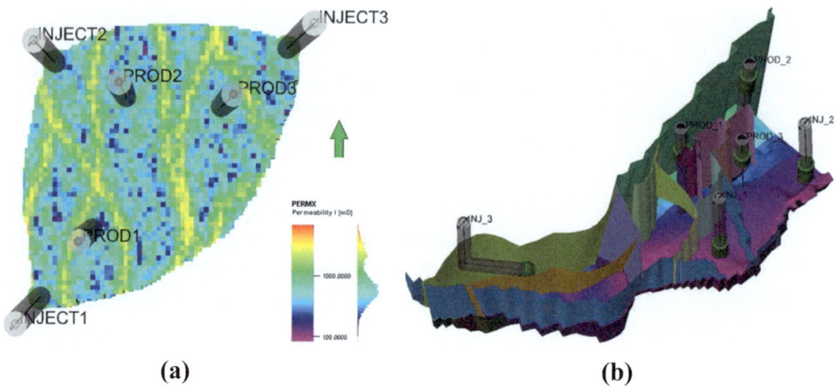

Fig. 46 Model illustration and well locations for **a** the Egg Model and **b** the Gullfaks Model (Matthew et al. 2023)

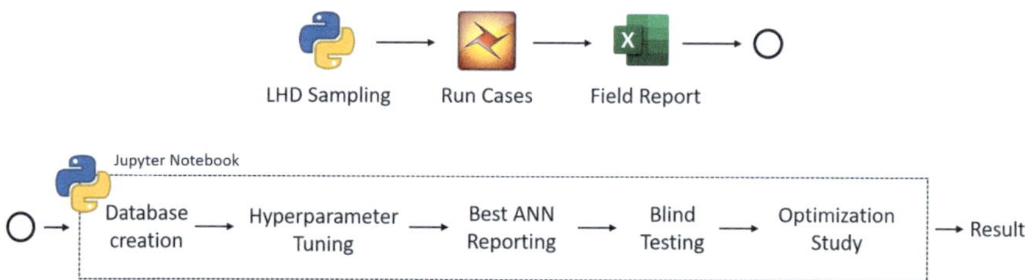

Fig. 47 Workflow of proxy building and optimization study (Matthew et al. 2023)

to improve the proxy building included mimicking the first-year behavior which was unique (proxies were segmented based on injection phase), and:

- For the Egg model: the proxy was then split into 1st year, 2nd–5th year, and 6th–10th year.
- For the Gullfaks model: the proxy was then split into 1st year, 1st–2.5th year, 2.5th–5th year, 5th–7.5th year, and 7.5th–10th year. Also, segmentation based on low half-cycle lengths (3 and 6 months) and high half-cycle lengths (9 and 12 months) were proposed.

The optimized parameters to find the best topology of the ANN for proxy segmentations were the learning rate, number of hidden layers, and number of nodes of each layer with an epoch of 100. Backpropagation was used to minimize the loss function (the difference between the predicted and the simulated values). The optimum hyperparameters were then retrained with 1500 epochs to obtain the best performance. The detailed hyperparameters are listed in Table 29.

Table 30 shows the results of proxy models during training and validation which are promising, except for the FCO_2PR, likely due to the error obtained at early times (first year).

Table 29 Hyperparameters in ANN design (Matthew et al. 2023)

Parameter	Value
Learning function	Adam
Learning rate	10^{-6}–10^{-2}
Number of hidden layers	1–3
Number of nodes	2–20
Activation function	ReLU
Batch size	64
Dropout	–
Training ratio	75
Validation ratio	25
Loss function	MSE
Epochs	100, then 1500

Table 30 APRE of each model for training and validation phases (Matthew et al. 2023)

Parameter	APRE, Egg Model (%)			APRE, Gullfaks Model (%)		
	Min	Average	Max	Min	Average	Max
FOPR	−2.307	−0.129	3.227	−2.364	0.042	1.443
FCO_2PR	−9.882	0.842	10.962	−2.871	1.270	11.166
FOPT	−2.435	−0.268	2.449	−2.533	−0.125	1.429
FCO_2PT	−3.252	1.052	6.114	−6.048	−0.909	2.778

For illustration, the proxy and simulator results (selected randomly) are compared in Fig. 48 for the Gullfaks model.

After training, 12 runs were used for blind validation for which the results are shown in Table 31.

A random blind test is shown in Fig. 49 for the Gullfaks Model.

The blind-validated proxy was then applied in the optimization study using the NSGA-II algorithm. Forty Pareto solutions from the Egg model proxy and 32 Pareto solutions from the Gullfaks model proxy were obtained (Fig. 50). We selected the solution closest to the average values of both functions.

The difference between the proxy and the simulator results (using the selected Pareto optimum) is shown in Table 32 for the selected cases.

The computational time needed to sample, build, and run the proxy models is shown in Table 33. The results of the optimization process are shown in Table 34.

In this study, the proxy models produced accurate results that were further improved by sampling and segmenting the behavior of the reservoir model (depending on the complexity of the reservoir model). The average error was less than 2% (compared with simulation results).

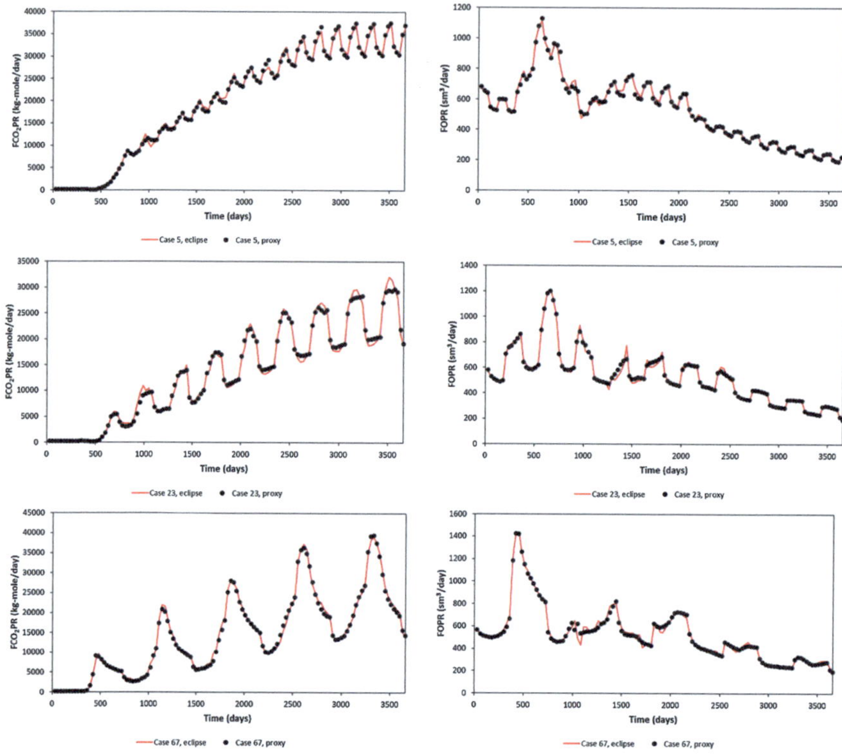

Fig. 48 Proxy results and comparison with the simulation output, for different injection patterns (3 months, 6 months, and 12 months) (Matthew et al. 2023)

Table 31 APRE of each model for the blind test phase (Matthew et al. 2023)

Parameter	APRE, Egg Model (%)			APRE, Gullfaks Model (%)		
	Min	Average	Max	Min	Average	Max
FOPR	−1.078	0.646	2.85	−0.7	0.511	3.93
FCO_2PR	−3.538	2.944	12.433	−1.823	2.006	10.055
FOPT	−1.575	0.168	2.162	−0.728	0.194	3.405
FCO_2PT	−1.783	1.74	5.81	−2.191	−0.071	5.612

Smeaheia CO_2 Storage (Fig. 51, composed of three prospects: Alpha, Beta, and Gamma) stores CO_2 in an aquifer hosted by the Horda Platform, east of the Troll Field (the North Sea) as part of the Northern Lights' full-scale CCS project (Equinor and Gassnova 2021).

50-year numerical simulations of the Smeaheia black-oil model take more than 10 h; so, optimization studies for a safe and efficient CO_2 storage design lead to high computational costs. We implemented two DNNs to construct grid-based proxy

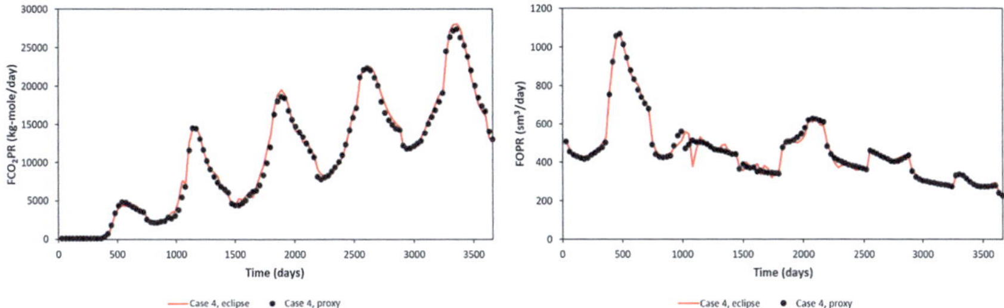

Fig. 49 Proxy results and comparison with the simulation output (Gullfaks Model), blind test. (Matthew et al. 2023)

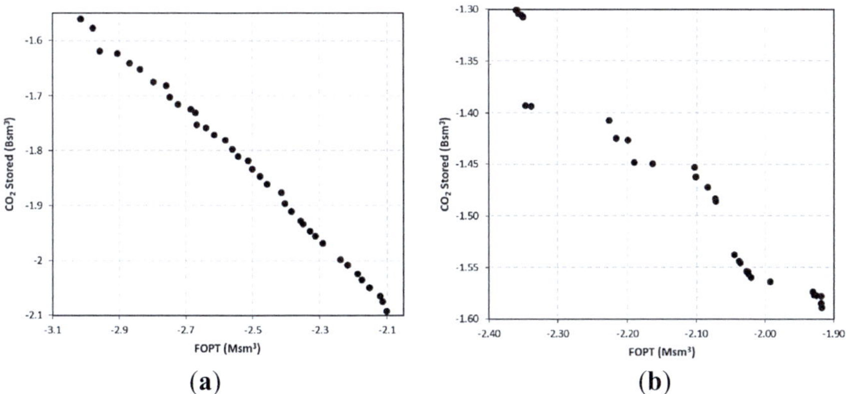

(a)   **(b)**

Fig. 50 Pareto front for **a** the Egg model and **b** the Gullfaks model (Matthew et al. 2023)

Table 32 Error analysis (between the proxy model and the simulation results) (Matthew et al. 2023)

Parameter	Egg Model			Gullfaks Model		
	PETREL	Proxy	Relative error (%)	PETREL	Proxy	Relative error (%)
FOPR *	–	–	0.26	–	–	1.04
FOPT (Msm^3)	2.49	2.51	1.05	2.08	2.1	1.19
FCO_2PR *	–	–	1.23	–	–	0.45
FCO_2PT (Msm^3)	78.84	79.43	0.75	95.91	94.83	−1.12

* The relative error shows the average error between the simulation and proxy results for each timestep. The values of FOPR and FCO_2PR are not specified as they comprise a set of data points

models with a non-cascading approach (Amiri 2022). The study used an updated numerical model (with over 1.5 million cells with 200 m × 200 m lateral resolution and 100 layers with 1.6 m thickness) and formulation of a viable injection strategy for Smeaheia CO_2 Storage. The proxy models were developed to reproduce simulator's

Table 33 Proxy development statistics (Matthew et al. 2023)

Process		Egg	Gullfaks
Sampling	LHS	68	97
	Sampling time	9 h 43 m	12 h 35 m
	Run time per case	9 m 25 s	8 m 45 s
	Memory space per case	97 MB	85 MB
Proxy development	FOPR	2 ANNs	8 ANNs
	FCO_2PR	5 ANNs	9 ANNs
	Total time needed	1 h 23 m	2 h 49 m
Proxy performance	Run time	8.75 s	9.16 s
	Memory space	14 KB	13 KB

Table 34 Optimization study statistics (Matthew et al. 2023)

		Egg	Gullfaks
Pareto optima		40	32
Time needed		2 h 25 m	2 h 23 m
Runs needed		1030	1030
Eclipse equivalent time		6 d 17 h 39 m	6 d 5 h 38 m
Results	Half-cycle (months)	3–6	3–9
	Gas rate (Msm^3/d)	1.98–2	1.95–2
	Water rate (sm^3/d)	3000–9000	3000–9000

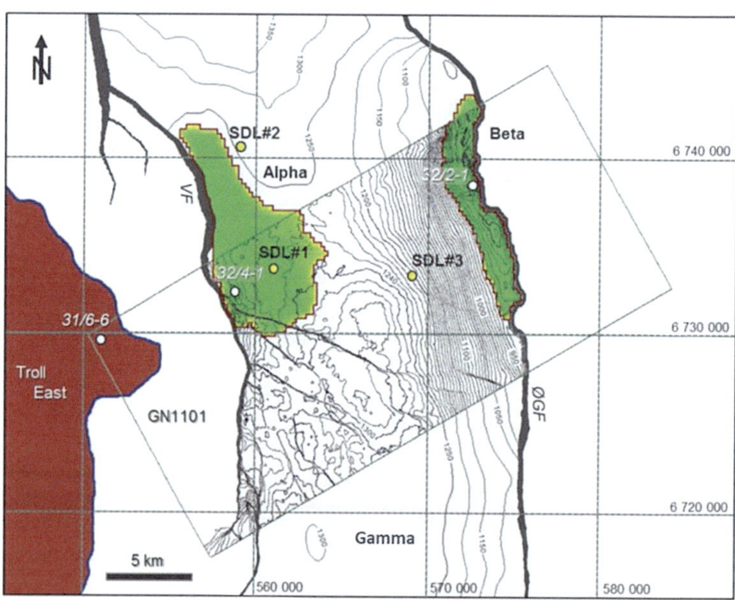

Fig. 51 Smeaheia's prospects location (Equinor and Gassnova 2021)

dynamic distribution of CO_2 saturation and pressure across all grid blocks, and to substitute numerical models for long-term simulation and optimization studies. Proxy models demonstrated a significant runtime improvement (about 300 times faster) and an accuracy of 99% compared to the simulated data. Genetic Algorithm (GA) was employed to optimize the injection and storage of CO_2, constrained by a maximum caprock fracture pressure (maximum pressure of 150 bar at the top of zone Alpha) and the CO_2 plume migration path, to avoid any leakage through caprock fracturing and available faults. The optimal rate and period of injection in zone Alpha were determined for a safe storage preventing the CO_2 plume from reaching zone Beta due to leakage risk.

For data generation, seven rates between 1.3 Mt/year and 5.5 Mt/year were selected, with a 6-month time step for a 50-year simulation study. Using a non-cascading approach, the input features for training the ANN were selected as cell index (i, j, k), cell coordinate (X, Y, Z), porosity, permeability, distance to injection well and production wells, injection rate, saturation at 0 time-step, pressure at 0 time-step, tier model at 0 time-step (grid blocks surrounding individual cells in the current layer), time step. Table 35 represents the performance of CO_2 saturation and pressure models, showing the successful training of the models. The best model structure had 3–5 hidden layers activated by the ReLU function in the hidden layers and Sigmoid for the output layer.

Table 36 shows the blind validation performance at the 100th time step for a CO_2 injection time frame of 50 years with a rate of 3.45925 million Sm^3/day.

Optimization study indicated the injection rate of 4.683495 million Sm^3/day for 50 years could store over 160 Mt CO_2 safely. A numerical simulation case was run with the optimum CO_2 injection rate and time (Fig. 52), which demonstrated the CO_2 saturation distribution in the year 2100 without any evidence of CO_2 migration toward zone Beta. This case was used to verify the optimization accuracy in Fig. 53. Predicted pressure and CO_2 saturation (by proxy and numerical model) were compared after 50

Table 35 Proxy models accuracy evaluation. Data extracted from Amiri (2022)

	Training	Validation	Testing
CO_2 saturation model			
MSE	7.91E−05	8.14E−05	8.13E−05
MAE	4.2E−03	4.2E−03	4.2E−03
Pressure model			
MSE	2.65E−07	2.66E−07	2.64E−07
MAE	3.53E−04	3.54E−04	3.53E−04

Table 36 Blind validation accuracy results (for the 100th time step). Data extracted from Amiri (2022)

MAE of the CO_2 saturation model	3.32E-04
MAE of the pressure model	8.44E-02

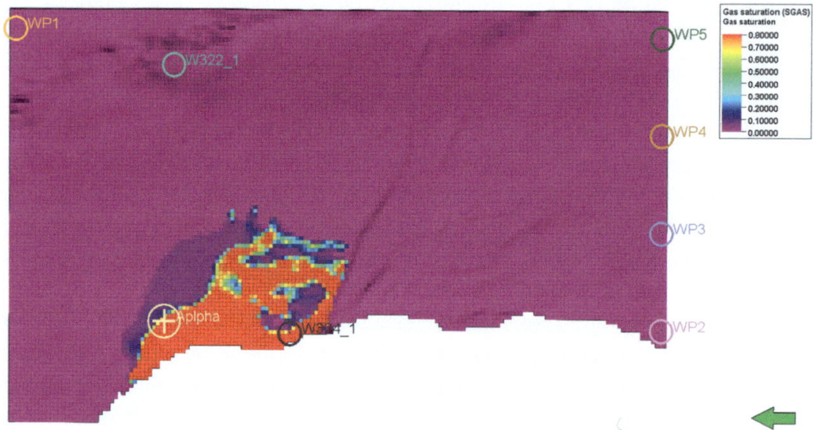

Fig. 52 CO_2 saturation distribution in the year 2100 (28 years after stopping the 50 years injection at the rate of 4.683495 million Sm^3/day) (Amiri 2022)

years of injection. Results demonstrated the model's accuracy, and it was concluded that the non-cascading grid-based proxy modeling is appropriate for long-term proxy modeling and optimization. It could simulate each dynamic parameter in only 30 s.

Development of proxy models for CO_2 injection and storage in Svelvik CO_2 Field Lab (a small-scale field laboratory, about 50 km southwest of Oslo) was studied (Wiranda 2022; Wiranda et al. 2023). The field lab consists of an injection well (for water and CO_2 injection at a depth of 64–65 m) surrounded by four monitoring wells covering an area of approximately 300 m $\times$ 150 m (Grimstad et al. 2020) (Fig. 54).

An injection campaign was performed in 2019, and the data gathered consisted of cross-well geophysical monitoring, injection rates, and pressure for the injection well and the monitoring wells. A numerical model was developed, history matched, and simulations were performed, after modifying the mud layer (50.7m–61.2m) to a non-continuous mud layer that contains sand lobes to explain the observed CO_2 migration to the upper layers (~38m depth). (Fig. 55).

A fine-grid simulation model with a grid size of $1 \times 1 \times 0.5$ m was used to capture the CO_2 movement, which took about seven hours to simulate the injection. Proxy models were then developed to reduce the simulation time (to seconds) and to further study the design of the next injection campaign. This was done using a prediction scenario of 1-week injection and 1-week shut in. Eighteen cases with different CO_2 injection rates were simulated and used for training and validation. The proxy models used were response surface (linear and bilinear) and universal kriging proxies with the first and the second cycle CO_2 injection rates as input. The outputs of the proxies were CO_2 dissolved in water, CO_2 ingas phase, injector bottom hole pressure, and average field pressure. The proxies reproduced the numerical simulation results with very good accuracy as shown in Figs. 56 and 57. The results were promising for both the training and validation cases with R^2 greater than 0.99 for all cases. The

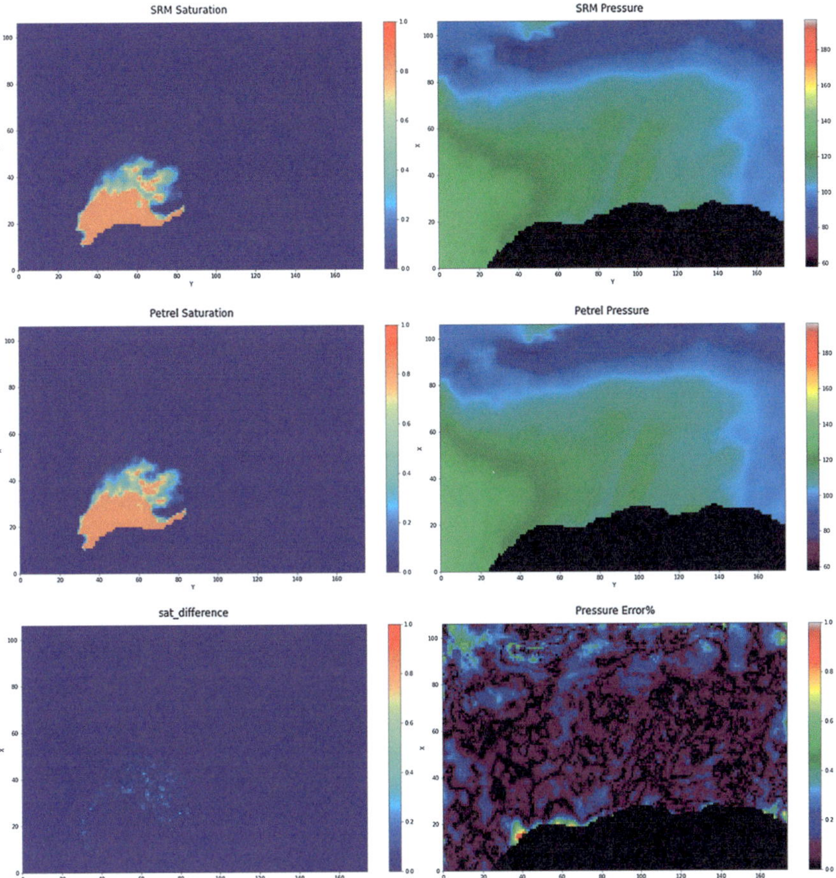

Fig. 53 Pressure and CO_2 saturation distribution in the first layer and error maps after 50 years of injection with the rate of 4.683495 million Sm^3/day (optimum case) (Amiri 2022)

universal kriging proxy (minimizing the error through the Gaussian process) showed better accuracy than the response surface proxy.

Diffusivity coefficient of CO_2 in brine is an important parameter in the design and monitoring of CO_2 storage projects. We established two explicit correlations for predicting this parameter under various operating conditions (Nait Amar and Jahanbani Ghahfarokhi 2020). To this end, two machine learning techniques, group method of data handling (GMDH) and gene expression programming (GEP) were implemented with pressure, temperature, and viscosity of the solvent as input parameters, using a comprehensive experimental database of 92 data points (Table 37). Dependency of CO_2 diffusivity on salinity was emulated by considering the solvent viscosity while establishing the correlations and the paradigms.

After rearrangements, the best correlation based on GEP was expressed as

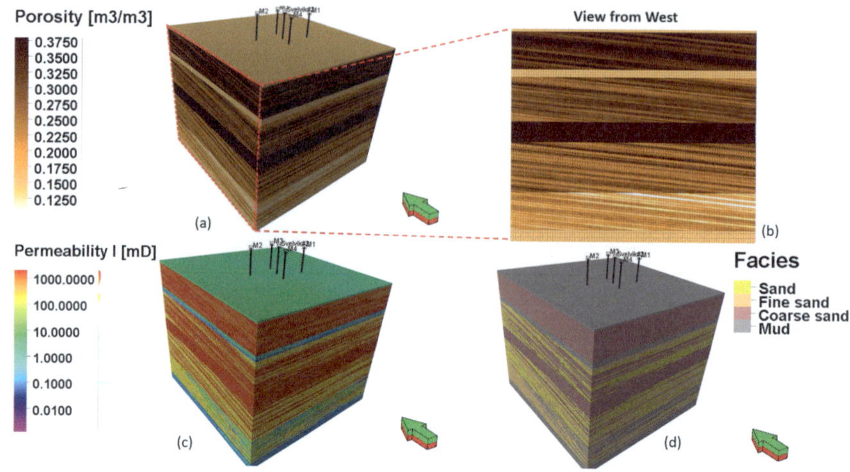

Fig. 54 Svelvik geological model **a**, **b** porosity; **c** permeability; **d** facies (Wiranda 2022)

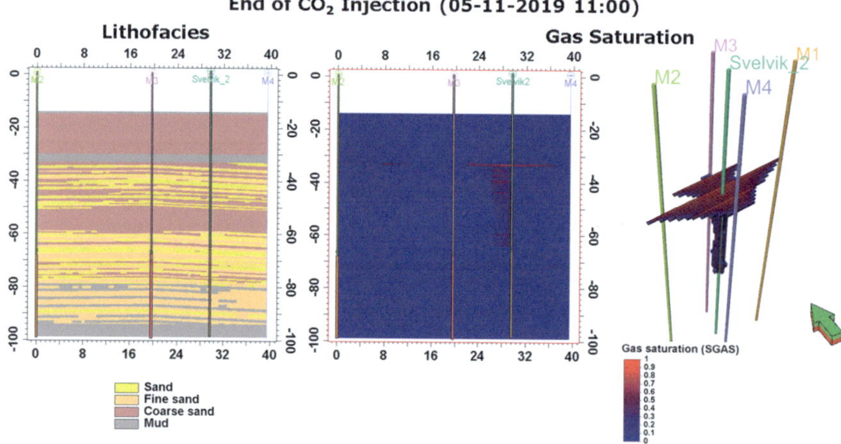

Fig. 55 CO₂ plume at the end of the injection campaign (Wiranda 2022)

$$D_c = 10^{-9} \times \left[\frac{A_1}{\mu} + \frac{A_2}{\sqrt{\mu}} + A_3 \times \sqrt{T} + A_4 \times P + A_5 \right] \qquad (5)$$

The terms A_1, A_2, A_3, A_4, and A_5 were defined as

$$A_1 = -0.0001564 \times P^3 + 0.01113 \times P^2 + 0.02935 \times P - 2.83 \times \sqrt{P} + 8.362$$

$$A_2 = 0.02426 \times (P + T) - 2.466 \times \sqrt{P}$$

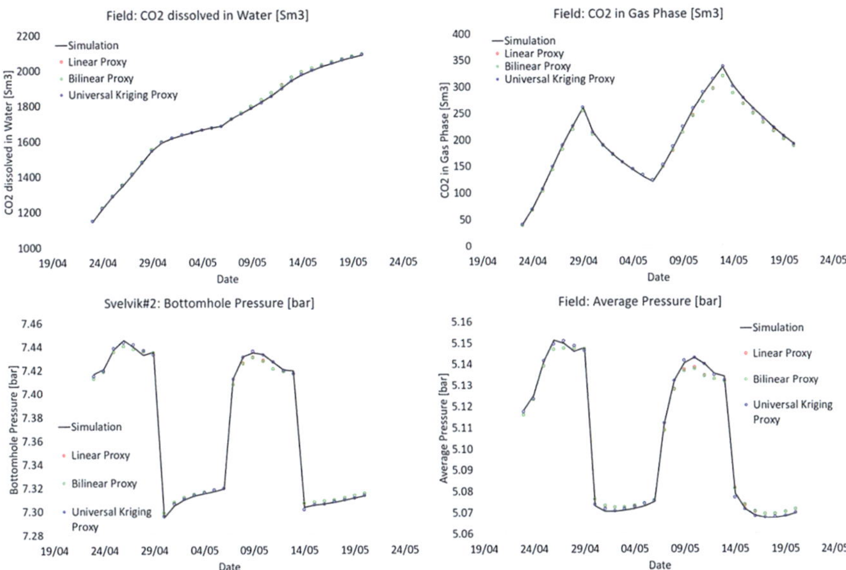

Fig. 56 Simulation and proxy results (Wiranda 2022)

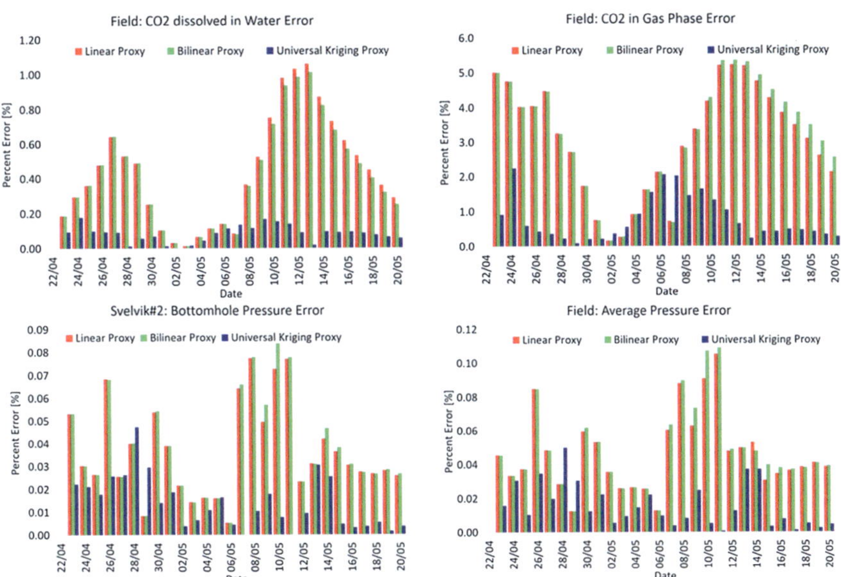

Fig. 57 Proxy models accuracy evaluation (Wiranda 2022)

Table 37 Summary of the gathered data (Nait Amar and Jahanbani Ghahfarokhi 2020)

	Max	Avg	Min	SD
P (MPa)	49.30	9.64	0.1000	14.8030
T (K)	473.15	317.76	273	39.8
Viscosity (mPa s)	1.9500	0.9003	0.1390	0.4720
Diffusivity coefficient ($\times 10^{-9}$ m^2/s)	16.1000	3.3522	0.3100	3.0874

$$A_3 = 0.4583 + 0.123 \times P$$

$$A_4 = 6.832 \times 10^{-5} \times P^2 + 0.003955 \times P + 19.34 \times \frac{1}{\sqrt{P}} - 3.801$$

$$A_5 = \frac{30.95 \times \ln(P)}{5.629 \times (P - \mu)} - \frac{0.0006024 \times T \times \sqrt{\mu}}{P} - 4.259 \times \ln\left(P^2 \times T\right) - 7.945$$

Figure 58 illustrates the cross-plots of the proposed correlations. The two cross-plots demonstrate promising agreement between the correlations and real data.

Table 38 further evaluates the correlation's performance. The proposed correlations showed promising results with an overall AARD of 8.0404% and 4.3015% for GMDH and GEP, respectively. Besides, Decision Trees (DTs) and Random Forest (RF) were considered for comparison with the best correlation. Analysis revealed that GEP outperformed GMDH, DTs, RFs, and the previous predictive models, with an overall coefficient of determination (R^2) of 0.9979.

The developed models were then compared with an intelligent paradigm based on hybrid genetic algorithm and mixed Kernels-based support vector machine (Feng et al. 2019) which outperformed the available empirical models for predicting the diffusivity coefficient with a global predictive AARD of 7.91%. The comparison revealed the superiority of the newly proposed GEP correlation, as it outperformed both the prior intelligent model and the empirical paradigms.

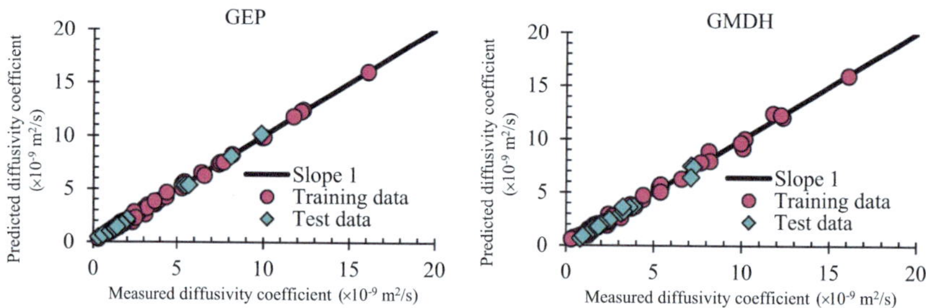

Fig. 58 Cross-plots of the established GEP and GMDH correlations for diffusivity coefficient (Nait Amar and Jahanbani Ghahfarokhi 2020)

Table 38 Performance analysis of the implemented models. Data extracted from Nait Amar and Jahanbani Ghahfarokhi (2020)

		GEP	GMDH	DTs	RF
Training data	AARD (%)	3.8584	8.6269	4.2969	6.3627
	R^2	0.9980	0.9943	0.9980	0.9973
	RMSE ($\times 10^{-9}$ m^2/s)	0.1427	0.2479	0.1598	0.1647
Test data	AARD (%)	6.0035	5.6292	8.8426	9.0015
	R^2	0.9978	0.9874	0.9924	0.9940
	RMSE ($\times 10^{-9}$ m^2/s)	0.1245	0.2271	0.2532	0.2764
All data	AARD (%)	4.3014	8.0404	5.1862	6.8790
	R^2	0.9979	0.9937	0.9969	0.9966
	RMSE ($\times 10^{-9}$ m^2/s)	0.1391	0.2440	0.1785	0.1870

The trends of the GEP outputs were logical with respect to the independent variables. A trend analysis is depicted in Fig. 59 by presenting the real measurements and the predictions of GEP as function of pressure for the whole database. As shown, there is overlap regardless of the operating conditions. Temperature was found to be the most influencing parameter in the prediction of diffusivity coefficient by GEP correlation.

Thermal conductivity of CO_2 is another important thermophysical parameter to be included in heat transfer modeling and design. We used soft computing approaches

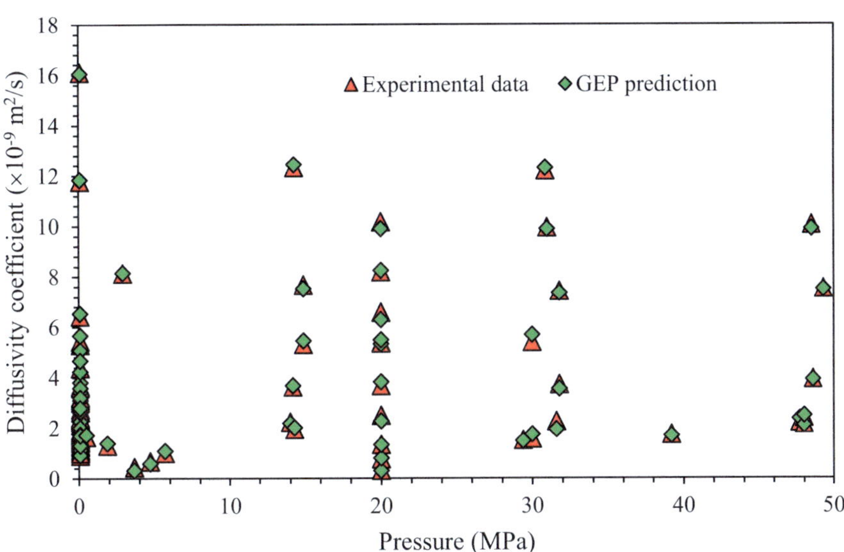

Fig. 59 Comparison of the diffusivity coefficient from measurements and generated by GEP correlation (Nait Amar and Jahanbani Ghahfarokhi 2020)

to develop new models for predicting the CO_2 thermal conductivity (Nait Amar et al. 2020b) using a massive database including 5890 experimental points covering a wide range of operational conditions (pressure: 0.097–209.763 MPa and temperature: 217.931–961.05 K). A detailed description of the collected data is reported in Table 39. Temperature and pressure were the input parameters for establishing the models for the thermal conductivity of CO_2.

Models were developed using multilayer perceptron (MLP) optimized by Levenberg–Marquardt Algorithm (LMA), Bayesian Regularization (BR), Resilient Backpropagation (RB), and Scaled Conjugate gradient (SCG), and also radial basis function neural network (RBFNN) coupled with PSO.

The evaluation of the ANN models including MLP optimized with LMA, BR, SCG, and RB, and RBFNN-PSO is shown in Table 40. The two best models, i.e., MLP-LMA and MLP-BR were then linked under committee machine intelligent systems (CMIS) using weighted averaging and group method of data handling (GMDH). The results are depicted in the table.

Figure 60 illustrates the structure of the proposed CMIS-GMDH. Furthermore, the mathematical formulation of the CMIS is shown below:

$$\lambda_{CO_2} = -0.016363 + 2.187283 \times N - 1.186711 \times M_1 - 0.817671 \times M_1$$
$$\times N + 0.396172 \times N^2 + 0.421492 \times M_1^2 \tag{6}$$

where M_1 refers to the MLP-LMA model and N is expressed as follows:

$$N = 0.015045 + 0.443216 \times M_2 + 0.5565212 \times M_1 - 0.029334 \times M_1 \times M_2$$
$$+ 0.013144 \times M_2^2 + 0.016189 \times M_1^2 \tag{7}$$

where M_2 represents the MLP-BR model.

The results showed that CMIS-GMDH was the most accurate model with an overall AARD% value of 0.8379% and a total coefficient of determination (R^2) of 0.9997. Figure 61 illustrates the cross-plot of CMIS-GMDH for the training and testing sets. The predictions for both training and testing showed the approach is accurate.

Comparisons with prior correlations showed that the CMIS-GMDH could outperform all the existing models for predicting CO_2 thermal conductivity. A trend analysis

Table 39 Summary of the gathered data (Nait Amar et al. 2020b)

Data	Max	Avg	Min	Standard deviation (SD)
T (K)	961.0500	421.5979	217.9310	150.1531
P (MPa)	209.7630	20.0119	0.0970	22.9104
Thermal conductivity (mW/ m.K)	198.7900	59.8571	10.8200	41.9814

Table 40 Performance evaluation of the implemented models (Nait Amar et al. 2020b)

		AARD (%)	R^2	RMSE	SD
Training data	MLP-LMA	1.2213	0.9991	1.0608	0.0003
	MLP-BR	1.2845	0.9987	1.2495	0.0004
	MLP-SCG	2.3963	0.9897	3.4930	0.0018
	MLP-RB	2.7941	0.9915	3.1677	0.0021
	RBF-PSO	2.6195	0.9896	3.5534	0.0019
	CMIS-LINEAR	1.2347	0.9986	1.1779	0.0003
	CMIS-GMDH	0.8372	0.9997	0.6966	0.0002
Test data	MLP-LMA	1.5759	0.9987	1.2025	0.0005
	MLP-BR	1.4581	0.9979	1.5636	0.0004
	MLP-SCG	2.7468	0.9918	3.2603	0.0023
	MLP-RB	2.8965	0.9789	5.2664	0.0026
	RBF-PSO	2.9461	0.9881	3.7283	0.0025
	CMIS-LINEAR	1.1537	0.9881	0.9127	0.0002
	CMIS-GMDH	0.8407	0.9997	0.6836	0.0002
All data	MLP-LMA	1.2922	0.9990	1.0906	0.0004
	MLP-BR	1.3193	0.9986	1.3183	0.0004
	MLP-SCG	2.4664	0.9900	3.4477	0.0019
	MLP-RB	2.8146	0.9887	3.6843	0.0022
	RBF-PSO	2.6848	0.9893	3.5891	0.0021
	CMIS-LINEAR	1.2185	0.9991	1.1248	0.0003
	CMIS-GMDH	0.8379	0.9997	0.6941	0.0002

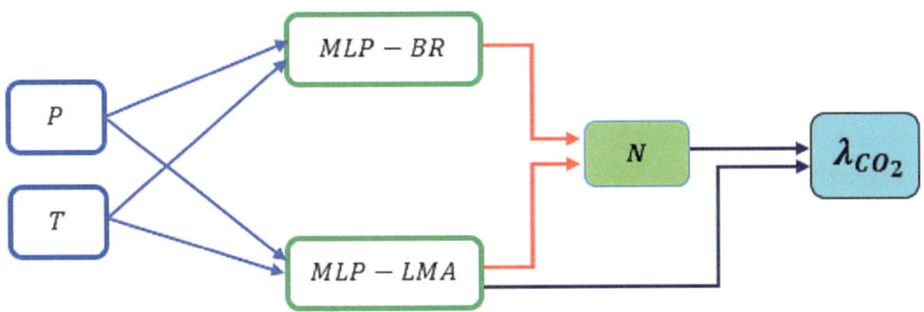

Fig. 60 CMIS-GMDH model proposed (Nait Amar et al. 2020b)

of the outcomes of the CMIS-GMDH with respect to pressure and temperature also demonstrated an excellent fit with the experimental data under different conditions.

Hydrogen is used for clean energy production for energy transition. Underground hydrogen storage (UHS) is a challenging task in the context of conserving energy for later use. The interfacial tension (IFT) of the H_2–brine system is an important parameter in the successful design and implementation of UHS. We implemented machine

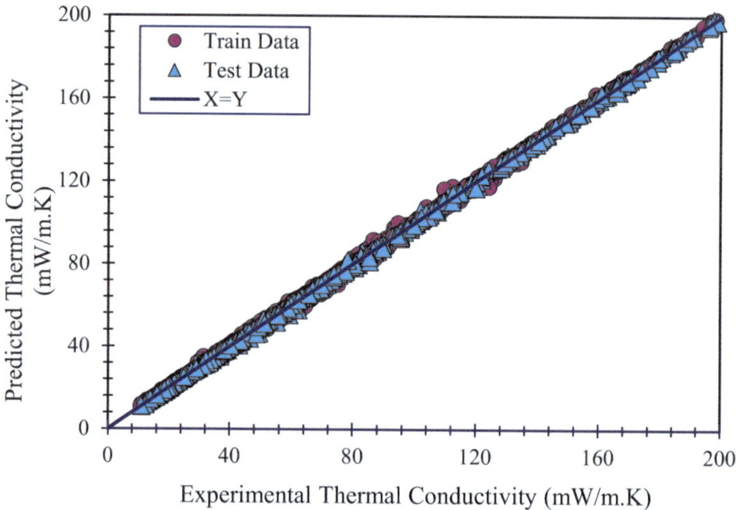

Fig. 61 Cross-plot of the CMIS-GMDH (Nait Amar et al. 2020b)

learning (ML) techniques (Ng et al. 2022a), such as genetic programming (GP), gradient boosting regressor (GBR), and multilayer perceptron (MLP) optimized with LM and Adam algorithms to establish ML-based models to predict the IFT of the H_2–brine system using molality, pressure, and temperature as input. The models were trained and tested using experimental IFT data consisting of 107 measurements, illustrated in Table 41.

Table 42 evaluates the performance and Fig. 62 illustrates the cross-plots of the developed models. The proposed models provided excellent estimations of the IFT. MLP-LMA outperformed the other models in the predictions (using all the data points) with R^2 and AAPRE values of 0.9997 and 0.1907%, respectively.

To determine the proper architecture of the models, trial and error approach was implemented. For MLP-LMA, the best topology consisted of a network with one hidden layer of 12 neurons, while the suitable activation functions were Tansig and pure line for the hidden and output layers of the MLP-LMA model, respectively.

The trend analysis of the MLP-LMA (best-performing model) demonstrated the physical coherence of the predictions. At four different molality values, the trend of IFT values was captured well by the MLP-LMA by changing the temperature and

Table 41 Summary of the database (Ng et al. 2022a)

	Molality, m (mol kg^{-1})	Pressure, P (MPa)	Temperature, T (K)	IFT (mN.m^{-1})
Minimum	0.00	0.50	298.03	42.90
Maximum	4.95	45.20	448.35	80.77
Average	1.37	17.87	353.94	63.21

Table 42 Performance evaluation of the implemented paradigms (Ng et al. 2022a)

		AAPRE	RMSE	R^2
MLP-LMA	Training	0.1557	0.1593	0.9998
	Testing	0.3344	0.2625	0.9995
	Total	0.1907	0.1842	0.9997
MLP-Adam	Training	0.3353	0.3116	0.9990
	Testing	1.0622	0.9509	0.9935
	Total	0.4848	0.5129	0.9979
GBR	Training	0.0940	0.0722	1.0000
	Testing	0.6559	0.4917	0.9979
	Total	0.2095	0.2320	0.9996
GP	Training	0.7862	0.5295	0.9964
	Testing	0.6353	0.2137	0.9975
	Total	0.7560	0.4663	0.9966

pressure. It was deduced that temperature was the most influencing parameter for the IFT. The MLP-LMA was also compared with an empirical correlation (Hosseini et al. 2022) and outperformed it.

Accurate prediction of wax deposition is important to avoid issues with the flow assurance during hydrocarbon production. Important parameters related to wax deposition (viewed as Solid–Liquid Equilibrium descriptors) are (Benamara et al. 2020): (1) wax appearance temperature (WAT) which is defined as the threshold temperature below which paraffin crystals are formed; (2) wax disappearance temperature (WDT) which means the temperature at which the last deposited paraffin is removed; and (3) the amount of deposited wax also known as the weight percent of deposited wax.

We established intelligent models to estimate the weight percent of the deposited wax under different production conditions (Nait Amar et al. 2022), where the input parameters included temperature, pressure, specific gravity, and compositions. To this end, two multilayer perceptron (MLP) models were trained and optimized with Levenberg–Marquardt algorithm (MLP-LMA) and Bayesian Regularization algorithm (MLP-BR) using a comprehensive database of 88 experimental measurements.

The MLP models were trained with three hidden layers, and Tansig as the activation function. The number of neurons in each hidden layer was 12, 11, and 9 for the first, second, and third hidden layers, respectively.

Figure 63 shows cross-plots of the two developed MLP models. Tables 43 and 44 show the performance evaluation of these models. The resulting models showed very good performance, where MLP-LMA exhibited the best performance with an overall root mean square error of 0.2198 and a coefficient of determination (R^2) of 0.9971. It also outperformed the prior approach developed by Kamari et al. (2013).

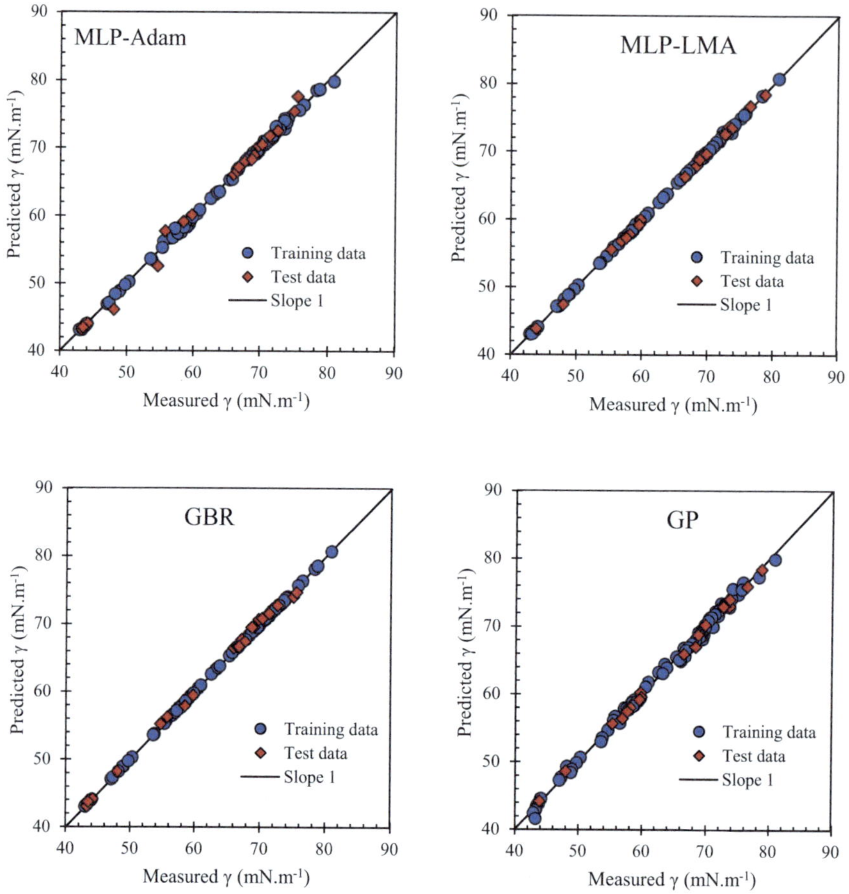

                                                                                    A. Jahanbani Ghahfarokhi

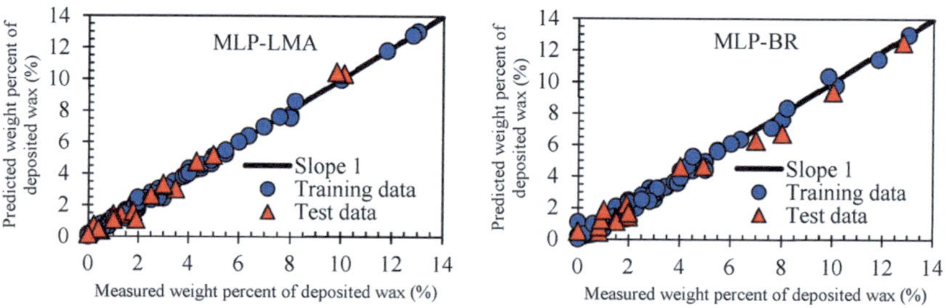

Fig. 62 Cross-plots of the implemented models for predicting IFT of the H$_2$–brine system (Ng et al. 2022a)

Fig. 63 Cross-plots of the MLP models for predicting the weight percent of deposited wax (Nait Amar et al. 2022)

Table 43 Performance evaluation of the MLP models (Nait Amar et al. 2022)

		MLP-LMA	MLP-BR
Training	RMSE	0.1840	0.2967
	R^2	0.9982	0.9947
Test	RMSE	0.3589	0.5178
	R^2	0.9944	0.9914
All	RMSE	0.2198	0.3420
	R^2	0.9974	0.9940

Table 44 Performance comparison (Nait Amar et al. 2022)

	RMSE	R^2
MLP-LMA	0.2198	0.9974
Kamari et al. (2013)	0.32	0.989

A trend analysis was also performed (Fig. 64) to verify that the MLP-LMA model followed the trend of the weight percent of deposited wax (%) as a function of temperature. The result showed excellent correspondence with the measured values for different temperature values.

Interwell connectivities between injector-producer well pairs and production estimates are vital parameters in reservoir management. There are techniques to determine the interwell connectivity such as the Capacitance–Resistance Model (CRM) (Sayarpour et al. 2009). This method is based on time-dependent material balance where the reservoir is acting as a tank for which injection and production rates are considered as input and output, respectively. The original CRM requires injection/production rate and producer BHP to train the model. Interwell connectivity and time constant are considered two unknown parameters which can be estimated using nonlinear regression. The main advantages over numerical simulators are simplicity, speed, and the limited number of input data. The original model was developed for slightly compressible fluid systems assuming a linear productivity index. So, a modified CRM (referred to as M-CRM) for gas-oil systems was developed considering density changes with pressure and non-linear productivity index (Yousefi et al. 2021). We then applied the M-CRM (as a physical approach) and a statistical method using least square support vector machine (LSSVM) and multiple linear regression (MLR) to obtain the interwell connectivities and production rates in two immiscible gas injection cases. The workflow of the study is shown in Fig. 65.

Streamline simulation was used as the reference tool for validation of the results. Both LSSVM-MLR and M-CRM were applied to investigate interwell connectivity and rate calculations. Unknowns such as interwell connectivities, time constant, and productivity index were determined and then production rate was estimated in the M-CRM using GA. In the statistical approach, production rate was calculated by LSSVM and then using MLR, interwell connectivities were optimized by GA. Finally, the results were compared with streamline simulations and the validity of the methods was determined.

                                                                A. Jahanbani Ghahfarokhi

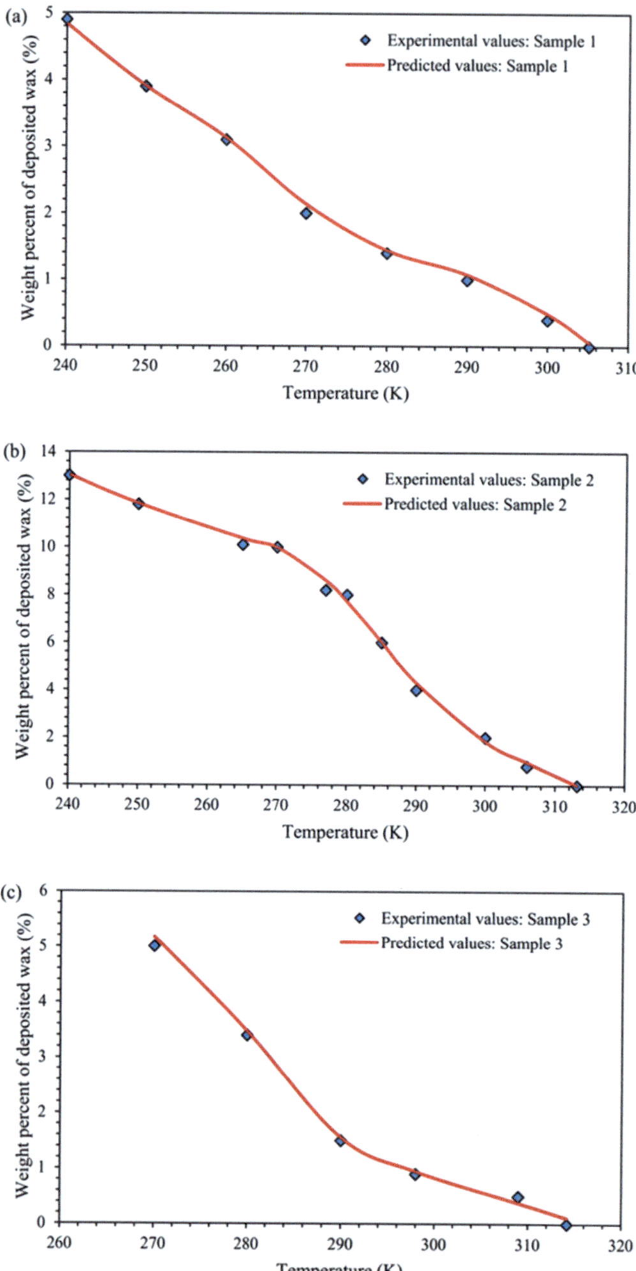

Fig. 64 Trend analysis (Nait Amar et al. 2022)

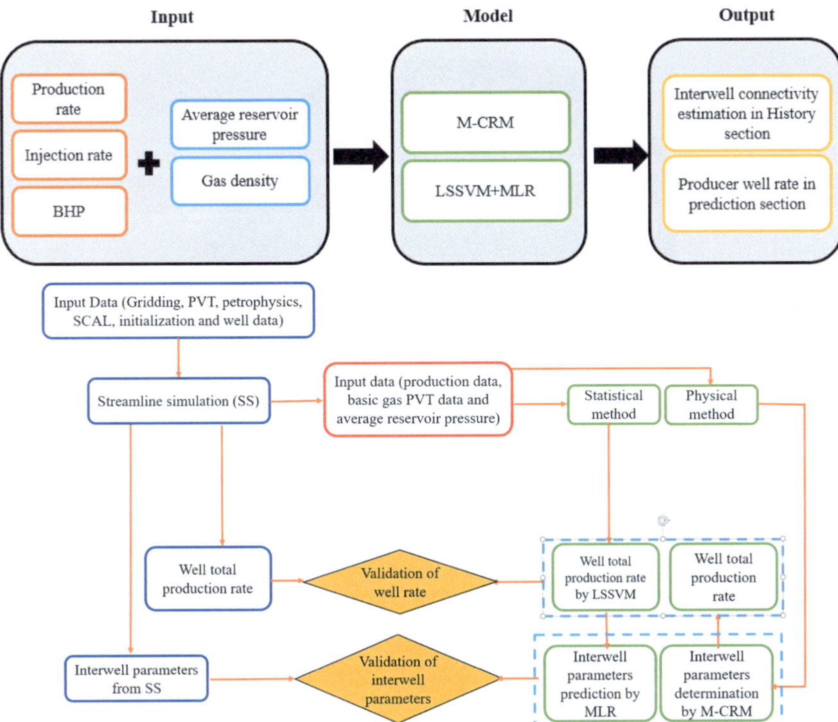

Fig. 65 The workflow used in the study (Yousefi et al. 2021)

The first case study was a simple case (a synthetic model) with one flow barrier and one high permeability streak, and the second case was a sector model of an oil reservoir with immiscible gas injection. Here, we only present the results for the second case. The 3D model and horizontal permeability distribution are illustrated in Fig. 66. Dynamic producer and injector data were imported using the oil rate control for producers and gas rate control for injectors. This analysis was performed using 181 data points of which 145 points were used for training and the remaining for testing.

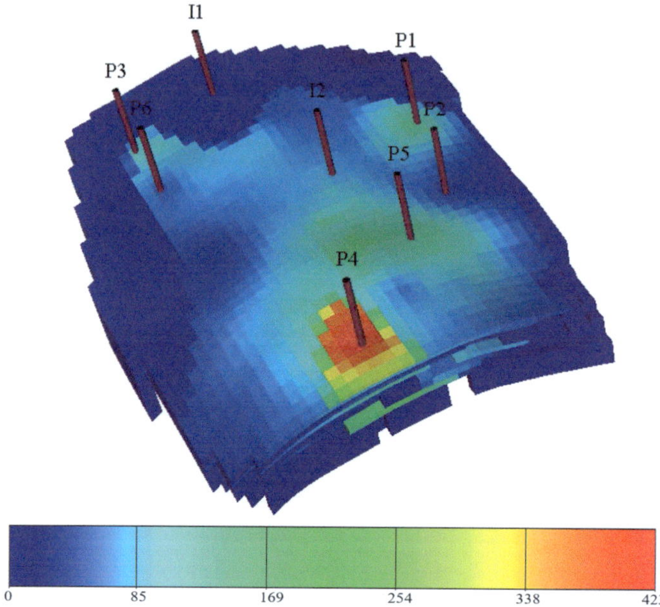

Fig. 66 3D geometry and horizontal permeability distribution of the sector model (Yousefi et al. 2021)

Figure 67 shows the cross-plots comparing production rates of LSSVM and M-CRM against streamline simulations during the testing phase (36 points).

Table 45 presents the interwell connectivities and production rates calculated by both methods. It was observed that the LSSVM results were aligned well with simulations for production rate predictions (1.19% relative error), while M-CRM also showed acceptable results with 5.20% relative error. M-CRM was more reliable for predictions of interwell connectivities compared to MLR (11.95% vs. 46.2% relative error). The results of the study showed that both methods were reliable in terms of validity, speed, and flexibility. The order of speed (time) in both M-CRM

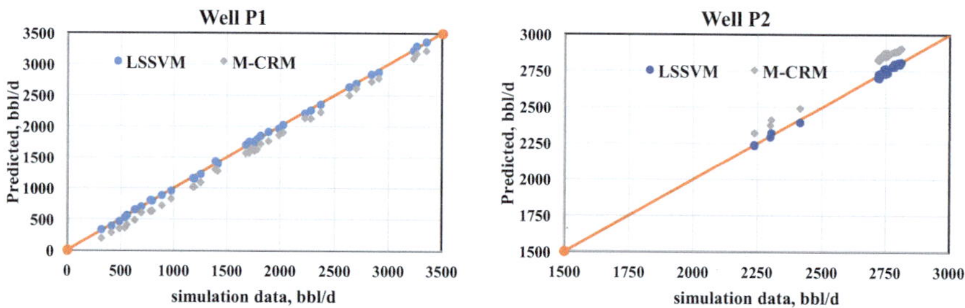

Fig. 67 Well production rates for LLSVM and M-CRM compared with streamline simulations in the testing phase (Yousefi et al. 2021)

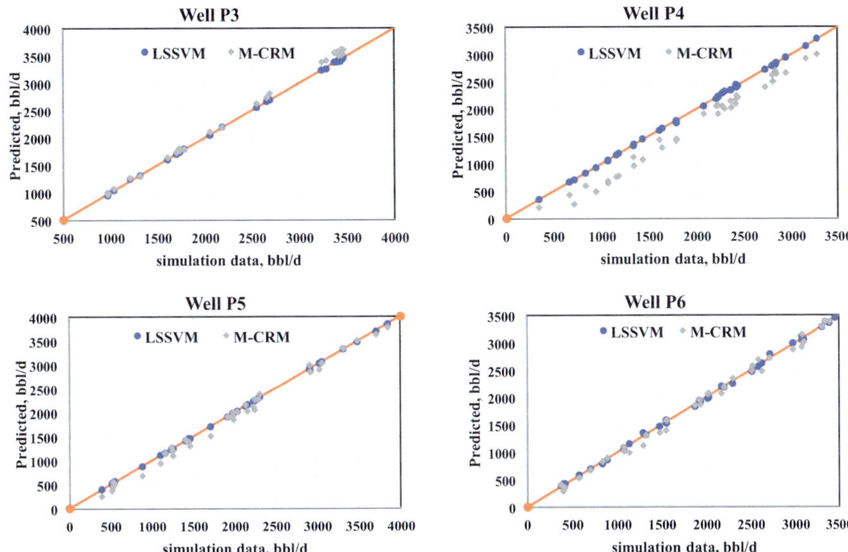

Fig. 67 (continued)

Table 45 Calculation results; Streamline vs. M-CRM and LSSVM (Yousefi et al. 2021)

	Well pair		Streamline well factors	M-CRM	Statistical	M-CRM		Statistical	
				Well factor	Well factor	MAPE (%)	CC	MAPE (%)	CC
Interwell connectivity	I1	P1	0.141	0.120	0.086	11.95	0.99	46.20	0.81
		P2	0.025	0.023	0.012				
		P3	0.548	0.554	0.407				
		P4	0.011	0.009	0.007				
		P5	0.004	0.000	0				
		P6	0.271	0.278	0.488				
	I2	P1	0.115	0.106	0.084				
		P2	0.362	0.353	0.215				
		P3	0.005	0.007	0.006				
		P4	0.201	0.177	0.161				
		P5	0.252	0.242	0.399				
		P6	0.065	0.070	0.136				
Total rate production (average for all producers)						5.20	0.99	1.19	1.00

and the statistical methods was a couple of minutes. However, the simulation study in the streamline approach took hours.

Acknowledgements This book chapter was written mainly based on a collection of over twenty research articles and the work done at the Department of Geoscience and Petroleum at NTNU, during the past 5 years in the area of proxy model development. The author would like to thank all the authors/co-authors of the studies discussed in this chapter. Appreciation for support also goes to:

- BRU21—NTNU Research and Innovation Program on Digital Automation Solutions for the Oil and Gas Industry (www.ntnu.edu/bru21).
- CEORS Gemini Centre at the Department of Geoscience and Petroleum at NTNU (a strategic collaboration between NTNU and SINTEF on CO_2 EOR and Storage, https://www.ntnu.edu/ceors).

References

Amiri B (2022) A fast and accurate investigation into CO_2 storage challenges by making a proxy model on a developed static model with the application of artificial intelligence/machine learning. Politecnico di Torino. http://webthesis.biblio.polito.it/id/eprint/23043. http://creativecommons.org/licenses/by-nc-nd/3.0/

Bahrami P, Sahari Moghaddam F, James LA (2022) A review of proxy modeling highlighting applications for reservoir engineering. Energies 15:5247. https://doi.org/10.3390/en15145247. https://creativecommons.org/licenses/by/4.0/

Benamara C, Gharbi K, Nait Amar M, Hamada B (2020) Prediction of wax appearance temperature using artificial intelligent techniques. Arab J Sci Eng 45:1319–1330

Bratvold RB, Bickel JE, Lohne HP (2009) Value of information in the oil and gas industry: past, present and future. SPE Reserv Eval Eng 12

Chaturvedi A (2021) Well control optimization by coupling smart proxy models with genetic algorithm. Norwegian University of Science and Technology. NTNU Open: https://hdl.handle.net/11250/2781628

Equinor (2018) Disclosing all Volve data [WWW Document]. https://www.equinor.com/en/news/14jun2018-disclosing-volve-data.html. Accessed 28 June 2021

Equinor and Gassnova (2021) Smeaheia dataset. https://co2datashare.org/dataset/smeaheia-dataset. https://co2datashare.org/view/license/26af9426-203f-4993-9d41-2e1bf191ceaf

Feng Q, Cui R, Wang S, Zhang J, Jiang Z (2019) Estimation of CO_2 diffusivity in brine by use of the genetic algorithm and mixed kernels-based support vector machine model. J Energy Resour Technol 141:41001

Fonseca RM, Rossa ED, Emerick AA, Hanea RG, Jansen JD (2020) Introduction to the special issue: overview of OLYMPUS optimization benchmark challenge. Comput Geosci 24

Grimstad A-A, Sundal A, Hagby KF, Ringstad C (2020) Modelling medium-depth CO_2 injection at the Svelvik CO_2 field laboratory in Norway. SSRN Electron J. https://doi.org/10.2139/ssrn.3365967

Hong A, Bratvold RB, Lake LW (2019) Fast analysis of optimal improved-oil-recovery switch time using a two-factor production model and least-squares Monte Carlo algorithm. SPE Reserv Eval Eng 22:1144–1160

Hosseini M, Fahimpour J, Ali M, Keshavarz A, Iglauer S (2022) H2−brine interfacial tension as a function of salinity, temperature, and pressure; implications for hydrogen geo-storage. J Pet Sci Eng 213:110441

Jahanbani Ghahfarokhi A, Chaturvedi A (2023) Well control optimization using smart proxy models, vol 2023. European Association of Geoscientists and Engineers, pp 1–5. https://doi.org/10.3997/2214-4609.202332027

Jahanbani Ghahfarokhi A, Ng CSW, Nait Amar M (2022) Artificial intelligence/machine learning for sustainable utilization of the subsurface. In: EAGE GET 2022, vol 2022. European

Association of Geoscientists and Engineers, pp 1–5. https://doi.org/10.3997/2214-4609.202 221116

Jansen JD et al (2014) The egg model - a geological ensemble for reservoir simulation. Geosci Data J 1:192–195

Kamari A, Khaksar-Manshad A, Gharagheizi F, Mohammadi AH, Ashoori S (2013) Robust model for the determination of wax deposition in oil systems. Ind Eng Chem Res 52:15664–15672

Krevor S et al (2023) Subsurface carbon dioxide and hydrogen storage for a sustainable energy future. Nat Rev Earth Environ 4:102–118

Lake LW, Lotfollahi M, Bryant SL (2019) CO_2 enhanced oil recovery experience and its messages for CO_2 storage. Sci Carbon Storage Deep Saline Form Process Coupling across Time Spat Scales 15–31. https://doi.org/10.1016/B978-0-12-812752-0.00002-2. Used with permission from Elsevier

Matthew DAM (2021) Proxy modeling for CO_2-EOR design study: water alternating gas and storage. Norwegian University of Science and Technology. NTNU Open: https://hdl.handle.net/ 11250/2786742

Matthew DAM, Jahanbani Ghahfarokhi A, Ng CSW, Nait Amar M (2023) Proxy model development for the optimization of water alternating CO_2 gas for enhanced oil recovery. Energies 16:1–19. https://doi.org/10.3390/en16083337. http://creativecommons.org/licenses/by/4.0/

Mohaghegh SD (2017) Data-driven reservoir modeling. In: Society of petroleum engineers, vol 53

Nait Amar M, Jahanbani Ghahfarokhi A (2020) Prediction of CO_2 diffusivity in brine using white-box machine learning. J Pet Sci Eng 190. https://doi.org/10.1016/j.petrol.2020.107037. Used with permission from Elsevier

Nait Amar M, Zeraibi N, Jahanbani Ghahfarokhi A (2020a) Applying hybrid support vector regression and genetic algorithm to water alternating CO_2 gas EOR. Greenh Gases Sci Technol. https:// doi.org/10.1002/ghg.1982. http://creativecommons.org/licenses/by/4.0/

Nait Amar M, Jahanbani Ghahfarokhi A, Zeraibi N (2020b) Predicting thermal conductivity of carbon dioxide using group of data-driven models. J Taiwan Inst Chem Eng. https://doi.org/10. 1016/j.jtice.2020.08.001. http://creativecommons.org/licenses/by/4.0/

Nait Amar M, Jahanbani Ghahfarokhi A, Ng CSW, Zeraibi N (2021) Optimization of WAG in real geological field using rigorous soft computing techniques and nature-inspired algorithms. J Pet Sci Eng. https://doi.org/10.1016/j.petrol.2021.109038. http://creativecommons.org/licenses/by/ 4.0/

Nait Amar M, Jahanbani Ghahfarokhi A, Ng CSW (2022) Predicting wax deposition using robust machine learning techniques. Petroleum 8:167–173. https://doi.org/10.1016/j.petlm. 2021.07.005. http://creativecommons.org/licenses/by-nc-nd/4.0/

Negahban S, Pedersen KS, Baisoni MA, Sah P, Azeem J (2010) An EoS model for a Middle East reservoir fluid with an extensive EOR PVT data material. In: Society of petroleum engineers - 14th Abu Dhabi international petroleum exhibition and conference 2010, ADIPEC 2010, vol 1, pp 289–303

Ng CSW, Jahanbani Ghahfarokhi A, Nait Amar M, Torsæter O (2021a) Smart proxy modeling of a fractured reservoir model for production optimization: implementation of metaheuristic algorithm and probabilistic application. Nat Resour Res 30:2431–2462. https://doi.org/10.1007/ s11053-021-09844-2. http://creativecommons.org/licenses/by/4.0/

Ng CSW, Jahanbani Ghahfarokhi A, Nait Amar M (2021b) Application of nature-inspired algorithms and artificial neural network in waterflooding well control optimization. J Pet Explor Prod Technol. https://doi.org/10.1007/s13202-021-01199-x. http://creativecommons. org/licenses/by/4.0/

Ng CSW, Jahanbani Ghahfarokhi A (2022a) Adaptive proxy-based robust production optimization with multilayer perceptron. Appl Comput Geosci 16:100103. https://doi.org/10.1016/j.acags. 2022.100103. http://creativecommons.org/licenses/by/4.0/

Ng CSW, Jahanbani Ghahfarokhi A (2022b) Fast well control optimization using machine learning based proxy models. In: EAGE conference on digital innovation for a sustainable future, vol

2022. European Association of Geoscientists and Engineers, pp 1–5. https://doi.org/10.3997/2214-4609.202272009

Ng CSW, Djema H, Nait Amar M, Jahanbani Ghahfarokhi A (2022a) Modeling interfacial tension of the hydrogen-brine system using robust machine learning techniques: Implication for underground hydrogen storage. Int J Hydrogen Energy 47:39595–39605. https://doi.org/10.1016/j.ijhydene.2022.09.120. http://creativecommons.org/licenses/by/4.0/

Ng CSW, Jahanbani Ghahfarokhi A, Nait Amar M (2022b) Well production forecast in Volve field: application of rigorous machine learning techniques and metaheuristic algorithm. J Pet Sci Eng 208. https://doi.org/10.1016/j.petrol.2021.109468. http://creativecommons.org/licenses/by/4.0/

Ng CSW, Nait Amar M, Jahanbani Ghahfarokhi A, Imsland LS (2023a) A survey on the application of machine learning and metaheuristic algorithms for intelligent proxy modeling in reservoir simulation. Comput Chem Eng 170:108107. https://doi.org/10.1016/j.compchemeng.2022.108107. http://creativecommons.org/licenses/by/4.0/

Ng CSW, Jahanbani Ghahfarokhi A, Nait Amar M (2023b) Production optimization under waterflooding with long short-term memory and metaheuristic algorithm. Petroleum 9:53–60. https://doi.org/10.1016/j.petlm.2021.12.008. http://creativecommons.org/licenses/by/4.0/

Ng CSW, Jahanbani Ghahfarokhi A, Wiranda W (2023c) Fast well control optimization with two-stage proxy modeling. Energies 16. https://doi.org/10.3390/en16073269. http://creativecommons.org/licenses/by/4.0/

Ng CSW, Jahanbani Ghahfarokhi A (2023) Optimizing initiation time of waterflooding under geological uncertainties with value of information: application of simulation-regression approach. J Pet Sci Eng 220:111166. https://doi.org/10.1016/j.petrol.2022.111166. http://creativecommons.org/licenses/by/4.0/

Olson RS, Bartley N, Urbanowicz RJ, Moore JH (2016) Evaluation of a tree-based pipeline optimization tool for automating data science. In: GECCO 2016 - proceedings of the 2016 genetic and evolutionary computation conference (2016). https://doi.org/10.1145/2908812.2908918

Sayarpour M, Zuluaga E, Kabir CS, Lake LW (2009) The use of capacitance-resistance models for rapid estimation of waterflood performance and optimization. J Pet Sci Eng 69:227–238

Schiozer DJ, De Souza Dos Santos AA, De Graça Santos SM, Von Hohendorff Filho JC (2019) Model-based decision analysis applied to petroleum field development and management. Oil Gas Sci Technol 74

Wiranda W (2022) Simulation of pre-ACT injection and development of proxy model for Svelvik CO_2 field laboratory. Norwegian University of Science and Technology. NTNU Open: https://hdl.handle.net/11250/3034264

Wiranda W, Jahanbani Ghahfarokhi A, Ringstad C, Grimstad A-A (2023) Simulation of CO_2 injection and development of proxy models for Svelvik CO_2 field lab. In: InterPore2023

Yousefi SH, Rashidi F, Sharifi M, Soroush M, Jahanbani Ghahfarokhi A (2021) Interwell connectivity identification in immiscible gas-oil systems using statistical method and modified capacitance-resistance model: a comparative study. J Pet Sci Eng 198:108175. https://doi.org/10.1016/j.petrol.2020.108175. Used with permission from Elsevier

Comparison of Three Machine Learning Approaches in Determining Total Organic Carbon (TOC): A Case Study from Marcellus Shale Formation, New York State

Danijela Dimitrijevic and Constantin Cranganu

Abstract Total Organic Carbon (TOC) contained by subsurface source rock units is an ideal parameter for predicting the potential production of gas and oil shales because it primarily relates to the organic matter content. When discussing drilling operations for gas shale, generating an accurate TOC depth distribution is one of the most important steps in the process of estimating the gas abundance. TOC is measured customarily by retrieving core samples from the boreholes and analyzing them in specialized labs. Unfortunately, TOC measurements are not consistently recorded in drilling operations. Also, the TOC measurements on samples from a borehole are discrete rather than continuous. One way to solve these issues is to employ the capabilities of machine learning (ML) approaches. We present a novel comparative study of the performance of three different ML methodologies in generating TOC content of the Marcellus Shale in New York State: Multilayer Perceptron Neural Networks (MLPNN), Genetic Algorithms (GA), and Support Vector Machines (SVM). Estimating and evaluating the intelligent models of predicted TOC divides the data set of the well logs into three steps: training, validation, and application step. The input data are gamma ray, neutron porosity, and bulk density logs, with the expected output being TOC log.

We built three models using MLPNN, SVM, and GA methods, with both normalized and non-normalized input data. Evaluation of statistical model performance indicates that the best method for creating the synthetic TOC log is GA using normalized data, with Regression (R) of 0.89, Normalized Mean Square Error (nMSE) of 0.35, Mean Square Error (MSE) of 1.0, and Mean Absolute Error (MAE) of 0.48.

This work was supported in part by PSC-CUNY Award # 69548-00 47

D. Dimitrijevic · C. Cranganu (✉)
Earth and Environmental Sciences Department, Brooklyn College, The City University of New York, New York, NY, USA
e-mail: cranganu@brooklyn.cuny.edu

D. Dimitrijevic
e-mail: d.danijela@icloud.com

C. Cranganu (ed.), *Artificial Intelligent Approaches in Petroleum Geosciences*,
https://doi.org/10.1007/978-3-031-52715-9_2

Modeling with normalized input data provided better results than modeling with non-normalized data.

Keywords Marcellus shale · Total Organic Carbon (TOC) · Multilayer Perceptron Neural Networks (MLPNN) · Genetic Algorithms (GA) · Support Vector Machines (SVM) · Well logs

1 Introduction

The recent development of exploration and exploitation of unconventional hydrocarbon resources brought up the necessity of estimating Total Organic Carbon (TOC) in source rocks. TOC is a vital parameter for prediction of potential production of oil and gas shales, like Marcellus Shale, because this parameter is a prime indicator of organic matter richness. TOC is measured directly on sidewall cores and drilling cuttings, using dedicated geochemical laboratories. However, the necessary rock samples cannot be obtained from all intervals in any source rock play. Consequently, TOC estimations from well logs have become increasingly more important because the continuing exploration and development of gas shales, oil shales, and tight oil reservoirs requires detailed information about the richness of their organic content.

In recent years, models that can predict the total organic carbon were developed by applying machine learning approaches in many studies: Liu et al. (2013) and Tan et al. (2015) quantified TOC variation from well logs using support vector regression (Liu et al. 2013), while Zhu et al. (2019) combined unsupervised learning with semi-supervised learning to obtain the same TOC prediction. A list of various machine learning techniques, including support vector machines (SVM), random forest (RF), and decision tree (DT), used in predicting TOC from well logs (formation resistivity, spontaneous potential, sonic transit time, bulk density, neutron porosity, gamma ray and spectrum logs of thorium, uranium, and potassium) is found in Siddig et al. (2021). More recent studies describe a super learner approach to predict total organic carbon using stacking machine learning models based on well logs (Goliatt et al. 2023; Asante-Okyere et al. 2023), a TOC prediction using a gradient boosting decision tree method (Zhang et al. 2023) or employing hybrid machine learning models for estimating TOC from mineral constituents in core samples of shale (Saporetti et al. 2023).

We present a comparative study of the performance of three different ML methodologies in generating TOC content of the Marcellus Shale in the State of New York:

1. Artificial Neural Networks (ANNs), version Multilayer Perceptron Neural Networks (MLPNNs)
2. Genetic Algorithms (GAs)
3. Support Vector Machines (SVMs).

Estimating and evaluating the intelligent models of predicted TOC divides the data set of the well logs into three steps:

1. Training
2. Validation
3. Application.

The input data are gamma ray (GR), neutron porosity (NPHI), and bulk density (RHOB) logs with the expected output being TOC log.

Each model's quality is assessed using various error parameters.

It is important to note that this comparison of three machine learning techniques in predicting TOC values was not performed as of today.

2 Methods and Procedures

Estimating and evaluating the ML-generated models of predicted TOC divide the data set of the well logs into three steps: *training, validation,* and *application.* The input data are gamma ray, porosity, and density logs, with expected output being TOC.

The first step, *training,* presents the data, trains the model by combining input values taken from input logs, and creates a modeled output. The second step, *validation,* focuses on a given model and seeks the best performance approach to determine the influence of the training on the model. In the third step, *application,* an output, synthetic log of TOC variations is generated. The model quality is assessed using various error parameters.

We built three models using MLPNN, SVM, and GA methods, with both normalized and non-normalized input data.

Our comparative study of three popular ML approaches aimed to assess the intrinsic amount of uncertainty generated by each method, to identify the main sources of uncertainty, and to make recommendations for future uncertainty management. We analyzed, characterized, estimated, and managed various types of uncertainty produced by applying the three ML methods to solving certain complex problems in the hydrocarbon industry.

The higher aim of our endeavor was locating the *sweet spots* before starting hydraulic fracturing and horizontal drilling in oil and gas shale formations. A *sweet spot* is characterized by a relatively high concentration of hydrocarbons and represents a coveted target for exploitation companies (Ter Heege et al. 2015). When trying to identify and locate a *sweet spot* area in underexplored shale formations, like Marcellus and Utica in New York state, limited data of petrophysical properties hampers the location efforts and require more extensive prospection and exploration work.

Among the parameters that define the organic, rock, and mechanical qualities of a shale formation viewed as a prime target for hydraulic fracturing, TOC is the prime indicator of organic matter content of subsurface rocks. Thus, it represents an ideal parameter for prediction of potential production of gas and oil shales (Aldrich and Seidle 2018).

Core-measured petrophysical properties of reservoir rocks, like sonic velocity or TOC, are usually considered to be the most direct and accurate method. But core measurements have several shortcomings: high costs associated with extracting, handling, and lab works, often incomplete borehole evaluation due to limited number of core samples, long times of coring, etc.

Because there is not a wireline log capable of measuring TOC in a borehole, this petrophysical reservoir property requires extensive coring of the wells and lab measurements of extracted rock samples. Consequently, the available TOC data in underexplored areas are scarce. So, to identify the sweet spots for future hydrocarbon production in frontier regions, there is a need to perform petrophysical characterization and modeling before drilling, hydraulic fracturing, and hydrocarbon production have started. The Marcellus and Utica gas shales in New York state are prime targets of our proposed research. A limited number of cores from those two shale formations were collected and their TOC was measured in a specialized lab. Those direct measurements will help constrain the results of our ML modeling.

2.1　Artificial Neural Networks (ANNs)

ANNs are computer models that attempt to simulate specific functions of the human brain, which is a highly organized system consisting of a large number of interconnected and specialized biologic units, called *neurons*. This kind of research originated in the 1940s, when McCulloch and Pitt (1943) and Pitts and McCulloch (1947) developed a computational model of the brain.

The current model of ANN is organized as a series of layers $L_1, ..., L_n$ of computing units (Cranganu 2007). The first layer contains *the input units*, while the last layer is made up of *the output units*. The connections between these two layers are assured by a *perceptron*—a computational unit consisting of some parallel structures with nonlinear processing nodes (artificial neurons) that are connected by fixed, variable, probabilistic, or fuzzy weights (interconnections). A perceptron may contain one or more "hidden" layers with a variable number of artificial neurons in each layer (Fig. 1).

The perceptron analyzing the input information $x_1, ..., x_n$, relayed by the first layer, is defined by $n + 1$ numbers: weights $w_1, ..., w_n$, plus a bias b. The weights establish a relationship between the inputs and outputs of each neuron in the ANN. With adequate training and verification, ANN becomes able to predict unknown parameters (like TOC, in Fig. 1), with (very) good accuracy. ANNs are recognized for their ability to solve nonlinear complex problems (Cranganu 2007; Wang et al. 2012; Ashena and Thonhauser 2015).

In the simplest case, illustrated in Fig. 1, we may describe a neuron k using the following pair of equations (Haykin 1999; Cranganu 2007; Taravat et al. 2015):

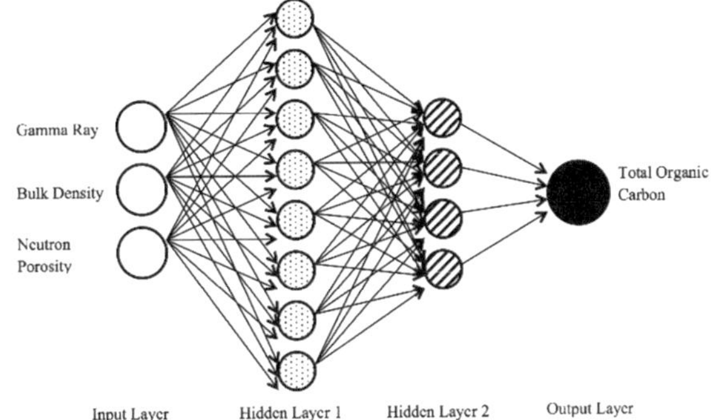

Fig. 1 Artificial neural network architecture used in modeling TOC in Marcellus shale formation of New York State

$$u_k = \sum_{j=1}^{n} w_{kj} x_j \tag{1}$$

and

$$y_k = f(u_k + b_k) \tag{2}$$

where x_1, x_2, ...,x_n are the input signals; w_{k1}, w_{k2}, ..., w_{kn} are the synaptic weights of neuron k; u_k is the *linear combiner output* due to the input signals; b_k is the bias; $f(.)$ is the *activation function*; and y_k is the output signal neuron. The activation function $f(.)$ is often the sigmoid or logistic function, the hyperbolic tangent, or the Gaussian function, but other functions have also been used (heaviside, threshold, or piece-wise linear).

Multilayer perceptron neural networks (MLPNNs) are the most used version of ANN due to their topology, which allows multiple connection weights, outputs, and more complex neurons (Simovici 2015). MLPNNs are also favored for their fast operation, easy implementation, and smaller training step requirements (Orhan et al. 2011). In this research, MLPNN with an error back-propagation (BP) algorithm was used. The algorithm is based on the error-correcting learning rule and consists of two passes: a forward pass and a backward pass (Haykin 1999). In the forward pass, the inputs are transmitted through the network, and output is generated so that synaptic weights are fixed. In the backward pass, an error signal is calculated by comparing the output generated value with the actual response. Depending on the error, the synaptic weights are adjusted in the backward pass. MLPNN networks consist of one or more layers of neural networks (Yan et al. 2006).

Using only one hidden layer is sufficient for almost all nonlinear problems. However, adding more layers improves the model's performance. ANNs with two or more "hidden" layers provide more accurate results (Panchal et al. 2011). Therefore, in this research, we will use an ANN model with two "hidden" (Fig. 1).

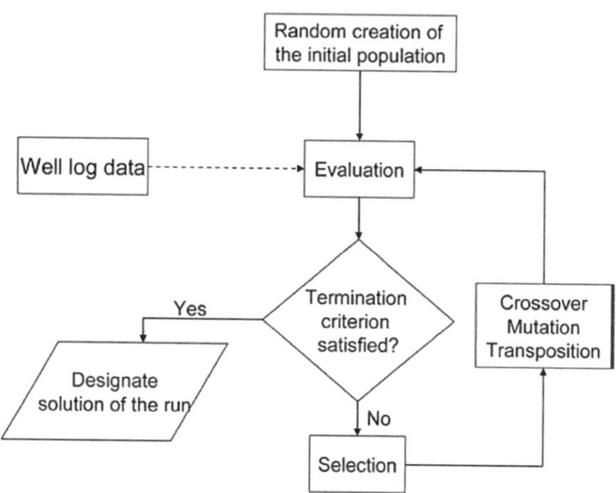

Fig. 2 Flow chart indicating the main steps in using genetic algorithms to simulate TOC values (after Cranganu and Bautu 2010)

2.2 Genetic Algorithms (GAs)

Introduced by Holland (1992), genetic algorithms (GAs) have become the most used and the best-known class of evolutionary algorithms (Luchian et al. 2015). They are part of the stochastic optimization methods used to simulate the process of natural evolution (survival of the fittest, Charles Darwin). GA works best when the parameter space is large, not perfectly smooth, noisy, has multiple local optima, or is not well understood. It is also advisable to use GA when one needs lots of quite good solutions, rather than one very good solution.

The main requirements for using GA for a given problem are (Cranganu and Bautu 2010; Luchian et al. 2015): (1) a solution for the problem that can be coded; this codification is called a chromosome or individual and (2) the existence of a fitness function, which gives a score to each individual. The genetic search starts with a population of randomly generated solutions and uses the fitness function and the genetic operators of reproduction, crossover, and mutation to improve the solutions, searching for better solutions. New generations of solutions are evaluated and the whole cycle repeats until the best solution is obtained (Cranganu and Bautu 2010; Luchian et al. 2015). The architecture of the GA model used in this research is depicted in Fig. 2.

2.3 Support Vector Machines (SVMs)

SVM is a learning algorithm largely used in regression, classification, and outlier detection (Simovici 2015). This ML technique analyzes data, recognizes the pattern, and then classifies them into two categories; each input data will fall into one category or another according to the condition of the classifier. Depending on which category the data falls into, the SVM will create the training model that will predict the desired output for one category.

The mathematical model of SVM was developed by Vapnik and collaborators (Cortes and Vapnik 1995; Vapnik et al. 1997) at AT&T Bell Laboratories for supervised classification analysis. SVMs have quickly become very popular, being nowadays one of the state-of-the-art classification algorithms. By using Vapnik's ε-insensitive loss functions, SVM can solve nonlinear regression problems, if kernel functions are employed. The most popular kernel functions: polynomial, radial, and sigmoid (Cranganu and Breaban 2013) were used to assess the accuracy and robustness of SVM in nonlinear environments.

We also investigated using Support Vector Regression (SVR) (Drucker et al. 1996; Cranganu and Breaban 2013; Al-Anazi and Gates 2015), which is just an SVM for regression analysis.

Generally, the regression problem, which seeks functions that model data with minimal errors, is formulated as follows: If there exists a set of *training* data $\{(x^{(1)}, y^{(1)}), (x^{(2)}, y^{(2)}), ..., (x^{(n)}, y^{(n)})\}$, where each example i consists of a vector of numerical features $x^{(i)} \in R^d$ and a numerical variable $y^{(i)} \in R$, then a function (called *regression function*) $f : R^d \rightarrow R$ has to be derived such that for each example i, $f(x^{(i)}) \approx y^{(i)}$, i.e., f predicts as close as possible the output $y^{(i)}$ based on the input $x^{(i)}$ (Cranganu and Breaban 2013).

Regression, either linear (Fig. 3), logistic, nonlinear, Cox, etc., is basically an optimization problem and the objective/cost function is most frequently formulated as the sum (over all training examples) of squared deviations of the predicted values from the real/training values and is to be *minimized*:

$$\sum_{i=1}^{n} (f(x^{(i)}) - y^{(i)})^2 \tag{3}$$

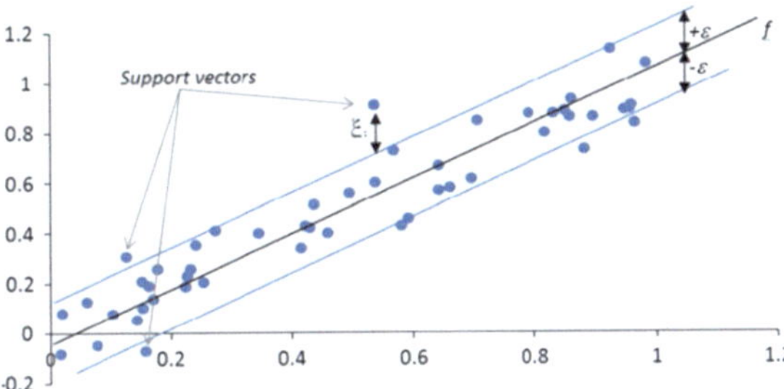

Fig. 3 Illustration of ε–SVR: linear regression function f built by penalizing only deviations that exceed ε; the data items that deviate from f by a value larger than ε are called support vectors and their deviations from the ε-boundary correspond to the slack variables (ξ_i and $\xi_i{}^*$) (From Cranganu and Breaban 2013)

Among many applications of SVM/SVR in petroleum geosciences, it is of interest for this research the simulation of missing logs (Cranganu and Breaban 2013) and the identification of the organic gas-rich Marcellus Shale in the Appalachian Basin (Wang et al. 2014).

2.4 A Three-Step Approach to Modeling

Three main steps will be performed in modeling with each artificial intelligence method (Cranganu 2007; Cranganu and Bautu 2010; Cranganu and Breaban 2013; Cranganu and Bahrpayema 2015):

- *Supervised training* of the model, using available predictors, such as gamma ray (GR), bulk density (RHOB), and neutron porosity (NPHI) extracted from logs recorded in a single borehole;
- *Confirmation and validation of the trained model* by blind-testing the results using log data in a well containing both the predictors and the target (TOC) values;
- *Application of the predicted model* to wells containing only the predictors and obtaining the synthetic (modulated) values TOC.

2.5 Metrics of Uncertainty Quantification and Approaches to Uncertainty Management

Each of the studied ML methods will perform petrophysical model simulations on the same data set.

The simulation results will then be benchmarked for the strengths and weaknesses of each method with respect to the following parameters.

2.5.1 Accuracy of the Models

The main quantity we like to predict is TOC based on gamma ray (GR), neutron porosity (NPHI), and bulk density (RHOB), extracted from logs recorded in a single borehole. We will evaluate *prediction accuracy* of a given training model we used measures like Mean Squared Error (MSE), Normalized Mean Squared Error (nMSE), Mean Absolute Error (MAE), and Regression correlation coefficient (R).

By using all of these statistical errors, the best method for predicting the values of TOC can be determined. Many researchers chose one or a combination of two or three statistical errors to present their models. The reason for using four parameters in this research is to detect more errors, which is more productive and beneficial in determining the best model performance.

The MSE is estimated by finding the difference between the desired target output and the real data:

$$\text{MSE} = \frac{1}{n} \sum_{i=0}^{n} (Oi - Ti) \tag{4}$$

where Oi is the desired output for the training data, Ti is the targeted prediction value, and n is the number of the points in the data set (Sharifi and Mohebbi 2012).

The MSE is widely used for the evaluation of models because it captures the error that estimators make. MSE is built on mathematical properties that make it easier to compute the gradient differences between the desired and real data. The MSE is relevant when the measurement noise variance is nearly uniform. Using the MSE when the measurement noise is highly heterogeneous leads to models being highly sensitive to single data points. In such cases, we will use the *MAE* to fit robust models.

The nMSE estimates deviations between the measured and the computed values:

$$n\text{MSE} = \frac{\sum_{i=0}^{n}(Ti - Oi)^2}{\sum_{i=0}^{n}(Ti - O)^2} \tag{5}$$

where Oi is the desired output, Ti is the targeted prediction value, and O is the mean value of the desired output. The primary distinction between the nMSE and other statistical parameters is that the nMSE sums the deviation instead of calculating the differences between measured and computed values. Low values of nMSE indicate that the model is performing well. However, high values on nMSE do not mean that the model is absolutely wrong.

The *MAE* calculates the mean error value from differences between all data through the formula:

$$\text{MAE} = \frac{1}{n} \sum_{i=0}^{n} |ei| \tag{6}$$

where ei is the model error for all n values.

The smaller the value of MAE, the more accurate the model is. Compared to the MSE, the MAE requires a more complex tool, which is a linearly programmed program to compute the gradient between the computed and the measured data. The problem with *MAE* is that sometimes it is hard to differentiate between the sizes of the errors.

The *Regression correlation coefficient R* measures the strength of the linear correlation between two variables, the real and the predicted data. Correlation between these two variables can be computed as

$$R = \frac{\frac{1}{n}\sum_{i=0}^{n}(Ti - T)(Oi - O)}{\sqrt{\frac{1}{n}\sum_{i=0}^{n}(Ti - T)^2}\sqrt{\frac{1}{n}\sum_{i=0}^{n}(Oi - O)^2}} \tag{7}$$

where O is the mean value of the desired output; T is the mean value of the targeted prediction value; Ti is the targeted prediction value; and Oi is the desired value for all n values. Ideally, the regression correlation coefficient R between the measured and the predicted data should be greater than 0.8 or 80%.

2.5.2 The Role of Normalization

In data analysis, performing data normalization is a very popular pre-processing step. One common use is to speed up the convergence of the gradient descent algorithm that constitutes the basis of many machine learning tasks. Data normalization is not mandatory in all regression algorithms. For example, symbolic regression with genetic programming (Cranganu and Bautu 2010) is not influenced by normalization. However, in ANN and SVM, data normalization may be crucial for the performance of the trained model (Cranganu 2007; Cranganu and Breaban 2013). We will seek to determine the role of normalization as an uncertainty mitigation procedure across the three intelligent methods investigated.

The bottom line of our uncertainty quantification and management approach is: The ML models need and can be designed to be less sensitive to various unknowns. They will succeed if they are reliable (doing the job we asked to do in specific circumstances), accurate, and robust (achieve the goals in the presence of uncertainty). We can mitigate not the existence, but the ultimate impacts of uncertainty.

2.6 Study Area

The Marcellus Shale has a large energetic potential as a natural gas source in the United States (Arthur et al. 2009). An important relationship exists between the amount of organic matter in the source rock and the volume of natural gas it contains: the more organic matter is present, the more likely natural gas will be found. Organic matter richness is measured by the TOC parameter. Study of the TOC measurements is crucial to understanding the production potential of present and future volumes of natural gas in the Marcellus Shale.

The Marcellus Shale, a Middle Devonian black shale is a widespread deposit in the Appalachian Basin (Fig. 4). When anaerobic degradation of these deposits occurs, the natural gas is formed (Kargbo et al. 2010). The Marcellus Shale gas is mostly thermogenic, containing enough heat and pressure to produce primarily dry natural gas. In the last few years, the Marcellus Shale has become a valuable resource due to its large amount of gas reserves in the United States.

The well logging and TOC discrete data in Marcellus Shale formation of New York state, used in this research, are provided by the Empire State Oil and Gas Information System of the State of New York (ESOGIS).

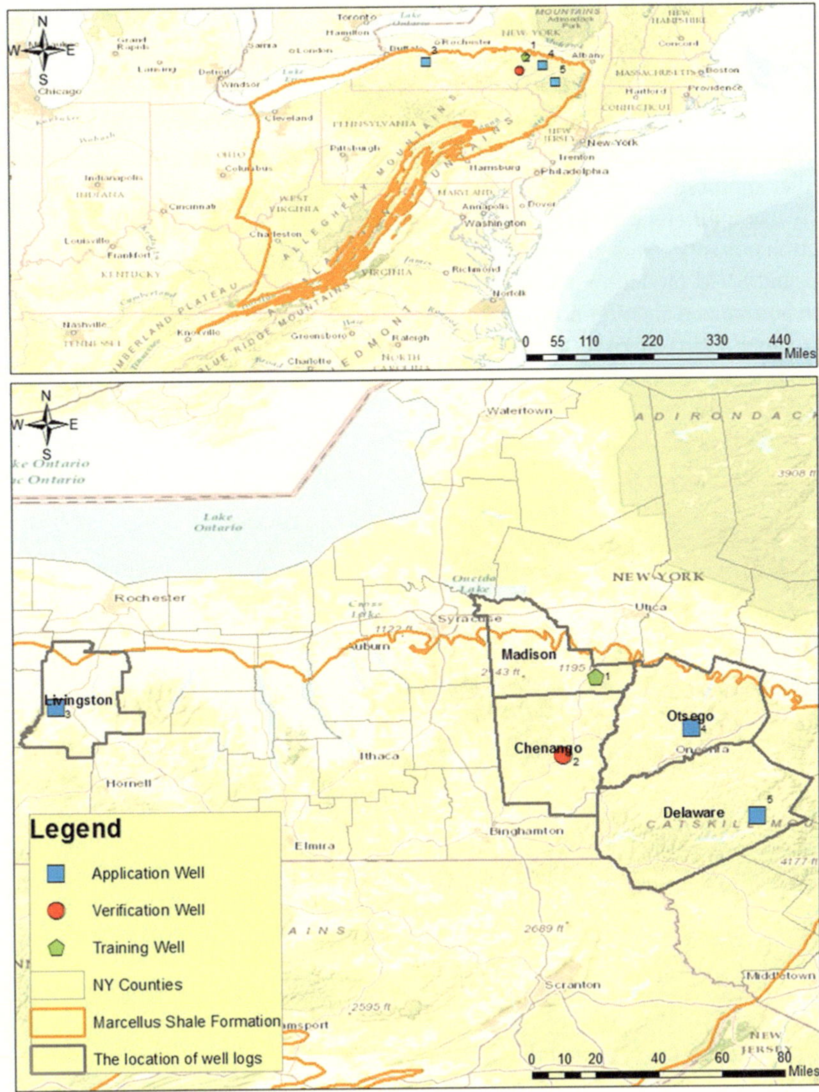

Fig. 4 Distribution of Marcellus Shale formation in the Appalachian Basin (up) and location of well logs used in this research (bottom) (modified from http://pubs.usgs.gov/fs/2011/3092/)

3 Results

3.1 Training and Verification

The supervised *training step* is implemented to build a model that can be used to predict appropriate values of TOC.

To exemplify the training step, three input well logs (gamma ray, bulk density, and neutron porosity), and an output, TOC, from Larkin 1 well logs were used for ANN, GA, and SVM modeling. As an example, Fig. 5 shows the results of training using both normalized and non-normalized input data for modeling using ANN. Similarly, for using SVM (Fig. 6).

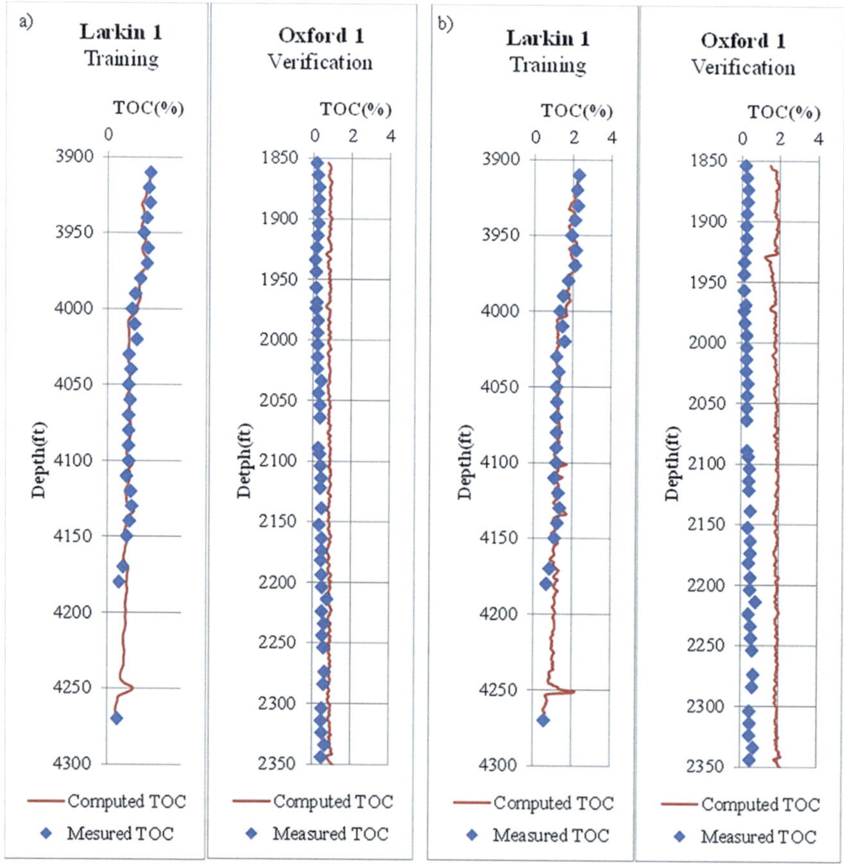

Fig. 5 Multilayer Perception Neural Network model for **a** normalized data of training well Larkin 1 and verification well Oxford 1; **b** non-normalized (raw) data of training well Larkin 1 and verification Oxford 1

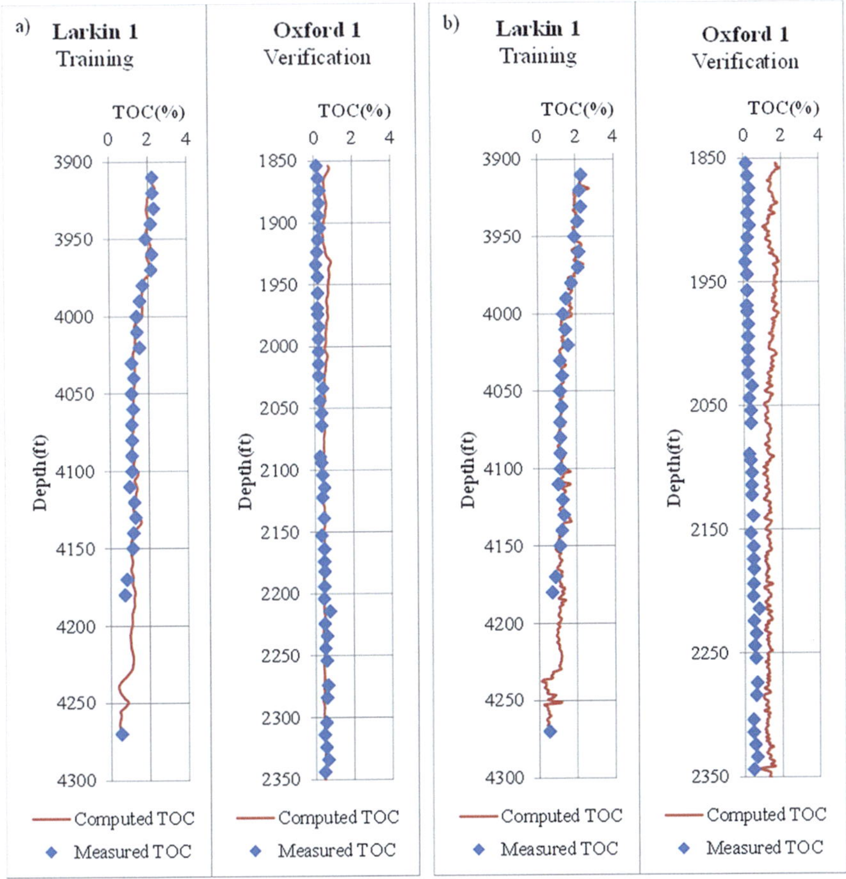

Fig. 6 Support Vector Machine model for **a** normalized data of training well Larkin1 and verification well Oxford 1; **b** non-normalized data of training well Larkin1 and verification Oxford 1

When the training of the model is completed, *validation* is applied on a different set of input data, selected from another well, and the computed outcome (TOC) is compared to the actual outcome, which is not known by the software.

If regression (R) between the actual and the predicted values is more than 0.8 and the nMSE is less than 5, we consider the model being validated. Otherwise, the model would need to be modified again until the desired results are achieved.

The validation well for all three ML methods was Oxford 1 (Figs. 5 and 6).

3.2 Application of the Validated Models

In this final step, the trained and validated models are applied to other wells with few measured TOC data. We applied the validated models of the three ML algorithms in the following wells: Hilts 20,617-T, Holdridge Charles 1, and Hoose 1 (Figs. 7, 8 and 9).

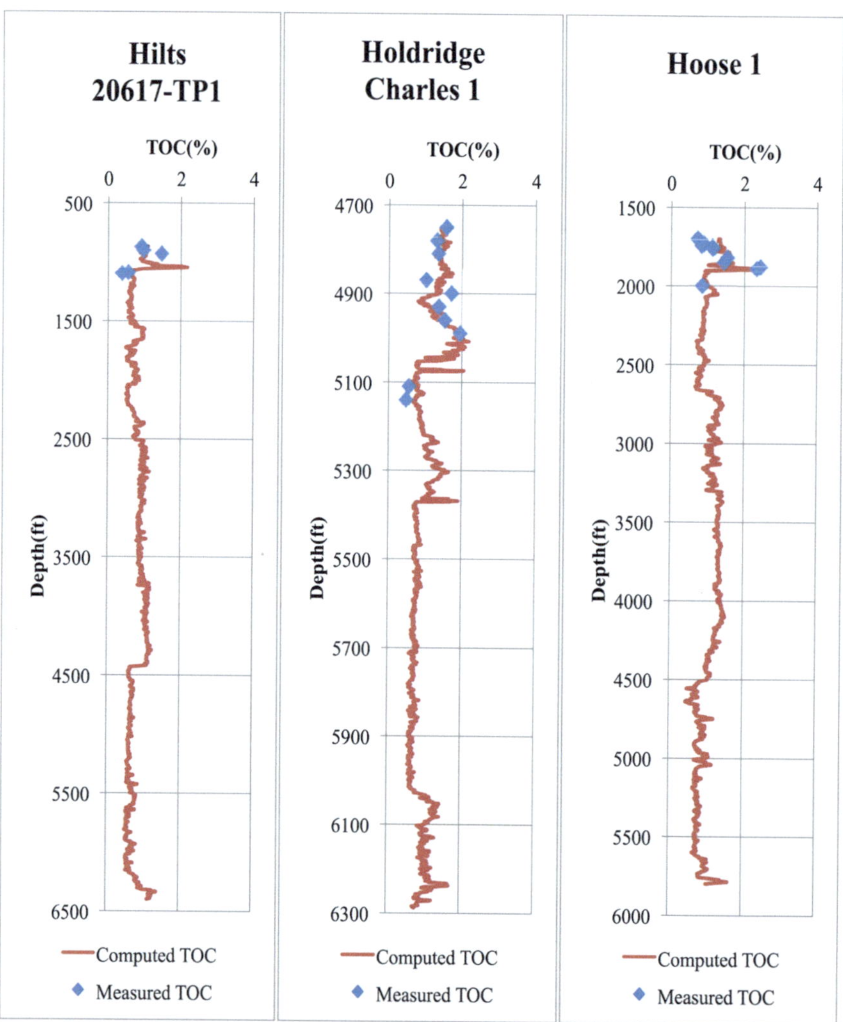

Fig. 7 Application of MLPNN model for predicting the TOC values

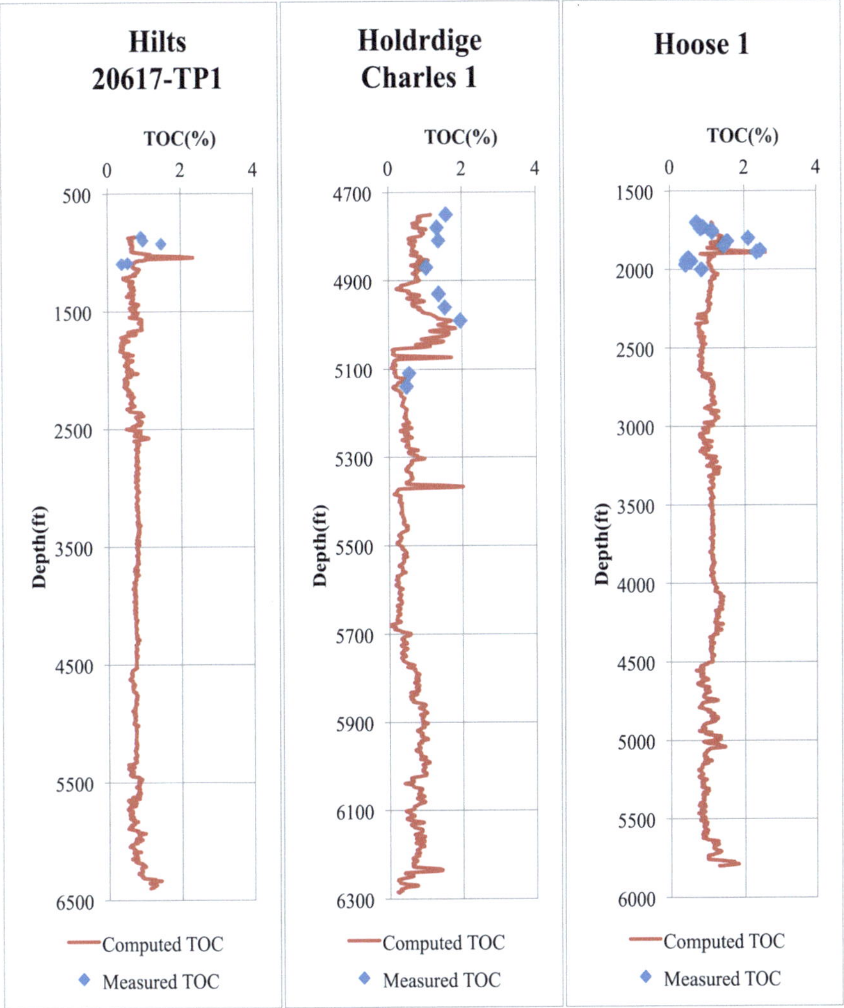

Fig. 8 Application of SVM model for predicting the TOC values

3.3 Error Analysis

For each of the three ML algorithms used, with both non- and normalized input values, we calculated the statistical four parameters discussed earlier (R, nMSE, MSE, and MAE). The results for training and validation (Verification) steps are presented in Table 1.

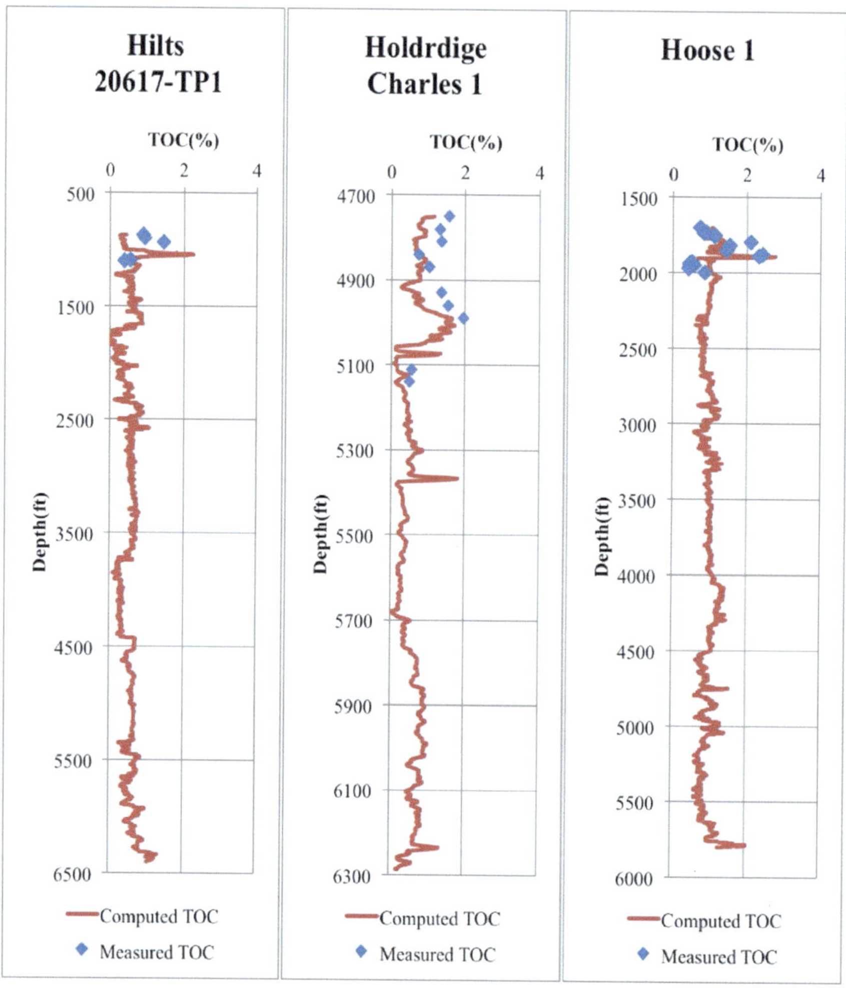

Fig. 9 Application of GA model for predicting the TOC values

4 Discussion

Normalized and non-normalized data are computed for the MLPNN, SVM, and
GA models, and three inputs of well logs are developed and tested using a three-step
approach: training, validation, and application. In the training step, a model is created
using data from the well Larkin 1 with input values from GR, NPHI, and RHOB well
logs, and the output values from the TOC well logs. The next step, validation, is
performed using the data from the new well log Oxford 1. If the model performance
satisfies statistical criteria (e.g., errors criteria-MSE, nMSE, MAE, and R), it is then
applied to additional well-log data with the known and unknown values of TOC.

Tabel 1 Error analysis for training and verification models, with normalized and non-normalized data, for three ML approaches

ERROR ANALYSIS

Method- MLPNN	Normalized data				Nonnormalized data			
Well nameType	R	nMSE	MSE	MAE	R	nMSE	MSE	MAE
Larkin 1Training	0.881	0.227	0.052	0.186	0.876	0.233	0.053	0.196
Oxford 1Verification	0.906	0.689	2.653	0.943	0.813	0.874	3.365	1.472

Method- GA	Normalized data				Nonnormalized data			
Well name Type	R	nMSE	MSE	MAE	R	nMSE	MSE	MAE
Larkin 1Training	0.905	0.184	0.210	0.157	0.837	0.069	0.210	0.157
Oxford 1Verification	0.881	0.511	1.970	0.801	0.909	0.675	2.598	1.499

Method- SVM	Normalized data				Nonnormalized data			
Well name Type	R	nMSE	MSE	MAE	R	nMSE	MSE	MAE
Larkin 1Training	0.895	0.207	0.047	0.146	0.892	0.049	0.141	0.146
Oxford 1Verification	0.864	0.729	2.847	0.891	0.855	0.485	1.870	1.025

4.1　Evaluation of Statistical Model Performance

Figure 9 shows the average of the correlation errors between the non-normalized and normalized data for the training well log Larkin 1, and verification well log Oxford 1, using all three models (MLPNN, SVM, and GA).

The results in Fig. 10 show that, for all three models, the use of normalized data yields more accurate results than the use of non-normalized data.

Analysis of the four statistical parameters suggests the following:

(a) *Mean Absolute Error (MAE)*: The *MAE* of the GA model provides better results, with an average error of 0.48, compared to the MLPNN 0.56 and SVM models which yield average errors of 0.56 and 0.52.

(b) *Normalized Mean Square Error (nMSE)*: The GA model, with an average error of 0.35, and the MLPNN model with an average error of 0.46, perform better than the SVM model, with an average error of 0.47.

(c) *Mean Square Error (MSE)*: The *MSE* for all the three models averages less than 1.5. The values of *MSE* errors are lower in the GA model, with an average error of 1, when compared to the MLPNN and SVM models which yield average errors of 1.35 and 1.45, respectively.

(d) *Regression (R)*: The MLPNN and GA models give slightly better predictions than the SVM model: 0.89 versus 0.88. However, data normalization is not as useful in predicting R for the SVM model.

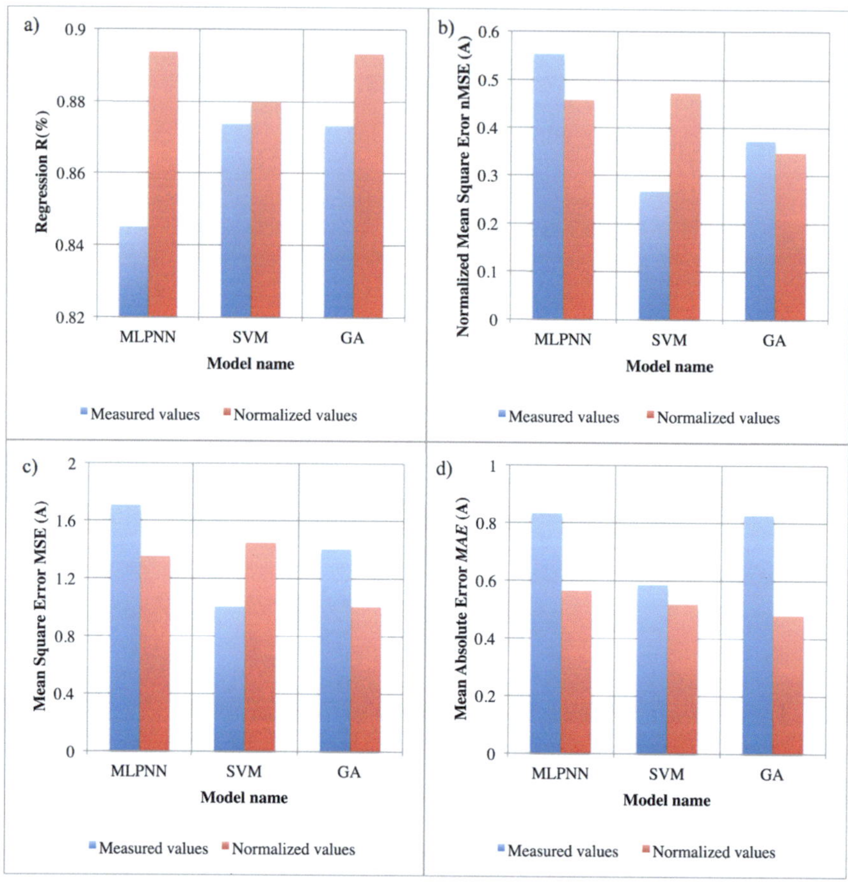

Fig. 10 Evaluation of statistical model performance: **a** Mean Square Error (MSE), **b** Normalized Mean Square Error (nMSE), **c** Mean Absolute Error (MAE), and **d** Regression (R) for MLPNN, SVM, and GA models. Each model values represent the average values of the training well logs Larkin 1 and verification well logs Oxford 1

Overall, the computed GA and MLPNN models proved to be somewhat better than the SVM model in determining the TOC in the Marcellus Shale.

5 Conclusions

Machine learning approaches can be used efficiently in estimating TOC missing values of Marcellus Shale from New York state, based on a suite of geophysical logs (gamma ray, neutron porosity, and density).

The detailed record of the well logs in determining the potential production of the gas reservoirs is not mandatory. Due to that, when total organic carbon value is absent from the record, it can be estimated using this method of intelligent modeling. Thus, this study represents a progressive and significant contribution to the field. Based on the performance of the models, this research can be highly beneficial because the exploration performed on the synthetic TOC log can drastically decrease companies' exploration costs, utilize their time better, and conserve human and natural resources.

Statistical analysis results show that models using normalized data performed better compared to those using non-normalized data.

The computed GA and MLPNN models suggest a slightly better prediction than the SVM model in determining the TOC in the Marcellus Shale formation of the New York state.

References

Al-Anazi AF, Gates ID (2015) On support vector regression to predict poisson's ration and young's modulus of reservoir rocks. In: Cranganu C, Luchian H, Breaban ME (eds) Artificial intelligent approaches in petroleum geosciences. Springer, pp 167–190

Aldrich JB, Seidle JP (2018) Sweet Spot" identification and optimization in unconventional reservoirs. AAPG Search and Discovery, Article #80644

Arthur JD, Bohm B, Layne M (2009) Hydraulic fracturing considerations for natural gas wells of the Marcellus Shale. Gulf Coast Assoc Geol Soc Trans 59:49–59

Asante-Okyere S, Marfo SA, Ziggah YZ (2023) Estimating total organic carbon (TOC) of shale rocks from their mineral composition using stacking generalization approach of machine learning. Upstream Oil Gas Technol, 11. Article 100089

Ashena R, Thonhauser G (2015) Application of artificial neural networks in geoscience and petroleum industry. In: Cranganu C, Luchian H, Breaban ME (eds) Artificial intelligent approaches in petroleum geosciences. Springer, pp 127–166

Cortes C, Vapnik V (1995) Support vector networks. Mach Learn 20:273–297

Cranganu C (2007) Using artificial neural networks to predict the presence of overpressured zones in the Anadarko Basin, Oklahoma. Pure Appl Geophys 164:2067–2081

Cranganu C, Bahrpayema F (2015) Use of active method to determine the presence and estimate the magnitude of abnormally pressured fluid zones: a case study from the Anadarko Basin, Oklahoma. In: Cranganu C, Luchian H, Breaban ME (eds) Artificial intelligent approaches in petroleum geosciences. Springer, pp 191–208

Cranganu C, Bautu E (2010) Using gene expression programming to estimate sonic log distributions based on the natural gamma ray and deep resistivity logs: a case study from the Anadarko Basin, Oklahoma. J Pet Sci Eng 70:243–255

Cranganu C, Breaban M (2013) Using support vector regression to estimate sonic log distributions: a case study from the Anadarko Basin, Oklahoma. J Pet Sci Eng 103:1–13

Cranganu C, Luchian H, Breaban ME (eds) (2015) Artificial intelligent approaches in petroleum geosciences. Springer, 290 p. ISBN 978-3-319-16530-1

Drucker H, Burges CJC, Kaufman L, Smola A, Vapnik V (1996) Support vector regression machines. In: Advances in Neural Information Processing Systems, NIPS 1996, vol 9. MIT Press, pp 155–161

Goliatt L, Saporetti CM, Pereira E (2023) Super learner approach to predict total organic carbon using stacking machine learning models based on well logs. Fuel 353. Article 128682

Haykin S (1999) Neural networks: a comprehensive foundation, 2nd edn. Prentice-Hall Inc., Upper Saddle River, New Jersey, p 842

Holland JH (1992) Genetic algorithms. Sci Am 267(1):66–73

Kargbo DM, Wilhelm RG, Campbell DJ (2010) Natural gas plays in the Marcellus Shale: challenges and potential opportunities. Environ Sci Technol 44(15):5679–5684

Liu C, Chen Z, Hu K, Liu C (2013) Quantifying total organic carbon (TOC) from well logs using support vector regression. CSPG/CSEG/CWLS Geo Convention: AAPG Search and Discovery, 90187. http://www.geoconvention.com/archives/2013/281GC2013QuantifyingTotalOrganicCarbon.pdf

Luchian H, Bautu A, Bautu E (2015) Genetic programming techniques with applications in the oil and gas industry. In: Cranganu C, Luchian H, Breaban ME (eds) Artificial intelligent approaches in petroleum geosciences. Springer, pp 101–126

McCulloch WS, Pitt WH (1943) A logical calculus of the ideas immanent in nervous activity. Bull Math Biophys 5:115–133

Orhan U, Mahmut H, Mahmut O (2011) EEG signals classification using the K-means clustering and a multilayer perceptron neural network model. Expert Syst Appl 38(10):13475–13481

Panchal G, Amit G, Parath S, Panchal D (2011) Determination of over-learning and over-fitting problem in back propagation neural network. Int J Soft Comput (IJSC) 2(2):40–45

Pitts W, McCulloch WS (1947) How we know universals; the perception of auditory and visual forms. Bull Math Biophys 9:127–147

Saporetti CM, Fonseca DL, Oliveira LC, Pereira E, Goliatt L (2023) Hybrid machine learning models for estimating total organic carbon from mineral constituents in core samples of shale gas fields. Mar Pet Geol 143. Article 105783

Sharifi A, Mohebbi A (2012) Introducing a new formula based on an artificial neural network for prediction of droplet size in venturi scrubbers. Braz J Chem Eng 29(3):549–558

Siddig O, Ibrahim AF, Elkatatny S (2021) Application of various machine learning techniques in predicting total organic carbon from well logs. Comput Intell Neurosci 2021. Article ID 7390055

Simovici D (2015) Intelligent data analysis techniques—machine learning and data mining. In: Cranganu et al (eds) Artificial intelligent approaches in petroleum geosciences. Springer, pp 1–52

Tan M, Song X, Yang X, Wu Q (2015) Support-vector-regression machine technology for total organic carbon content prediction from wireline logs in organic shale: a comparative study. J Nat Gas Sci Eng 26:792–802

Taravat A, Proud S, Peronaci S, Del Frate F, Oppelt N (2015) Multilayer perceptron neural networks model for meteosat second generation SEVIRI daytime cloud masking. Remote Sens 7(2). https://doi.org/10.3390/rs70201529

Ter Heege J, Zijp M, Nelskamp S, Douma L, Verreussel R, Ten Veen J, de Bruin G, Peters R (2015) Sweet spot identification in underexplored shales using multidisciplinary reservoir characterization and key performance indicators: example of the Posidonia Shale Formation in the Netherlands. J Nat Gas Sci Eng 27(2): 558–577

Vapnik V, Golowich S, Smola A (1997) Support vector method for function approximation, regression estimation, and signal processing. In Mozer MC, Jordan MI, Petsche T (eds) Advances in neural information processing systems, vol 9. MIT Press, Cambridge, MA, pp 281–287

Wang G, Cheng G, Carr TR (2012) The application of improved Neuro Evolution of Augmenting Topologies neural network in Marcellus Shale lithofacies prediction. Comput Geosci 54:50–65

Wang G, Carr T, Ju Y, Li C (2014) Identifying organic-rich Marcellus Shale lithofacies by support vector machine classifier in the Appalachian basin. Comput Geosci 64:52–60

Yan H, Jiang Y, Zheng J, Peng C, Li Q (2006) A multilayer perceptron-based medical decision support system for heart disease diagnosis. Expert Syst Appl 30(2):272–281

Zhang H, Wensheng W, Hao W (2023) TOC prediction using a gradient boosting decision tree method: a case study of shale reservoirs in Qinshui Basin. Geoenergy Sci Eng 221. Article 111271

Zhu L, Zhang C, Zhang C, Zhang Z, Nie X, Zhou X, Liu W, Wang X (Oct 2019) Forming a new small sample deep learning model to predict total organic carbon content by combining unsupervised learning with semisupervised learning. Appl Soft Comput 83:105596

Gated Recurrent Units for Lithofacies Classification Based on Seismic Inversion

Runhai Feng

Abstract As a qualitative indicator, subsurface lithofacies is an important parameter that can characterize hydrocarbon reservoirs for the degree of compartmentalization. In order to account for the geological dependency between data samples along the vertical direction, the feed-backward Recurrent Neural Networks is applied to classify the sequential lithofacies in the subsurface. Particularly, Gated Recurrent Units (GRU) is used, which can be dedicated to learning how to update or reset hidden states (in this case, lithofacies), such that the information flow through the system is regulated. Operating on the output layer, the *softmax* function is able to map the probability values over various possible lithofacies, and the associated uncertainty could be analyzed subsequently. In addition, the statistical Hidden Markov Models (HMM) is applied to benchmark the performance of GRU, in which the embedded transition matrix could enforce the conditional probability between different lithofacies. The designed GRU and HMM are applied to a synthetic model of the Book Cliffs and a real dataset from the Vienna Basin. Instead of using well logs, elastic rock properties from a non-linear inversion scheme are proposed as inputs for the classification purpose, which could help to overcome the location limitations of cored wells, and 2D sections of reservoir lithofacies are then obtained.

Keywords Recurrent Neural Networks · Rock properties · Lithofacies classification · Seismic inversion

1 Introduction

During the process of reservoir characterization, different reservoir parameters are expected to be estimated, such as lithofacies, porosity, permeability, etc. As a qualitative indicator, lithofacies is a very important reservoir parameter, which could imply the degree of reservoir compartmentalization, as well as the rock-physical behaviors

R. Feng (✉)
Department of Geoscience and Engineering, Delft University of Technology, Delft 2628CN, The Netherlands
e-mail: r.feng@tudelft.nl

    97
C. Cranganu (ed.), *Artificial Intelligent Approaches in Petroleum Geosciences*,
https://doi.org/10.1007/978-3-031-52715-9_3

(Bosch et al. 2002; Garland et al. 2012; Zhou et al. 2016; Gan et al. 2019). In order to distinguish between different lithofacies, various methods have been developed. For example, Mukerji et al. (2001) predicted reservoir lithofacies and fluid types based on a statistical relation between rock-physical properties and seismic information. Using rock-physical models, Bosch et al. (2010) provided a quantitative description of reservoir properties in joint and simultaneous workflows.

Other than rock-physical templates being used, the non-linear relationship between rock properties and reservoir lithofacies could also be explored through advanced deep-learning methods. Qian et al. (2018) analyzed a seismic facies interpretation by deep convolutional autoencoders. Pires de Lima et al. (2019) used deep Convolutional Neural Networks (CNN) to aid the core description, which can achieve a high level of classification accuracy (~90%). Wei et al. (2019) developed a data padding strategy in CNN to characterize various rock facies.

However, these aforementioned methods can be regarded as pointwise, since every depth point that is fed into the classification system is treated as independent of each other. From a geological point of view, the intrinsic transitions of lithofacies along the vertical direction, such as the fining-upward trend in terms of grain size found in distributary channels, are missing (Feng et al. 2017). CNN uses convolutional filters to only consider the local information within the size of pre-defined filters, and thus it is more suitable for image segmentation at the pixel level (Goodfellow et al. 2016). Lithofacies are deposited according to geological rules, and the transition information in a long and short range cannot be accounted for by the convolutional filters in CNN. On the other hand, Recurrent Neural Networks (RNN) can employ the internal memory units to process series variables, and the connections between nodes form a directed graph (Goodfellow et al. 2016; Zhang et al. 2018). By means of a feedback loop, RNN allows information cycles, and both of the current state and what has learned from previous steps could be accounted for, which makes it suitable for time series analysis or tasks in the prediction of sequential variables, such as natural language processing. Imamverdiyev and Sukhostat (2019) developed effective deep-learning models for geological facies classification, in which 1D CNN, RNN, and support vector machine are compared with each other. Grana et al. (2020) proposed an RNN framework for the lithofacies classification. In this paper, the Gated Recurrent Units (GRU) (Cho et al. 2014), a gating mechanism in RNN, is applied to classify sequential lithofacies with temporal dependencies, and the data manipulation is processed at an intelligent level. Besides, Hidden Markov Models (HMM), that can implement the inherent orderings of lithofacies using a constructed transition matrix (Elfeki and Dekking 2001), is applied to benchmark the performance of the proposed deep learning method (Eidsvik et al. 2004; Lindberg and Grana 2015; Feng 2020a). In HMM, a Gaussian assumption is usually adopted to describe the relationship between rock properties and reservoir lithofacies, which might be insufficient when complex distributions exist.

The novelty of this paper is that both geological orderings and complex data distributions are considered by the designed neural system, which is expected to have a better classification performance, compared to traditional approaches. Additionally, the inverted rock properties from seismic data are proposed as the inputs, which typically cover larger extents of target reservoir in a 2D or 3D way (Feng et al. 2018a) and can exclude the location limitation of borehole logs that are usually sparse in the field.

2 Methodology

Being able to recognize sequential patterns, RNN is a deep-learning algorithm applicable for serial data analysis, such as speech and handwriting recognitions (Goodfellow et al. 2016). It considers time and sequential information and has a temporal dimension. The outputs of RNN are not only influenced by weights applied to the inputs but also by hidden state vectors that represent the context based on prior inputs (x_t)/outputs (y_t) (Fig. 1) (Goodfellow et al. 2016; Grana et al. 2020). Therefore, RNN can maintain a class of states in memory cells, allowing it to perform jobs such as prediction of sequential data that are beyond the capability of feed-forward CNN.

Theoretically, RNN is entitled to capture long-term dependencies. However, in practice, training such network may fail because of vanishing or divergent gradients from long products of operation matrices (Goodfellow et al. 2016). To address this problem, gating mechanisms are proposed such that hidden states in the system can be updated or reset to regulate the information flow. Compared to the long short-term memory (LSTM) (Gers et al. 1999), Gated Recurrent Units (GRU) (Cho et al. 2014) has fewer parameters, as no output gate is assigned (Fig. 2) and is selected for classification task of sequential lithofacies in this study.

With gating support, GRU can be dedicated to learn how to reset or update hidden states, or when to skip irrelevant temporal information. For a given input x_t at time step t, the reset gate variable r_t and update variable u_t can control how much information of the previous state h_{t-1} will be remembered or copied, and are computed as follows:

$$r_t = \sigma(W_{xr}x_t + W_{hr}h_{t-1} + b_r) \tag{1}$$

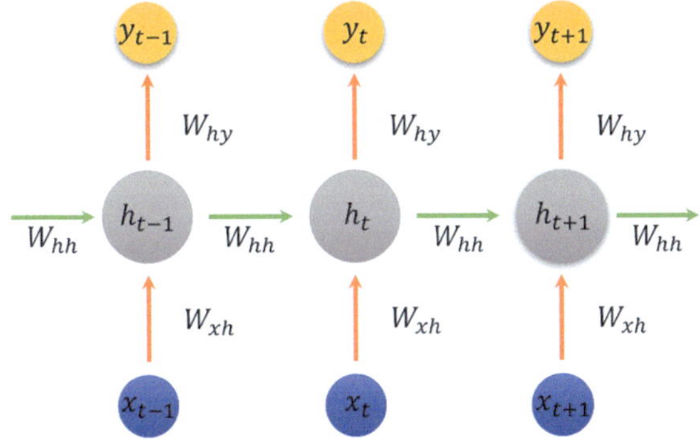

Fig. 1 Recurrent Neural Networks, with hidden state (h_t) that carries pertinent information at step t in the series. W_{xh}, W_{hh}, and W_{hy} are trainable weights associated with input (x_t)-hidden (h_t), hidden (h_{t-1})-hidden (h_t), and hidden (h_t)-output (y_t) connections, respectively

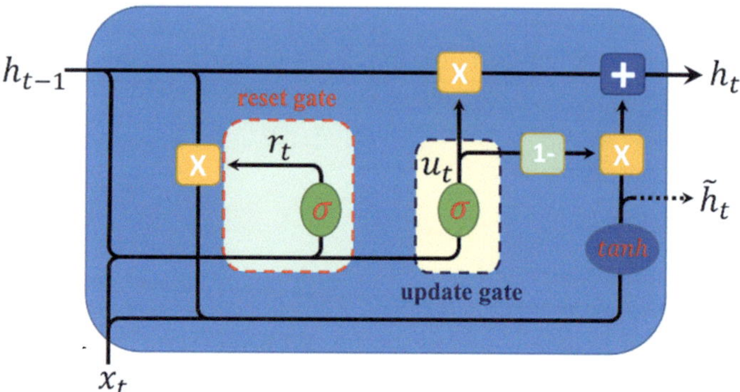

Fig. 2 Computational flow of Gated Recurrent Units

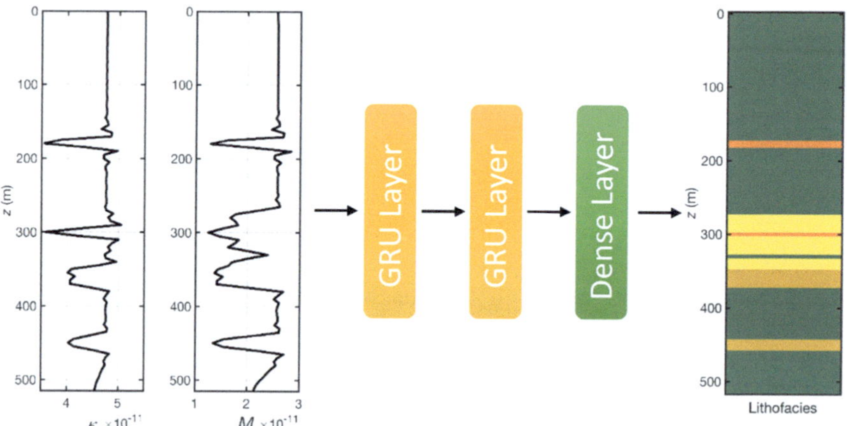

Fig. 3 Schematic view of the proposed GRU model for the lithofacies classification

$$u_t = \sigma(W_{xu}x_t + W_{hu}h_{t-1} + b_u) \tag{2}$$

where W_{xr}, W_{hr}, W_{xu}, W_{hu} and b_r, b_u are trainable weights and biases within each specific gate. σ is the *sigmoid* function with an output interval between 0 and 1.

By combining the reset gate with the effect of update gate, the new state h_t can be determined as

$$h_t = u_t \times h_{t-1} + (1 - u_t) \times \widetilde{h}_t \tag{3}$$

and

$$\widetilde{h}_t = \tanh(W_{xh}x_t + W_{hh}(r_t \times h_{t-1}) + b_h) \tag{4}$$

in which $\times$ represents an element-wise multiplication; W_{xh}, W_{hh} and b_h are trainable neural weights and biases for different states; tanh is the hyperbolic tangent activation function; and $\widetilde{h}_t$ can be considered as a candidate state, since the action of update gate has not been taken yet.

In this proposed GRU approach, the input layer is the inverted rock properties from seismic data in terms of compressibility (κ) and shear compliance (M). The network architecture includes two GRU layers with ten units for each layer. One dense layer is then added with *softmax* as the activation function, relating to the target lithofacies (Fig. 3). As a multiclass generalization/extension of the discrete Bernoulli distribution (Evans et al. 2000), *softmax* function can map the probability distributions over various possible labels after normalization (Eq. 5):

$$s_i = \frac{e^{z_i}}{\sum_{j=1}^{K} e^{z_j}} \tag{5}$$

in which, e^{z_i} is an exponential function for input vector z_i; K stands for the number of lithofacies; and s_i denotes the output probability that relates to state i (Goodfellow et al. 2016). Initial weights in these layers are randomly assigned using the Xavier initialization (Glorot and Bengio 2020), and initial biases are set to 0 before the network training. Nadam (Nesterov-accelerated Adaptive Moment Estimation) (Dozat 2016) is applied to update neural weights and biases to minimize the loss function that is a categorical cross-entropy.

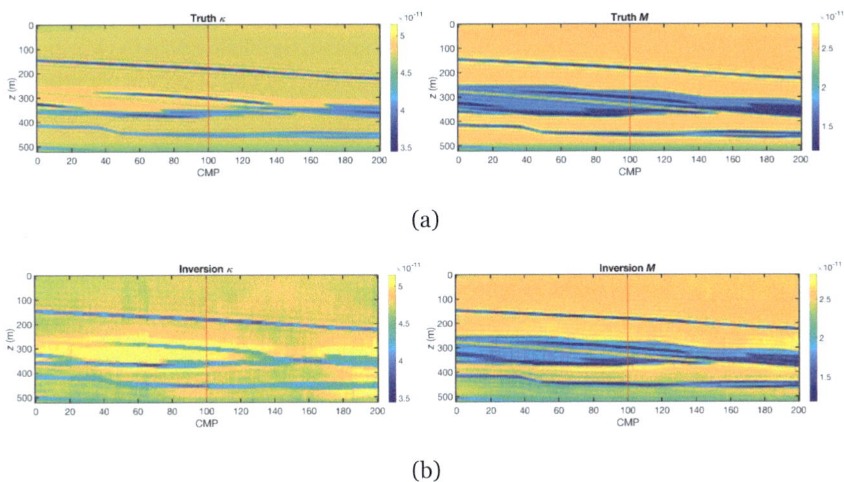

(a)

(b)

Fig. 4 True (**a**) and inverted (**b**) rock properties in terms of κ and M. The true rock properties at CMP 100 (red line in (**a**)) are used for the training of GRU system, and lithofacies are to be classified based on the input of seismic inversion results across the whole section. The unit for κ and M is square meter per newton (m^2/N)

3 Examples

3.1 Synthetic Case

The proposed methodology is first applied to a synthetic model of the Book Cliffs created by Feng et al. (2017). This geological and petrophysical model is based on the fluvio-deltaic Book Cliffs outcrops in Utah, USA, and has been added with more details by Feng et al. (2017) to emphasize the potential reservoir units. Elastic rock properties have been assigned within each lithofacies, which are generated using empirical rock equations based on different reservoir parameters, such as porosity and clay content (Feng et al. 2017). As a test, only a subset of the original 2D section has been selected, and Fig. 4 shows the true and inverted rock properties in terms of compressibility κ ($\kappa = 1/K$, with K being the bulk modulus) (Eq. 6), and shear compliance M ($M = 1/\mu$, with μ being the shear modulus) (Eq. 7):

$$\kappa = \frac{3}{3V_P^2\rho - 4V_S^2\rho} \tag{6}$$

$$M = \frac{1}{V_S^2\rho} \tag{7}$$

in which, V_P, V_S are the P- and S-wave velocities, respectively; ρ is the bulk density. κ and M are considered to be more closely related to different rock types, such that the sandstone has a large κ and a small M, since it is more easily to be compressed because of the higher porosity often found. In contrast, the shale could have a small κ and a large M, due to its weak rigidity. Moreover, κ and M can also indicate property changes in the time-lapse seismic inversion owing to their complementary property behaviors (Feng et al. 2017).

The seismic inversion approach used is a full-waveform scheme that is capable to fully explore the non-linear relationship between rock properties and seismic data (Feng et al. 2017). Compared to the truth (Fig. 4a), property values and geometrical structures in the inverted results (Fig. 4b) have been recovered quite well, which makes them as suitable inputs for lithofacies classification. For more details on the non-linear inversion scheme, please refer to Gisolf and Verschuur (2010) and Feng (2020b).

True rock properties at CMP 100 (Fig. 4a) are used to train the GRU network (Fig. 3). In total, there are four lithofacies identified at this well: Fine-grained sandstone (FS), Very fine-grained sandstone (VFS), Siltstone (SS), and Clay (Fig. 5a). Figure 5b shows the 90% confidence region of the Gaussian likelihood model in terms of κ and M, given each lithofacies. It can be seen that there is a large overlap between SS and VFS, and Clay could be easily separated from other lithofacies.

After training of the pre-designed GRU system shown in Fig. 3, the inversion results at CMP 100 (red lines in Fig. 4b) are used for the lithofacies classification (Fig. 6). Both inverted κ and M match the truth quite well, especially for M (Fig. 6a),

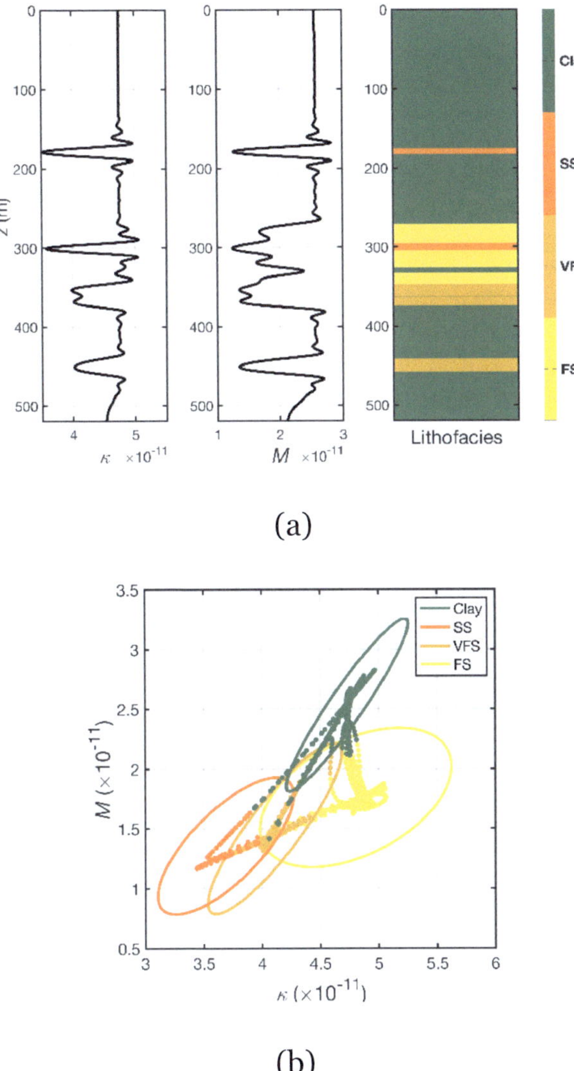

Fig. 5 **a** True rock properties and lithofacies at CMP 100 (red line in Fig. 4a) for training GRU. **b** 90% confidence region of bivariate Gaussian likelihood model given each lithofacies. Scattered points are the sample data, as colored by each lithofacies

since the PP (pressure-to-pressure) and PS (pressure-to-shear) seismic data have been used for the inversion. Compared to the reference, almost all lithofacies layers have been correctly classified by GRU, especially for the SS layer at 180 m (Fig. 6b). However, VFS is misclassified as SS between 355 and 370 m, due to the large overlap of their rock properties (Fig. 5b). The probabilities of each classified lithofacies calculated by the *softmax* function are also shown to assess the associated uncertainty, and most of the values are close to the bounds of probability interval, 0 or 1. Notice that the true rock properties and reference lithofacies profile in Fig. 6 are obtained after upscaling of the ones in Fig. 5a to match the seismic resolution.

With the section inversion results as inputs (Fig. 4b), the classified lithofacies by GRU are shown in Fig. 7b, together with the reference (Fig. 7a). Almost all of the lithofacies units have been correctly predicted, and they are close to the truth, with some misclassified SS in the lower part.

To analyze the classification uncertainty, probabilities of each lithofacies are displayed in Fig. 8, in which the variability is less fluctuated, since these values are either close to 0 (not likely) or 1 (likely).

For a comparison with GRU, the statistical HMM is used to classify lithofacies, in which the Viterbi path is utilized, and only a single highest probability value is computed (Rabiner 1989). The transition matrix in HMM that can describe the lithofacies successions is estimated from cored wells (Fig. 5a), and the emission function for the relationship between rock properties and reservoir lithofacies is Gaussian assumed (Fig. 5b). The classified lithofacies by HMM are shown in Fig. 9.

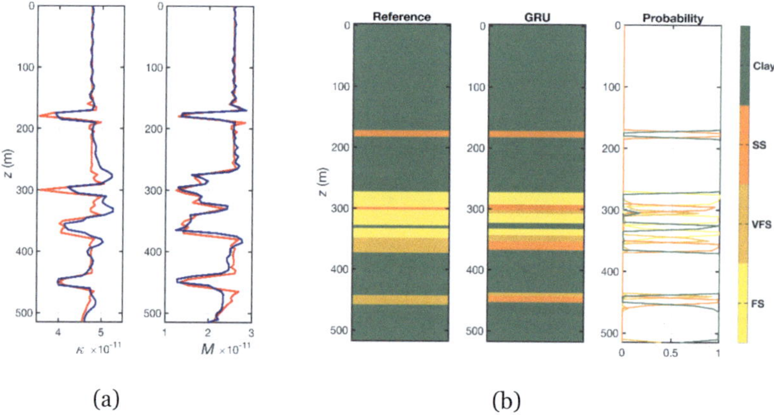

(a)   (b)

Fig. 6 **a** Inverted (blue curve) and true (red curve) rock properties at the well of CMP 100 (red lines in Fig. 4). **b** Classified lithofacies by GRU and their associated uncertainty

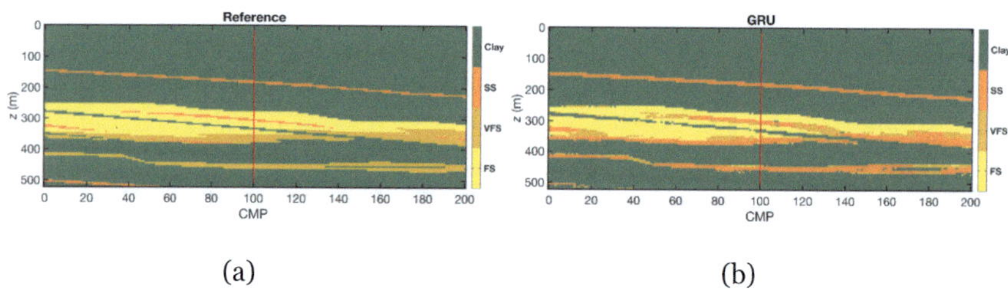

(a)   (b)

Fig. 7 Reference lithofacies (**a**) and classified results by GRU (**b**) of the whole cross section. The red line shows the CMP location for an inspection in Fig. 6b

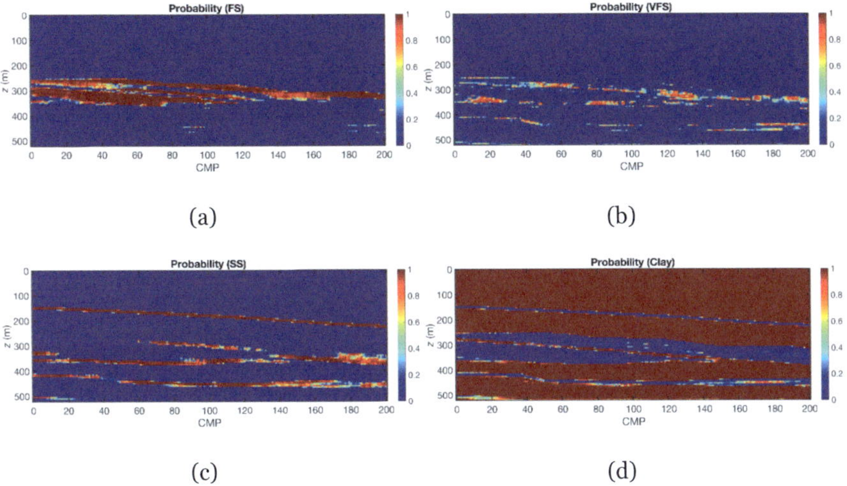

Fig. 8 Probability values of FS (**a**), VFS (**b**), SS (**c**), and Clay (**d**)

At the well location (CMP 100), compared to the results by GRU (Fig. 6b), the thin SS layer has also been predicted correctly around 180 m. Between 270 and 360 m, the result in HMM is more uniform, and Clay has been misclassified as FS between 460 and 520 m. To quantitatively analyze the categorical results by GRU and HMM at the well location, Matthews correlation coefficient (MCC) is computed, which is based on a multiclass confusion matrix, and its value range is between −1 (worst outcome) and 1 (perfect prediction) (Matthews 1975; Feng et al. 2018b). The MCC value is 0.7007 for GRU (Fig. 6b) and 0.6868 for HMM (Fig. 9a), which means that a higher accuracy by GRU has been achieved.

In the section result, the lateral continuity of SS layer around 180 m cannot be fully recovered by HMM, and there are some discontinuous VFS units being predicted (Fig. 9b). FS has been over-predicted by HMM, and the thin Clay layer in between FS is not classified around 300 m in depth (between CMP 0 and 140) (Figs. 7a and 9b). This might be caused by the large overlap between their distributions (Fig. 5b), and they cannot be fully explained by the Gaussian assumption used in HMM.

3.2 Field Case

In order to further test the capability of the proposed GRU, a field dataset from the Vienna Basin for the exploration of clastic reservoirs is used. As an extensional basin between the Eastern Alps and the Western Carpathians, a major part of the basin fills are shallow water sediments of marine to limnic, and fluviatile origins deposited in the Early to Middle Miocene age (Strauss et al. 2006). The main reservoirs in this basin are from Sarmatian and Badenian times (Strauss et al. 2006). Vintages of 3D

Fig. 9 Classified lithofacies by HMM at the well location (CMP 100) (**a**) and of the cross section (**b**). The red line shows the well location for training

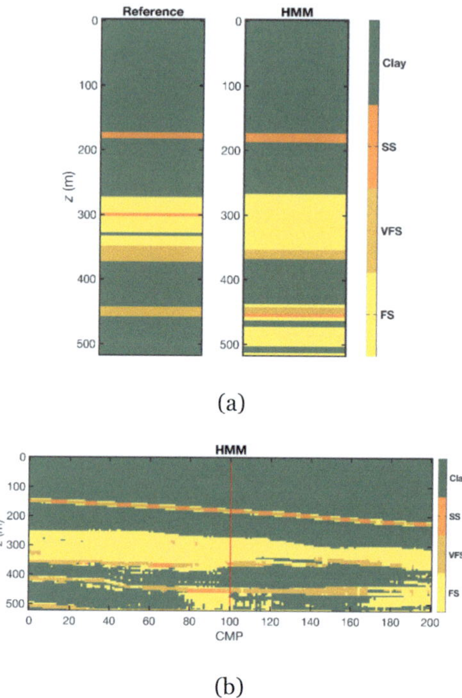

(a)

(b)

seismic surveys acquired in different years have been merged into a single dataset, namely Vienna Basin Super Merge (VBSM). These pre-stack seismic gathers are used as inputs for the non-linear inversion scheme (Feng et al. 2017, 2018a), and a single cross section of inverted κ and M is shown in Fig. 10, where there is a logging well drilled in the middle (CMP 70).

Figure 11a shows the rock properties and interpreted lithofacies at the well location (CMP 70) (Shale, Shaly Sand (SH_Sand), and Sand). Lithofacies are identified by petrophysicists based on different contents of clay minerals, as more clay minerals are found from the shale, while less clay minerals are contained in the sand. It can

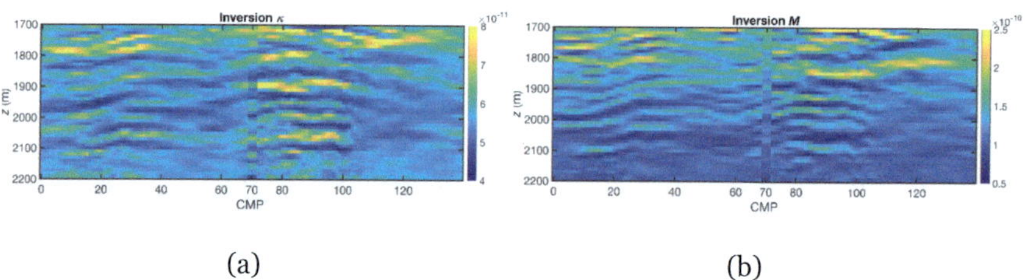

(a)

(b)

Fig. 10 Cross section of inverted rock properties in the Vienna Basin. **a** κ; **b** M. The logging properties at CMP 70 have been superposed on the inversions for a comparison

be seen from the confidence region and sample data in Fig. 11b that there are large overlaps between the three lithofacies, which would make the distinguish between them difficult.

Figure 12 displays the true and inverted rock properties at the well location (CMP 70). The quality of the inverted results is suboptimal, even though the match in the upper part is quite good. The lower part in the inversion matches the truth worse, which is attributed to the poorer seismic-to-well tie with an increase in depth. The inversion quality of M is worse than that of κ, since only PP data are available in the field.

After training based on the well-log data at CMP 70, the classified lithofacies with seismic inversions as inputs by HMM and GRU are shown in Fig. 13, in which the probability of each lithofacies is calculated by the *softmax* function in GRU. Both methods could recover the main Sand reservoir units in the upper part, while HMM predicts more Sand than GRU in the lower part. All the Shale layers have been overestimated by HMM and GRU. The MCC value is 0.1714 by GRU, which is higher than 0.0544 by HMM.

The confusion matrices showing the success and failure rates by classifiers for each lithofacies are displayed in Fig. 14. The correctly classified data samples and

Fig. 11 **a** Rock properties and interpreted lithofacies at the well location (CMP 70). **b** 90% confidence region of bivariate Gaussian likelihood function given each lithofacies. Colored points are the sample data

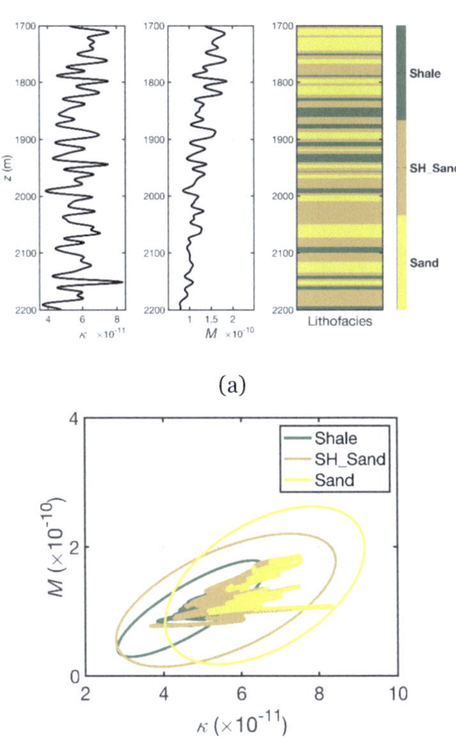

(a)

(b)

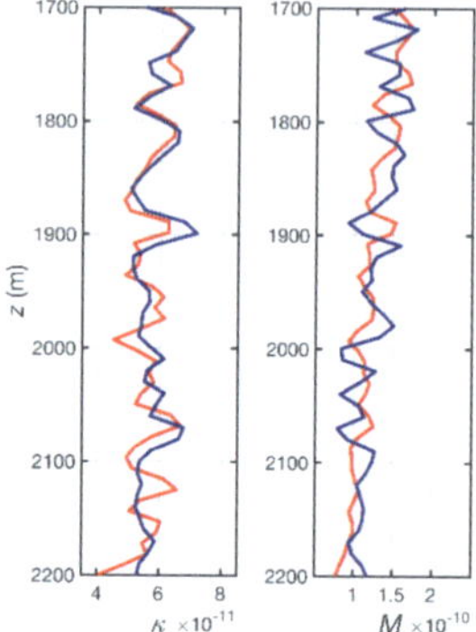

Fig. 12 True (red curve) and inverted (blue curve) rock properties at the well location (CMP 70). An upscaling is performed for rock properties shown in Fig. 11a to match the seismic scale

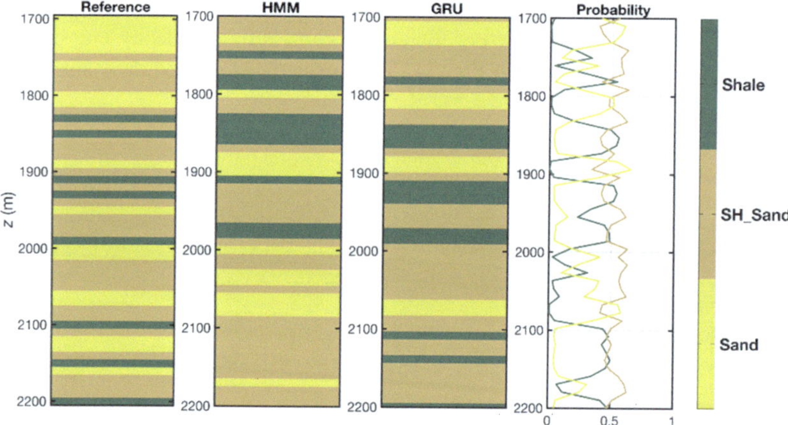

Fig. 13 Classified lithofacies by HMM and GRU at the well location (CMP 70). The reference profile is obtained by upscaling the true lithofacies at the well (Fig. 11a) to the seismic resolution

their related percentages are shown along the diagonal space, in which a score of 100% means a perfect classification. A suboptimal classification is obtained, if the classified samples are close to the diagonal, such that Sand is predicted as SH_Sand. The worst classification samples are those in the off-diagonal space, since the reservoir Sand has been misclassified as non-reservoir Shale or vice versa. GRU performs better than HMM for the classification of Sand and SH_Sand. For the classification of Shale, the same performance can be observed between them within the diagonal samples.

Based on the non-linear inverted rock properties of the cross-sectional seismic data (Fig. 10), the classified lithofacies by GRU are shown in Fig. 15, together with their corresponding probabilities. Most data samples have been classified as SH_Sand, which is also indicated by the reference lithofacies at the well location (Figs. 11a

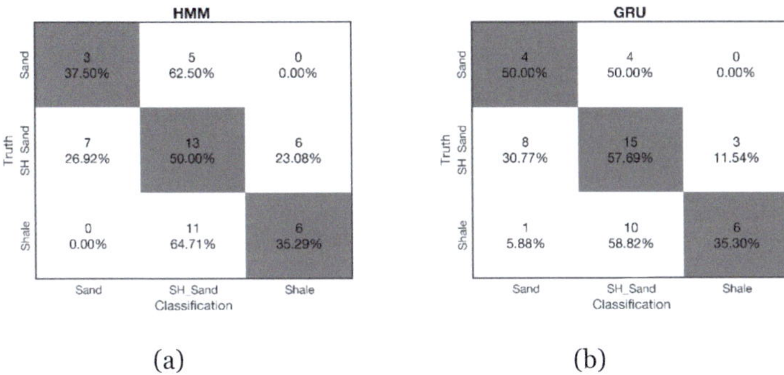

(a)   (b)

Fig. 14 Confusion matrices of HMM (**a**) and GRU (**b**), in which the gray color is associated with the diagonal cells where the classification is correct

and 13). Small Sand reservoir units are distributed across the section, separated by the Shale layers, which are conformable with the depositional environments of shallow marine and limnic origins with sand sheets and impermeable shales in between the Vienna Basin (Strauss et al. 2006).

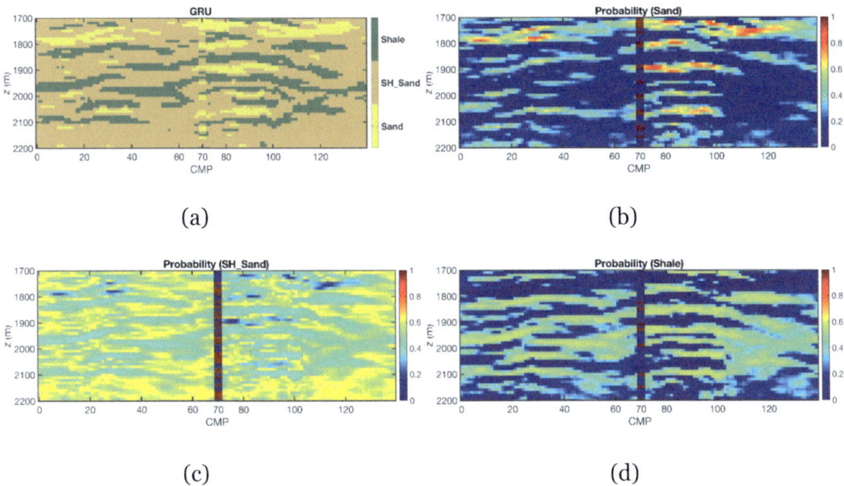

(a)   (b)

(c)   (d)

Fig. 15 **a** Classified lithofacies by GRU with inverted rock properties of the cross section as inputs (Fig. 10). Probability as calculated by *softmax* function in GRU for Sand (**b**), SH_Sand (**c**), and Shale (**d**). The reference lithofacies and their probability values (0 or 1) at the well location (CMP 70) are superposed on the sections

4 Discussion

In this paper, instead of using well logs, which have location limitations, the inversion results from seismic data are proposed as inputs for the classification purpose and a 2D section of reservoir lithofacies is obtained. The non-linear inversion scheme used in this paper is based on an integral representation of the full-elastic equations (Gisolf and Verschuur 2010), in which all internal transmission effects and internal multiples are considered, as well as the wave-mode conversions. The number of iterations will determine the order of multiples used in the inversion procedure. A good recovery of subsurface properties and geometries could be guaranteed (Gisolf and Berg 2010a, 2010b), which makes the inversion results as good candidate inputs for the lithofacies classification. The inverted κ (compressibility) and M (shear compliance) are deemed to relate more closely to lithofacies types (Feng et al. 2018a).

In the synthetic study, both PP and PS seismic data are used as inputs for the non-linear inversion scheme, which makes that the quality of κ and M are almost equally good (Figs. 4 and 6a). When only PP data are available such as in the field example, κ is determined better than M (Fig. 12). Another rock property—bulk density has strong connections with lithofacies and is only accessible when high-quality and wide-angle seismic gathers are available.

A comparison is made between GRU and HMM for the lithofacies classification since both methods could account for the data dependency along the vertical direction. In HMM, there is a Gaussian assumption used to draw the relationship between rock properties and reservoir lithofacies (Figs. 5b and 11b), which is not able to explain complex data distributions. Differently, the non-parametric GRU could relax the strong assumption of Gaussianity in HMM by fitting the non-linear relations between rock properties and reservoir lithofacies with the aid of synaptic neurons. Compared to the results by HMM, GRU can achieve a higher classification accuracy (larger MCC values), especially in the synthetic study for its ability to predict the SS layer around 180 m across the whole section (Fig. 7b).

To train the GRU system, logging data at the well location are used (CMP 100 in Fig. 5a and CMP 70 in Fig. 11a), and the seismic inversion results have been kept untouched during the training process, which can be considered as a blind test set. The training of GRU for 5000 epochs took ~2 min and ~6 min for synthetic and real studies, respectively, on a Xeon 3.70 GHz CPU (central processing unit). Afterward, the implementation of trained GRU on the remaining trace locations is very fast, which can greatly improve the computational efficiency, especially when a large volume of seismic inversion results is available, compared to other traditional rock physics-based methods.

Softmax function is applied to calculate the probability values of lithofacies, given the input rock properties. In the synthetic study, in total, there are 1300 data samples, which may not be representative enough for data distributions, especially for SS. Thus, the so-obtained probability values are less varied (Figs. 6b and 8), compared to the ones in the real case study (Figs. 13 and 15b–d). In the field example, there are 3417 data samples at the selected well location for the GRU training. It is important to note that the same training data in GRU are used for the estimation of parameters in HMM.

Design of network architecture and hyper-parameters tuning are important steps for a successful retrieval of reservoir lithofacies. A trial-and-error approach is usually adopted to select the hyper-parameters in GRU system for the classification task. In this research, a simple network architecture is used that can already provide a high accuracy, especially in the synthetic study. Other complex network structures could also be applied, which are still under investigation. To prevent the overfitting problem, a regularization technique combining batch-normalization and dropout is employed (Srivastava et al. 2014; Ioffe and Szegedy 2015). Moreover, since the proposed neural model is supervised, which means that a perfect classification performance can only be expected when lots of labelled examples are available, and the rock-physical features given lithofacies should be well-defined such that the intra-state variance is as low as possible, and the extra-state variance is wide enough.

As a regression process, reservoir porosity is to be predicted in the future, which could help to quantify the storage potential of hydrocarbon reservoirs. Furthermore, the implemented GRU system could be modified in order to account for the horizontal correlation between lithofacies (Feng et al. 2018a; Tan et al. 2019), which is missed in this study, since the current classification process is implemented trace-by-trace.

5 Conclusion

Gated Recurrent Units (GRU), a special form of Recurrent Neural Networks, was applied for the lithofacies classification, which is a qualitative indicator for hydro-carbon reservoirs. The designed GRU network could account for the spatial coupling between data points along the vertical direction. Therefore, the classification process could implicitly honor the geologically depositional rules. Meanwhile, the data manipulation has been taken to an intelligent level, instead of common Gaussian assumption being adopted, such as in Hidden Markov Models. Rather than using well-log data, which are only available at sparse locations in the field, results from a non-linear inversion scheme on seismic data are used as inputs in the classification process, and 2D sections are produced.

Acknowledgements This study is sponsored by the DELPHI Consortium. OMV is gratefully acknowledged for permission to publish this data.

Declaration of Interests The authors declare that they have no known competing financial interests or personal relationships that could have appeared to influence the work reported in this paper.

References

Bosch M, Zamora M, Utama W (2002) Lithology discrimination from physical rock properties. Geophysics 67(2):P573–P581

Bosch M, Mukerji T, Gonzalez EF (2010) Seismic inversion XE "Seismic inversion" for reservoir properties combining statistical rock physics and geostatistics: a review. Geophysics 75(5):A165–A176

Cho K, Merrienboer B, Gulcehre C et al (2014) Learning phrase representations using RNN encoder-decoder for statistical machine translation. arXiv:1406.1078

Dozat T (2016) Incorporating nesterov momentum into adam. ICLR workshop

Eidsvik J, Mukerji T, Switzer P (2004) Estimation of geological attributes from a well log: an application of hidden Markov chains. Math Geol 36(3):379–397

Elfeki A, Dekking M (2001) A Markov chain model for subsurface characterization: theory and applications. Math Geol 33(5):569–589

Evans M, Hastings N, Peacock B (2000) Statistical distributions. Wiley

Feng RH, Luthi SM, Gisolf A, Sharma S (2017) Obtaining a high-resolution geological and petrophysical model from the results of reservoir-orientated elastic wave-equation-based seismic inversion. Pet Geosci 23:376–385

Feng R, Luthi SM, Gisolf A, Angerer E (2018a) Reservoir lithology determination by hidden Markov random fields based on a Gaussian mixture model. IEEE Trans Geosci Remoting Sens 56(11):6663–6673

Feng R, Luthi SM, Gisolf D (2018b) Simulating reservoir lithologies by an actively conditioned Markov chain model. J Geophys Eng 15(3):800–815

Feng R (2020a) Lithofacies classification based on a hybrid system of Artificial Neural Networks and Hidden Markov Models. Geophys J Int

Feng R (2020b) Estimation of reservoir porosity based on seismic inversion results using deep learning methods. J Nat Gas Sci Eng. https://doi.org/10.1016/j.jngse.2020.103270

Gan L, Wang Y, Luo X et al (2019) A permeability prediction method based on pore structure and lithofacies. Pet Explor Dev 46(5):935–942

Garland J, Neilson J, Laubach SE, Whidden KJ (2012) Advances in carbonate exploration and reservoir analysis. Geol Soc Lond Spec Publ 370:1–15

Gers FA, Schmidhuber J, Cummins F (1999) Learning to forget: continual prediction with LSTM. In: Proceedings of the ICANN'99. IEEE, pp 850–855

Gisolf A, Van Den Berg PM (2010) Target oriented non-linear inversion of seismic data. In: 72nd annual international meeting of European association of geoscientists and engineers (expanded abstract), Barcelona, 14–17 Jun, 2010

Gisolf A, Van Den Berg PM (2010) Target-oriented non-linear inversion of time-lapse seismic data. In: 80th annual international meeting of society of exploration and geophysics (Expanded Abstract), Denver, 17–22 Oct, 2010

Gisolf A, Verschuur DJ (2010) The principles of quantitative acoustical imaging. EAGE Publications b.v., Houten

Glorot X, Bengio Y (2020) Understanding the difficulty of training deep feedforward neural networks. In: Proceedings of the 13th international conference on artificial intelligence and statistics, pp 249–256

Goodfellow I, Bengio Y, Courville A (2016) Deep learning. MIT Press

Grana D, Azevedo L, Liu M (2020) A comparison of deep machine learning and Monte Carlo methods for facies classification from seismic data. Geophysics 85(4):WA41–WA52

Imamverdiyev Y, Sukhostat L (2019) Lithological facies classification using deep convolutional neural network. J Petrol Sci Eng 174:216–228

Ioffe S, Szegedy C (2015) Batch normalization: accelerating deep network training by reducing internal covariate shift. In: Proceedings of the 32nd international conference on international conference on machine learning

Lindberg DV, Grana D (2015) Petro-elastic log-facies classification using the expectation-maximization algorithm and hidden Markov models. Math Geosci 47:719–752

Matthews BW (1975) Comparison of the predicted and observed secondary structure of T4 phage lysozyme. Biochimica et Biophysica Acta (BBA)—Protein Structure 405(2):442–451

Mukerji T, Jørstad T, Avseth P, Mavko G, Granli JR (2001) Mapping lithofacies and pore-fluid probabilities in a North Sea reservoir: seismic inversions and statistical rock physics. Geophysics 66(4):988–1001

Pires de Lima R, Suriamin F, Marfurt KJ, Pranter MJ (2019) Convolutional neural networks as aid in core lithofacies classification. Interpretation 7(3):SF27–SF40

Qian F, Yin M, Liu X, Wang Y, Lu C, Hu G (2018) Unsupervised seismic facies analysis via deep convolutional autoencoders. Geophysics 83(3):A39–A43

Rabiner LR (1989) A tutorial on hidden Markov models and selected applications in speech recognition. Proc IEEE 77(2):257–286

Srivastava N, Hinton G, Krizhevsky A et al (2014) Dropout: a simple way to prevent neural networks from over-fitting. J Mach Learn Res 15:1929–1958

Strauss P, Harzhauser M, Hinsch R, Wagreich M (2006) Sequence stratigraphy in a classic pull-apart basin (Neogene, Vienna Basin). A 3D seismic based integrated approach. Geol Carpathica 57(3):185–197

Tan X, Liu Y, Zhou X (2019) Multi-parameter quantitative assessment of 3D geological models for complex fault-block oil reservoirs. Pet Explor Dev 46(1):194–204

Wei Z, Hu H, Zhou H, Lau A (2019) Characterizing rock facies using machine learning XE "Machine learning" algorithm based on a Convolutional Neural Network and data padding strategy. Pure Appl Geophys 176:3593–3605

Zhang D, Chen Y, Meng J (2018) Synthetic well logs generation via recurrent neural networks. Pet Explor Dev 45(4):629–639

Zhou Z, Wang G, Ran Y et al (2016) A logging identification method of tight oil reservoir lithology and lithofacies: a case from Chang7 Member of Triassic Yanchang Formation in Heshui area. Ordos Basin NW China Pet Explor Dev 43(1):65–73

Application of Artificial Neural Networks in Geoscience and Petroleum Industry

Rahman Ashena and Gerhard Thonhauser

Abstract It has been shown that artificial neural networks (ANNs), as a method of artificial intelligence, have the potential to increase the ability of problem-solving to geoscience and petroleum industry problems, particularly in case of limited availability or lack of input data. ANN application has become widespread in engineering including geoscience and petroleum engineering because it has shown to be able to produce reasonable outputs for inputs it has not learned how to deal with. In this chapter, the following subjects are covered: artificial neural networks basics (neurons, activation function, ANN structure), feed-forward ANN, backpropagation and learning (perceptrons and backpropagation, multilayer ANNs and backpropagation algorithm), data processing by ANN (training, over-fitting, testing, validation), ANN and statistical parameters, an applied example of ANN, and applications of ANN in Geoscience and Petroleum Engineering.

Keywords Hide layer · Mean square error · Drill string · Average root mean square error · Bottom hole pressure

Nomenclature

α	Learning rate
ANN	Artificial neural network
AAPE	Average absolute percent error
APE	Average percent relative error
ARMSE	Average root-mean-square error
BB	Backpropagation
f	Activation or transfer function
Input$_i$	The input value corresponding to neuron i
Logsig	Logistic sigmoid activation/transfer function

R. Ashena (✉) · G. Thonhauser
Chair of Drilling and Completion Engineering, Petroleum Engineering Department, University of Leoben, Leoben, Austria
e-mail: rahman.ashena@unileoben.ac.at

© The Author(s), under exclusive license to Springer Nature Switzerland AG 2024
C. Cranganu (ed.), *Artificial Intelligent Approaches in Petroleum Geosciences*,
https://doi.org/10.1007/978-3-031-52715-9_4

m	Number of output neurons or nodes
MSE	Mean square error
O_{ANN}	Predicted output value by the artificial neural network
R	Pearson correlation coefficient
R^2	Squared Pearson correlation coefficient
SD	Standard deviation
T	Number of training samples from known data point given for training the network
V	Variance
$V_{expected}$	Expected real value (known or measured value of output)
W_i	The weight corresponding to link or connection i

1 Introduction

Artificial neural networks (ANNs) and their application in geoscience and petroleum industry are considered in this chapter. Application of ANNs has shown to be an effective tool to solve nonlinear complex engineering problems particularly when there are no straightforward analytical or even numerical solutions.

Frequently, when there is no analytical solution or approach to a complex or non-straightforward problem wherein the involved parameters and their exact relationship are not known clearly, ANN is applied. However, one can use ANN even in linearly behaving problems having an analytical solution so that its accuracy and functionality can be determined.

Indeed, a so-called flow of information in the ANN model takes place utilizing basic processing units which are artificial neurons connected to each other. In this way, processing of the data takes place through a network of neurons.

In petroleum industry applications to date, the commonly utilized ANN structure has been Feed-forward artificial neural network (FF-ANN) due to its simplicity compared to other neural structures. FF-ANNs are network structures in which the information or data will propagate only in one direction. This network has a learning ability to recognize the relationship between the inputs and outputs provided that adequate training data are supplied. The FF-ANNs typically consist of three layers including input layer, hidden layer, and output layer. There are other structures of ANNs which are not discussed in this chapter because of their limited usage in the petroleum industry. In terms of the number of hidden layers, perceptrons, and multilayer networks will be discussed.

Although it is possible to have more than one hidden layer in the network, normally a single layer is preferred in many applications. The number of neurons in the input and output layers is normally determined by the problem. However, the optimal number of neurons in the hidden layer (and even the number of hidden layers) must be determined by trial and error in order to obtain a proper network size with the highest possible performance.

The network learning or training is attained by adjustment of the weights corresponding to connections or links between the neurons to produce outputs with acceptable errors. After training, the network performance is tested in two stages (test and validation). If the network performance was successful in these three stages, the network would be recognized as a capable tool for simulation using new inputs and obtaining new results. Recognition and prevention of over-fitting will be discussed.

In most geoscience and petroleum engineering applications, the FNN will employ backpropagation as its training or learning algorithm. The algorithm is called backpropagation because the output error is propagated backward to the links between neurons in the previous layers during training. In this way, backpropagation helps to modify the weights of links in order to achieve a desired output. The work mechanism of backpropagation algorithm is adjustment of each weight of the network individually based on the selection of the path of the steepest gradient descent to minimize the error function (which is usually mean squared error or MSE). As each ANN performance is evaluated by the corresponding errors, knowledge about the relevant statistical model error parameters such as MSE is of great importance.

In this chapter, an attempt is made to give a comprehensive applied discussion of neural networks from basics to application. Thus, an applied example of ANN application in petroleum industry is given after presenting the basics. Then, a short overview of applications of neural network approach will be given.

2 Artificial Neural Network (ANN) Basics

2.1 ANN Structure in General

Artificial neural networks' structure and function are indeed an extremely simplified version or simulation of the biological human brain. ANNs have been developed as simplification of the mathematical models of biological neural networks (Fig. 1) by having some assumptions which include processing of information takes place in some processing elements called neurons (first assumption). In this simplification, the information is transferred between neurons using connection links (second assumption), and a specified weight is allocated to each link to be multiplied by the information or signal which is passing each link (third assumption). Then, each neuron allocates a desired bias or threshold value to be added to the sum to yield a net value (fourth assumption), the net value is given as input to an activation or transfer function (which is normally a nonlinear function) and in this way the output of the neuron would yield (fifth assumption). Simply, the function of the whole ANN structure is just the calculation of the output of all the neurons existing in the network.

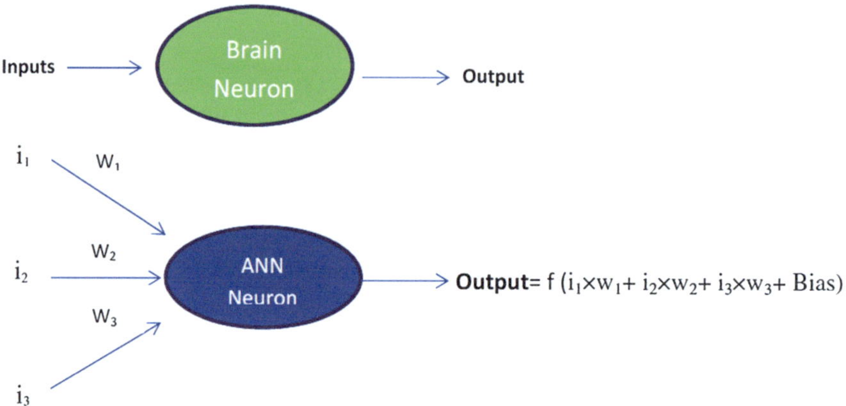

Fig. 1 Biological and artificial neuron similarity

2.2 Artificial Neurons

A typical neuron structure is shown in Fig. 2. As can be seen in this figure the inputs are multiplied by the corresponding weights, and then, their summation is found. In addition, a bias is added to the summation as an error correction. This value is called the net value. The net value (as the input) is passed through a function called the activation function (Fig. 3). The output of this function is indeed the output of the neuron which would be used as the input to the other neurons in the next layer. This is what actually happens inside each individual neuron.

Please note that the weights and inputs could be assumed as vectors and thus their multiplication is in reality considered as their inner product.

The weight allocated to each connection link, to be multiplied by each input of the neuron, is indeed representative of the importance of the input in the problem, and in this way, this input importance or strength is transferred to the next layer through the corresponding link. For example, in the bottom hole pressure (BHP) prediction downhole, there are several input parameters which are of importance and must be taken into consideration. A good trained ANN would allocate higher weights to be multiplied by the more important or stronger input parameter values.

2.3 Activation Function

It should be noted that the activation function (Fig. 3) takes the net as its input and its output is considered as the output of the neuron. Typical activation functions are as follows:

- Threshold function $\{f = 0$ when $x < 0$, 1 when $x \geq 0\}$,

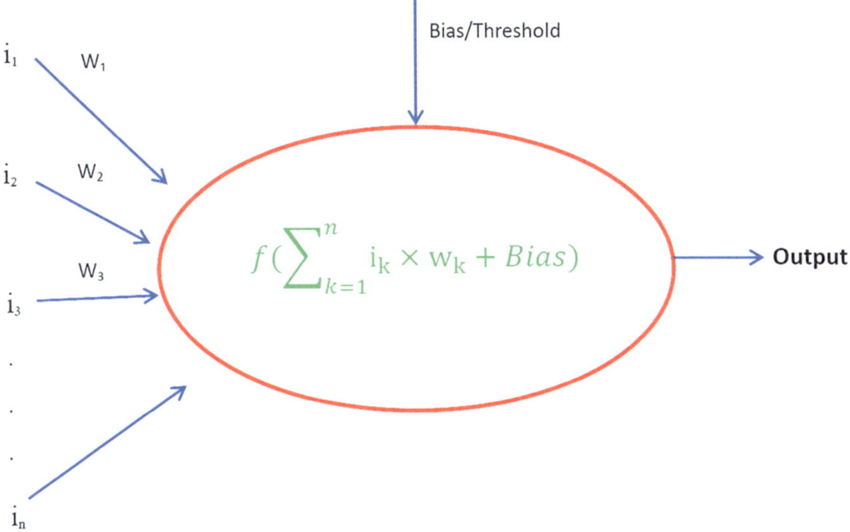

Fig. 2 A typical neuron structure and the neuron output by applying the activation function on net value

Fig. 3 Two popular sigmoid activation/transfer functions.
a logistic function $f(x) = \frac{1}{1+e^{-ax}}$ and
b hyperbolic tangent function $f(x) = \tanh(x)$

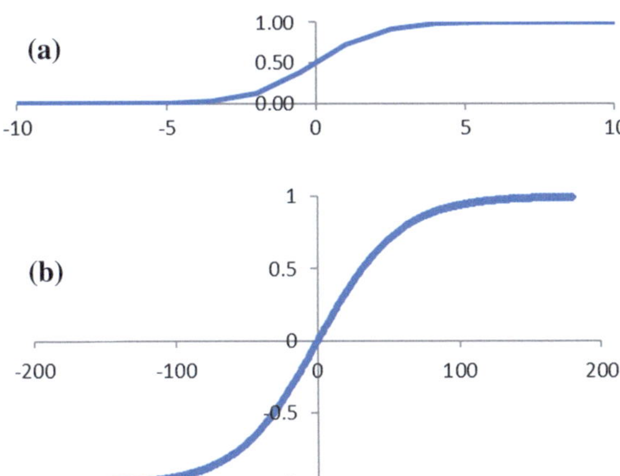

- Piecewise linear function $\{f = 0$ for $x \leq -0.5, f = 0$ for $-0.5 \leq x \leq 0.5$, and $f = 1$ for $x > 0.5\}$,
- Logistic sigmoid function $\left\{f(x) = \frac{1}{1+e^{-x}}\right\}$ which has values between 0 and 1,
- Sigmoid hyperbolic tangent function $\{f(x) = \tanh(x)\}$ with values between -1 and 1.

Among the activation functions, the sigmoid functions are very common. In Fig. 3, the common sigmoid functions have been illustrated, wherein *a* is a constant. The sigmoid functions are more frequently used because they are continuous and of course have positive smooth and derivative. The sigmoid functions, unlike some other functions, do have valid derivatives in all points. The derivative of the *logisticsigmoid* function is indeed smooth as shown below:

$$f'(x) = \frac{-e^{-ax}}{(1 + e^{-ax})^2} = af(x) \times \{1 - f(x)\} \tag{1}$$

The derivative of the *sigmoid hyperbolic tangentfunction* is also smooth as shown below:

$$f'(x) = 1 - \tanh^2(x) = \{1 - f^2(x)\} \tag{2}$$

It is just to note that the derivative of the activation or transfer function in ANN is of great importance in order to give more ability to the network. This is because the neuron activation function must be differentiable and continuous at each point. Consequently, depending on the purpose, the logistic sigmoid and sigmoid hyperbolic tangent functions are normally performing better than the rest.

It is also to be noted that in ANN with no hidden layers, no activation functions are applied and the output of each neuron is the net value (summation plus bias). These ANNs are usually used for simple problems.

3 ANN Structure and Feed-Forward Artificial Neural Networks (FF-ANNs)

As said above, ANNs structurally consist of some neurons which are linked to each other and cooperate to transform inputs into outputs in the best possible manner. However, on a larger scale, ANN structure is in turn made up of an input layer, one or several hidden layers, and also an output one. In each layer, there are one or several neurons.

There exists only one input and only one output layer all the time. However, there can exist a different number of hidden layers. They can be none, one, two, or even more. As a matter of fact, the number of hidden layers depends on the complexity of the problem.

The FF-ANN has been the first and simplest ANN yet introduced. In this type of ANN, as seen in Fig. 4, the flow or movement of information takes place from the input neurons, through the hidden neurons (if any) to the output neurons (only in the forward direction, without any cycles or loops). Simplicity of this network has helped it to be in common use in most of the petroleum engineering applications to date. FF-ANN is normally capable of learning the implicit governing relationship between the inputs and outputs.

The numbers of neurons in the input layer and output layer are normally determined by the problem. However, the number of neurons in the hidden layer has to be specified by the user. Based on the experience of the user, the optimal number

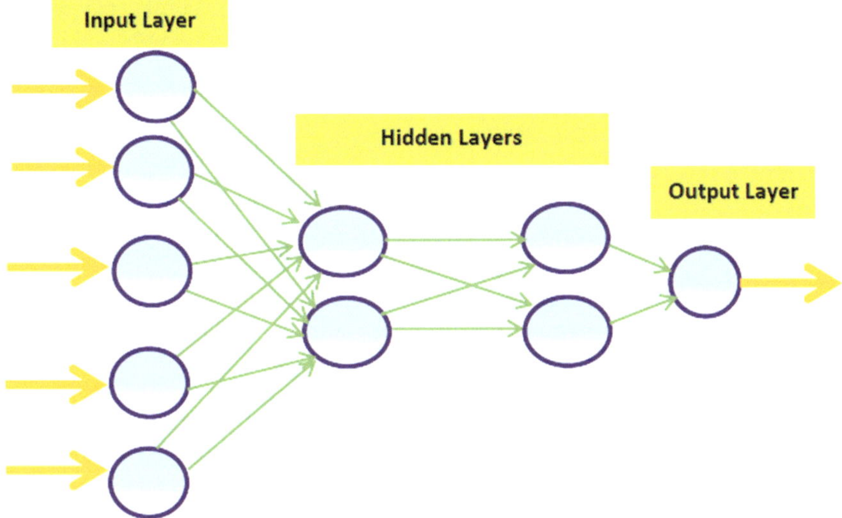

Fig. 4 Typical feed-forward artificial neural networks (FF-ANNs)

of neurons must be determined so that an efficient neural network is obtained. Yet, the only way to determine the optimal number of neurons in the hidden layer is performed by trial and error (based on the user's experience).

4 Backpropagation and Learning

In petroleum engineering application, FF-ANN employs backpropagation of errors in the training phase as their training algorithm (with a supervised learning method). Indeed, backpropagation gets its name from the fact that, during training, the output error is propagated backward to the links or connections between neurons in the previous layers. During this backpropagation of errors, the weights of the links between neurons are adjusted. This process is continued in an iterative manner. In this way, the weights corresponding to links are modified in order to obtain a desired output (with less error), or in other words better learning of the algorithm.

To obtain a more real understanding of the learning mechanism of backpropagation, the expected real output and the predicted output by ANN are compared, and an error function is evaluated. Before the training begins, some initial values are allocated to the links between neurons as the weights. Afterward, by the start of training process, a number of data points are fed to the network to be trained using these real examples. For simplicity, a perceptron[1] ANN (with no hidden layers) consisting of two input neurons and one output neuron is considered. The training data points have

---

[1] A neural network without hidden layer(s).

the frame of (Input$_1$, Input$_2$, V_{ex}). Commonly, the error function utilized to measure the deviation of the expected real output (V_{ex}) and the output by ANN is the mean squared error (MSE). The MSE is found by

$$\text{MSE} = (V_{ex} - O_{ANN})^2$$

where

V_{ex} Real expected output value corresponding to a specified input data point (a fixed value, already known)

O_{ANN} Evaluated value by whole ANN using specified input data points (the output value of the output neuron of the network).

In the mentioned perceptron above (Fig. 5), for instance, suppose a data point of (2, 1, 1) is taken into account for training the network. Please note that the values 2 and 1 are input-independent parameter values, and 1 is the dependent value. In this data point, the value of 1 is the expected real output value. If the mean squared error (MSE) values are plotted on the y-axis versus the possible output values computed by ANN (O_{ANN}) on the x-axis, a parabolic shape is resulted as shown in Fig. 6. The minimum of the parabola is corresponding to the global minimum of the error or MSE (the most favorable point with error equal to zero). The nearer the predicted output value by ANN (O_{ANN}) is to the expected real value (V_{ex}), the less the MSE would be.

Considering $O_{ANN} = \text{Input}_1 w_1 + \text{Input}_2 w_2$ (ignoring the bias for simplicity), the 3D map of the error surface (MSE) could be drawn considering the known data point of (2, 1, 1) as an example (Fig. 7).

The working mechanism of backpropagation algorithm is adjustment of each weight of the network individually based on the selection of the path of the steepest gradient descent to minimize the error function (which is usually MSE). In more details, backpropagation calculates the gradient descent or derivative of the error of the ANN prediction with respect to all the weights existing in the network. For

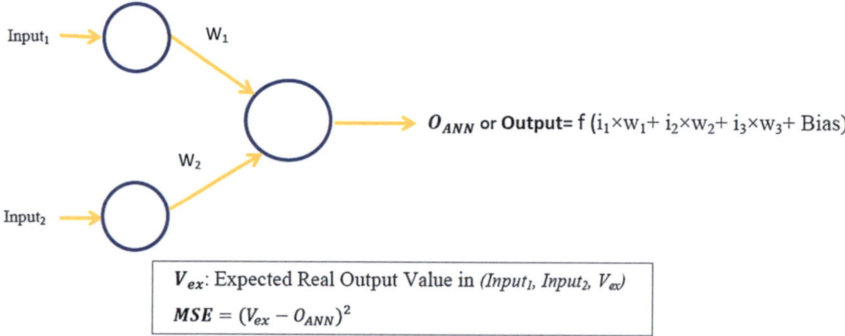

Fig. 5 Example perceptron with 2 input neurons and 1 output neuron for MSE consideration

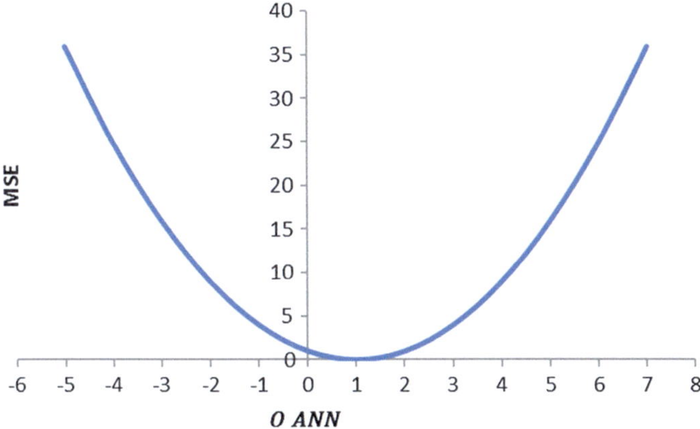

Fig. 6 Ideal 2D graph of (MSE) error versus predicted output by ANN (O_{ANN}) considering a known data point of (2, 1, 1) as an example. The value of 1 is the expected real output. Thus, $\mathrm{MSE} = (1 - O_{\mathrm{ANN}})^2$

Fig. 7 Ideal 3D map of error surface (MSE) versus x (w_1) and y (w_2) considering a data point of (2, 1, 1) for training a perceptron with 2 input neurons

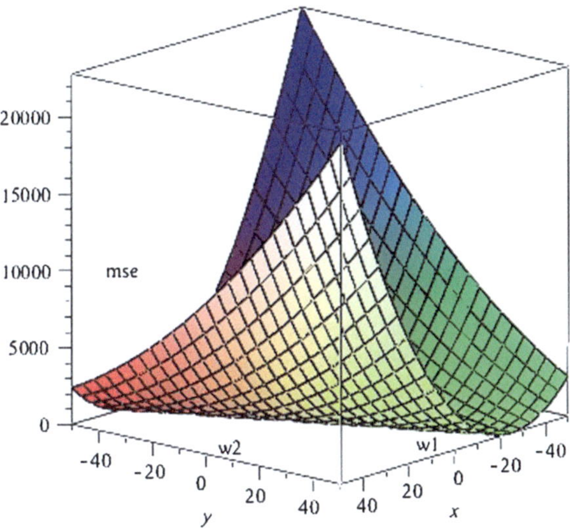

further simplicity of understanding, the learning mechanism by which backpropagation reduces MSE of the ANN (highest gradient descent) is analogous to the way a mountain climber can descend a hill just by selecting the steepest path down the hill at each point. The hill steepness and path that the climber has to select and go through at each point could be, respectively, considered as representative of the slope and gradient of the error surface at that point.

Ideally, it is assumed that only one global minimum exists in the error surface. But this is not necessarily the case (there may be a lot of local minima and also maxima

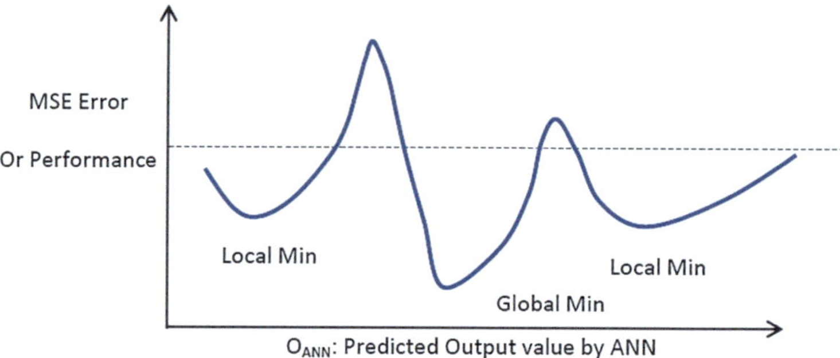

Fig. 8 Backpropagation (with gradient descent error mitigation mechanism) can only find the local minimum of error, which is not necessarily the global minimum of error

in the error surface as shown in Fig. 8). In reality, backpropagation learning algorithm (with gradient descent) would finally converge to an error minimum which is actually a local minimum of error. Undoubtedly, this local minimum of error may not be necessarily global at all. All optimizing algorithms such as genetic algorithm, ant colony, and particle swarm optimization have been designed to give the ANN backpropagation more capability to escape local error minima and reach global minimum of error.

If the initial point of the gradient descent process in backpropagation is somewhere between a local minimum and a local maximum, the training of the network would finally lead to the local minimum, which is not desirable to the user. This dependence of the ANNs which learn by backpropagation algorithm on the initial starting point is an important limitation to this algorithm (Fig. 8). Giving several different random initial values prior to each training is required to prevent trapping into local minima.

It is just to note that weights corresponding to the links of the network are the only variables that can be modified by the network to minimize the error. Still, to the authors' information, modification of ANN structure (number of neurons and hidden layers) with the objective of error reduction is not possible by the network itself, except by self-user's trial and error or automatic procedures.

As backpropagation is based on calculation of the MSE gradient with respect to all weights existing in the network, one of the requirements of the backpropagation is that differentiable activation functions should be used in neurons. This is the reason why sigmoid functions, as differentiable functions, are so popular in petroleum engineering and geoscience applications.

As said before, the MSE is found by

$$\text{MSE} = \frac{1}{2}(V_{\text{ex}} - O_{\text{ANN}})^2 \tag{3}$$

Please note that the ½ factor has been added so that the derivative has no coefficient: $\text{MSE}' = (V_{\text{ex}} - O_{\text{ANN}})$.

For perceptrons (ANNs with no hidden layers), the activation function is linear or O_{ANN} is simply the weighted sum of the inputs as follows:

$$O_{\text{ANN}} = \sum_{i=1}^{n} W_i \times \text{Input}_i + \text{Bias} \tag{4}$$

where

Input_i Input value to output neuron from neuron i

W_i Weight corresponding to link i (between input neuron i and the output neuron).

For multilayer ANNs, O_{ANN} is found after application of a nonlinear activation function as follows:

$$O_{\text{ANN}} = f(\text{Net}) = f\left(\sum_{i=1}^{n} W_i \times \text{Input}_i + \text{Bias} \right) \tag{5}$$

where

f Activation function (usually a sigmoid function).

Since backpropagation applies gradient descent method for error reduction (as said before), the derivate of MSE gradient with respect to weights in the network is calculated utilizing the chain rule of partial derivatives as follows:

$$\frac{\partial \text{MSE}}{\partial W_i} = \frac{\partial \text{MSE}}{\partial O_{\text{ANN}}} \times \frac{\partial O_{\text{ANN}}}{\partial \text{Net}} \times \frac{\partial \text{Net}}{\partial W_i} \tag{6}$$

where

$\frac{\partial \text{Net}}{\partial W_i}$ Rate of change of net value with respect to weight $i = \text{Input}_i$

Note $\text{Net} = \sum_{i=1}^{n} W_i \times \text{Input}_i$

$\frac{\partial O_{\text{ANN}}}{\partial \text{Net}}$ Rate of change of the output value from output neuron with respect to net value.

$= \frac{\partial f}{\partial \text{Net}} = O_{\text{ANN}} \times (1 - O_{\text{ANN}})$ (for logistic f)

$= \frac{\partial f}{\partial \text{Net}} = (1 - O_{\text{ANN}})^2$ for tanh

Note Activation functions (f) are normally considered as sigmoid functions (logistic or tangent hyperbolic sigmoid function):

$O_{\text{ANN}} = f(\text{Net}) = \frac{1}{1 + e^{-\text{Net}}}, f(\text{Net}) = \tanh(\text{Net})$

$\frac{\partial \text{MSE}}{\partial O_{\text{ANN}}}$ Rate of change of MSE with respect to the output value from output neuron.

$= V_{\text{ex}} - O_{\text{ANN}}$

Note $\text{MSE} = \frac{1}{2}(V_{\text{ex}} - O_{\text{ANN}})^2$

$\frac{\partial \text{MSE}}{\partial W_i}$ Rate of change of MSE with respect to weight i

$$= (V_{\text{ex}} - O_{\text{ANN}}) \times O_{\text{ANN}}(1 - O_{\text{ANN}}) \times \text{Input}_i \quad \text{(for logistic)} \tag{f}$$

$$= (V_{\text{ex}} - O_{\text{ANN}}) \times (1 - O_{\text{ANN}})^2 \times \text{Input}_i \quad \text{(for tanh) (weight corresponding to link from neuron } i)$$

Note $\frac{\partial \text{MSE}}{\partial W_i} = \frac{\partial \text{MSE}}{\partial O_{\text{ANN}}} \times \frac{\partial O_{\text{ANN}}}{\partial \text{Net}} \times \frac{\partial \text{Net}}{\partial W_i}$

To summarize the above calculations, the gradient of MSE with respect to W_i is equal to

$$\frac{\partial \text{MSE}}{\partial W_i} = (V_{\text{ex}} - O_{\text{ANN}}) \times f' \times \text{Input}_i \tag{7}$$

If the logistic sigmoid function is used for the activation function or f, we have

$$\frac{\partial \text{MSE}}{\partial W_i} = (V_{\text{ex}} - O_{\text{ANN}}) \times O_{\text{ANN}}(1 - O_{\text{ANN}}) \times \text{Input}_i \tag{8}$$

If the *sigmoid hyperbolic tangent function* is used for the activation function (f), we have

$$\frac{\partial \text{MSE}}{\partial W_i} = (V_{\text{ex}} - O_{\text{ANN}}) \times (1 - O_{\text{ANN}}^2) \times \text{Input}_i \tag{9}$$

ΔW_i is found by multiplying $\frac{\partial \text{MSE}}{\partial W_i}$ by the learning rate (α). Thus, for a multilayer ANN, the value of weight change is equal to

$$\Delta W_i = \alpha(V_{\text{ex}} - O_{\text{ANN}}) \times f' \times \text{Input}_i \tag{10}$$

For the logistic sigmoid activation function, we have

$$\Delta W_i = \alpha(V_{\text{ex}} - O_{\text{ANN}}) \times O_{\text{ANN}}(1 - O_{\text{ANN}}) \times \text{Input}_i \tag{11}$$

For the tangent hyperbolic sigmoid activation function, we have

$$\Delta W_i = \alpha(V_{\text{ex}} - O_{\text{ANN}}) \times (1 - O_{\text{ANN}}^2) \times \text{Input}_i \tag{12}$$

For a perceptron ANN, f is linear (or f' is equal to 1). Thus, the value of weight change is equal to

$$\Delta W_i = \alpha(V_{\text{ex}} - O_{\text{ANN}}) \times \text{Input}_i \tag{13}$$

Please note that in the above calculations, for simplicity, the value of weight change (stated above) has been evaluated using only one data point. It is also reminded, in the MSE function yet considered, it was assumed that only one neuron exists in the output layer. Generally, mean square error (MSE) can be evaluated using

$$\text{MSE} = \frac{1}{2} \sum_{k=1}^{T} \sum_{j=1}^{m} \left\{ (V_{\text{ex},j}(K) - O_{\text{ANN},j}(K)) \right\}^2 \tag{14}$$

where

T Number of training samples from known data point given for training the network

(1) $(\text{Input}_1, \text{Input}_2, \ldots, V_{\text{ex},1}, V_{\text{ex},2}, \ldots$
(2) $(\text{Input}_1, \text{Input}_2, \ldots, V_{\text{ex},1}, V_{\text{ex},2}, \ldots$

$\ldots$

(T) $(\text{Input}_1, \text{Input}_2, \ldots, V_{\text{ex},1}, V_{\text{ex},2}, \ldots$

m Number of output neurons or nodes

$V_{\text{ex},j}(K)$ Expected real value of output no. k $(\text{Input}_1, \text{Input}_1, \ldots, V_{\text{ex},1}(K)$

$O_{\text{ANN},j}(K)$ Predicted or estimated value by ANN.

4.1 *Perceptrons and Backpropagation Algorithm*

The ANN structure, in which there are no hidden layers, is called perceptron. A perceptron is indeed just like a simple neuron. Perceptrons are only applicable in linear simple problems. In perceptrons, only the simple bias/threshold activation function is utilized, namely by adding a bias/threshold value to the summation. A typical perceptron neural network (with no hidden layer) is shown in Fig. 9.

It is noted that training is usually performed in an iterative manner. An *epoch* is indeed the process of providing the network with the entire training data points, calculating the network output and error function, and modifying the network's weights for

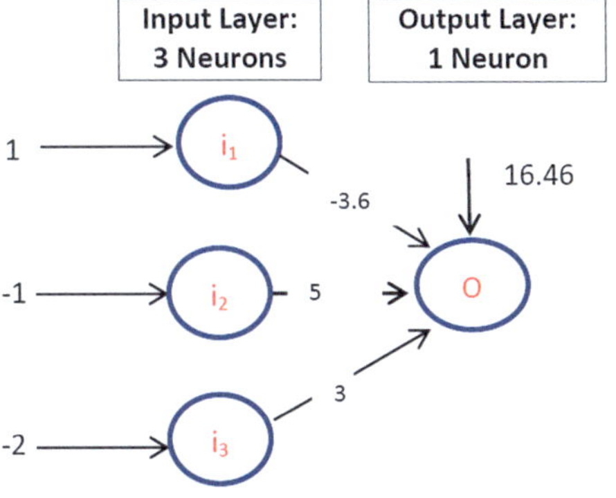

Fig. 9 A typical perceptron as an ANN example

the next epoch. An epoch is composed to several iterations (providing or presenting one data point to the network can be considered as an iteration). Based on complexity of the problem and the ANN, sometimes a large number of epochs are required for training the network.

Indeed, when the input data are presented to the network, the flow of information is forward (FF-ANN). However, as said previously, backpropagation is the backward error propagation of weight adjustments.

The calculations corresponding to the composing neurons of the perceptron in Fig. 9 are shown below.

In each epoch, the net value (summation plus the bias) and final output value of the neuron (O) are calculated by the perceptron by the ANN as follows:

$$\text{Output} = 1 \times (-3.6) + (-1) \times 5 + (-2) \times 3 + 16.46 = 1.86$$

Now, suppose that the real expected output value is equal to 1. If so, in each next epoch for each data point, the weight values are changed such that the output value gets nearer to the real expected value. This process continues until the nearest possible output values to the real expected values are obtained. In this simple ANN, the modified weights are evaluated as follows:

$$\Delta W_i = \alpha (V_{\text{expected}} - O_{\text{ANN}}) \times \text{Input}_i$$

$$W_{i,\text{modified}} = W_{i,\text{previous}} + \alpha (V_{\text{expected}} - O_{\text{ANN}}) \times \text{Input}_i \tag{15}$$

where

ΔW_i Magnitude of change of the weight corresponding to link i (between neuron i and output neuron)

α Learning rate (a constant)

V_{expected} Real expected output value corresponding to a specified input data point (already known)

O_{ANN} Evaluated value by whole ANN using specified input data points (the output value of the output neuron)

Input_i Input value to output neuron from neuron i

$W_{i,\text{previous}}$ Value of the weight in the previous epoch.

Thus, assuming α (called learning rate) to be equal to 1, the modified weights are as follows:

$$W_{1,\text{ mod }.} = -3.6 + 1 \times (1 - 1.86) \times 1 = -4.46$$
$$W_{2,\text{ mod }.} = 5 + 1 \times (1 - 1.86) \times (-1) = 5.86$$
$$W_{3,\text{ mod }.} = 3 + 1 \times (1 - 1.86) \times (-2) = 4.72$$

Thus, after training the network during one epoch using the data point {(1, −1, −2) as input and 1 as output}, the above-modified weight values would be used in the next epoch.

4.2 Multilayer ANNs and Backpropagation Algorithm

For more complex problems, normally the ANN should have hidden layers. These neural networks are called multilayer ANNs. Multi-layer ANNs could be considered as a developed or extended perceptron. The structure of a multilayer ANN with one hidden layer has been shown in Fig. 10. Application of multilayer ANN is not only restricted to very simple linear problems but can also utilized for complex or non-straightforward problems.

Unlike perceptrons, in multilayer ANN, the activation function is not just a bias/threshold, but activation functions (usually sigmoid functions) are usually applied. As could be seen in Fig. 10, the ANN structure is composed of one input layer (consisting of three neurons), one output layer (consisting of one neuron), and also one hidden layer (consisting of two neurons) as a typical example. For most engineering purposes, only one or two hidden layers at most are normally adequate to be used in the structure. The number of neurons in the hidden layers is crucial in the performance and functionality of the network. However, depending on the complexity of the problem, an optimization should be made not to use too many neurons in the hidden layers as this causes over-fitting or over-training.

For the multilayer ANN shown in Fig. 10, in each epoch, the net value and the output of each neuron in the hidden and output layers are calculated by the ANN as follows:

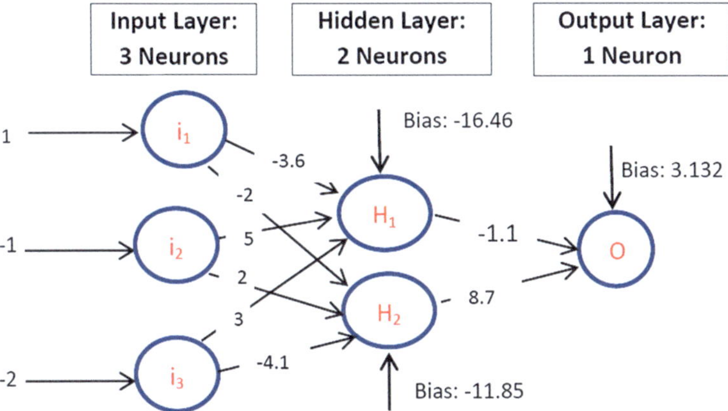

Fig. 10 A multilayer ANN with one hidden layer (a logistic sigmoid function has been considered as the activation/transfer function)

H$_1$: the net value and output of this neuron are found as follows:

$$\text{Net} = 1 \times (-3.6) + (-1) \times 5 + (-2) \times 3 + 16.46 = 1.86$$

$$\text{Output} = \frac{1}{1 + e^{-1.86}} = 0.86$$

H$_2$:

$$\text{Net} = 1 \times (-2) + (-1) \times 2 + (-2) \times (-4.1) - 11.85 = -7.65$$

$$\text{Output} = \frac{1}{1 + e^{-(-7.65)}} = 4.75 \times 10^{-4}$$

O: the net value and output of this neuron (which is the output of the ANN) are found as follows:

$$\text{Net} = 0.86 \times (-1.1) + 4.75 \times 10^{-4} \times 8.7 + 3.132 = 2.187$$

$$\text{Output} = \frac{1}{1 + e^{-2.187}} = 0.9$$

Now suppose that the real expected value is equal to 1. If so, in each next epoch for each data point, the weight values are changed or modified such that the output value gets nearer to the real expected value (here equal to 1). This process continues until the nearest possible output values to real expected values are obtained. In this multilayer ANN, the modified weights are evaluated as follows:

$$\Delta W_i = f' \times (V_{\text{expected}} - O_{\text{ANN}}) \times \text{Input}_i$$

Replacing the *logistic sigmoid activation function* derivative for the output neuron yields

$$\Delta W_i = \alpha O_{\text{ANN}}(1 - O_{\text{ANN}}) \times (V_{\text{expected}} - O_{\text{ANN}}) \times \text{Input}_i$$

The modified weight for this sigmoid function is as follows:

$$W_{\text{modified}} = W_{\text{previous}} + \alpha O_{\text{ANN}}(1 - O_{\text{ANN}}) \times (V_{\text{expected}} - O_{\text{ANN}}) \times \text{Input}_i \quad (16)$$

Replacing the *hyperbolic tangent sigmoid activation function* derivative for the output neuron yields

$$\Delta W_i = \alpha(1 - O_{\text{ANN}})^2 \times (V_{\text{expected}} - O_{\text{ANN}}) \times \text{Input}_i$$

The modified weight for the *hyperbolic tangent sigmoid function* is as follows:

$$W_{\text{modified}} = W_{\text{previous}} + \alpha(1 - O_{\text{ANN}})^2 \times (V_{\text{expected}} - O_{\text{ANN}}) \times \text{Input}_i \qquad (17)$$

where

O_{ANN} Evaluated value by whole ANN using specified input data points (the output value of the output neuron). This is the fixed value for each epoch

f' The derivative of the activation function corresponding to the output neuron

Input_i Input value of the neuron i

$W_{i,\text{previous}}$ Value of the weight corresponding to link i in the previous epoch

$W_{i,\text{modified}}$ Value of the weight corresponding to link i to be used for the next epoch.
   Input value is the value that enters each neuron. If the neuron is in the hidden layer, the *input* value is the **net** value (summation plus bias) of the neuron. In perceptrons (without hidden layer ANN) which are applicable to simple linear problems, the activation function is indeed $f(x) = \alpha x$. Therefore, applying the above relations to perceptrons would yield f' to be equal to 1 or α (15).
   Thus, assuming α (called learning rate) to be equal to 1, the modified weights are evaluated as follows:

$$W_{\text{modified}} = W_{\text{previous}} + \alpha O_{\text{ANN}}(1 - O_{\text{ANN}}) \times (V_{\text{expected}} - O_{\text{ANN}}) \times \text{Input}_i$$

$$W_{I1-H1} = -3.6 + 1 \times 0.9(1 - 0.9) \times (1 - 0.9) \times 1 = -3.591$$

$$W_{I2-H1} = 5 + 1 \times 0.9(1 - 0.9) \times (1 - 0.9) \times (-1) = 4.991$$

$$W_{I3-H1} = 3 + 1 \times 0.9(1 - 0.9) \times (1 - 0.9) \times (-2) = 2.982$$

$$W_{I1-H2} = -2 + 1 \times 0.9(1 - 0.9) \times (1 - 0.9) \times 1 = -1.991$$

$$W_{I2-H2} = 2 + 1 \times 0.9(1 - 0.9) \times (1 - 0.9) \times (-1) = 1.991$$

$$W_{I3-H2} = -4.1 + 1 \times 0.9(1 - 0.9) \times (1 - 0.9) \times (-2) = -4.118$$

$$W_{H1-O} = -1.1 + 1 \times 0.9(1 - 0.9) \times (1 - 0.9) \times 1.86 = -1.083$$

$$W_{H2-O} = 8.7 + 1 \times 0.9(1 - 0.9) \times (1 - 0.9) \times (-7.65) = 8.631$$

Thus, after training the network during one epoch using the data point $\{(1, -1, -2)$ as input and 1 as output$\}$, the above-modified weight values would be used in the next epoch. This process takes place for all the data points given to the network in the training stage.

It is noted that if the learning rate is too low, the learning process would be too slow. If the learning rate is too high, the weights and objective function would diverge and no good learning would occur. In linear problems, proper learning rates could be computed using the Hessian matrix (Bertsekas and Tsitsiklis 1996). It is also possible to adjust the learning rate while training. A number of proposals exist in the neural network literature to adjust the learning rate while training. However, most of them are not effective. Among these works, Darken and Moody (1992) could be named.

Please note that the ANN structures discussed above are feed-forward ANN (FF-ANN) with backpropagation algorithm for weight modification. This has been very popular in petroleum engineering practices. Thus, these types of neural networks are further discussed as follows.

5 Data Processing by ANN

Typically, processing of ANN data is performed in three sections including training, testing/calibration, and validation/verification phases. For this purpose, the number of the known data points comprising input and output values is divided into two categories. In most of the petroleum engineering practices performed by the authors, typically the number of available data points (inputs and output values available) is divided into two independent parts: (1) 60% of the data points for utilization in the training phase and (2) 40% of the data points for utilization in testing. Some users test the trained neural network in two stages: (a) 20% of the data points for first testing of the trained network and (b) 20% of the data for second testing purposes. Depending on the number of data points available and the experience of the users (which is very important), different users adopt different divisions of the data. Some may take 70% of the available data points for training and the rest for testing. The number of data points plays an important role in successful training. Fake data points could cause trouble for successful training particularly if their number is many.

Then, if testing gives promising results, the user can go for validation or verification phase. In this phase, some new input data is applied to the trained ANN to get the network outputs. The network outputs are thus considered to be very near to the real ones and could be used for instance in petroleum engineering.

Indeed, during the training phase, the desired network is developed. Then, it is tested. When the training and testing process has been finished successfully (after training and simultaneously testing the network by test data points), the network is applied to the validation data points in order to predict outputs using new input data.

5.1 Training

As one of the main similarities between artificial neural networks (ANNs) and biological neural networks, both have the ability to learn or to be trained. As said earlier, the output of a neuron is a function of the net value (which is the weighted sum of the inputs plus a bias). Indeed, after successful training, an ANN can have the capability of creating reasonable outputs using new inputs (during testing and validation phases). The more reasonably the neural networks respond to the new data, the neural networks have got higher functionality in terms of generalized prediction. Good training is thus of great importance to enhance the functionality of ANN.

In the beginning, the ANN allocates a random weight to each input value. Then, during the training process, when the inputs are introduced to the network, the weights (corresponding to the links between all the neurons of the network) would be adjusted or modified such that finally the ANN output(s) is/are very near to the expected real output value(s). Depending on how close the output created by ANN is to the expected real output, the weights between the neurons are modified such that if the same inputs are introduced to the network, the network would provide an output pattern closer to the expected real output.

Surely, if the difference between the created data (by ANN) and real data is considerable, more modification of the weights is performed during training. In this way, at any time in training, a kind of memory is attached to the neurons which store the weights in the past computations. Finally, if convergence is attained after training, we expect the ANN to give outputs which are in proximity to expected real outputs. In Fig. 9 (perceptron) and the corresponding calculations, it was previously shown how the weights were modified ($W_{1,\mathrm{mod}}$, $W_{2,\mathrm{mod}}$, and $W_{3,\mathrm{mod}}$) during training.

In more details, it could be stated that the training or the learning stage of the network is accomplished by summing up the errors at the output, creating a cost/risk function in terms of network inputs and weights, and minimizing the cost function with respect to the network inputs. The cost function is basically based on the mean squared error (MSE) of the outputs. The process of MSE error mitigation during training takes place in an iterative process during many iteration times. Each iteration, which is indeed the process of providing the network with inputs and modifying the network's weights, is called an epoch. Normally, a lot of epochs are required to train a typical ANN.

Training continues as said above until the created output value or pattern complied with the quality criteria based on statistical error values. In other words, the stopping criteria are based on the minimization of MSE (Cacciola et al. 2009). This type of training, wherein both inputs and actual outputs are supplied to the network (called Supervised Training) is usually more common in practice. In unsupervised training, only input values are supplied and ANN adjusts its own weights such that similar outputs are given out of the network while inserting similar inputs. Most of the neural network applications in the oil and gas industry are based on supervised training algorithms (Mohaghegh 2000).

5.2 Over-Fitting

In iterative backpropagation ANNs which are commonly used in petroleum industry, there is a serious problem which arises after too much training of the network. It is important to know when to stop training the network and go for testing and validation phases. As a matter of fact, too much training would cause over-fitting or over-training. Over-fitting is also called memorization as well because in this case, the network would indeed memorize the data points used in training and give a very accurate match with them, but it would lose its generalization capability which is required in testing and validation phases. Please note that over-fitting does not really apply to ANNs which are trained using non-iterative processes. However, it can be said that ANNs used in petroleum are normally iterative.

Indeed, during the initial phase of training, both the training and validation set errors show a decreasing trend with time or number of epochs. It is noted that training and validation set errors are, respectively, errors corresponding to training and valida-tion data points. Nevertheless, when the over-fitting of the ANN starts, the validation error suddenly starts to increase, while the training error still continues its decreasing trend (Fig. 11). The blue dots in Fig. 12 illustrate the train data points (X, Y). The blue and red lines in Fig. 12, respectively, show the true functional relationship and the ANN learned function. As can be seen, due to too much training, there is a large difference between the true functional curve and the ANN learned function in points other than train data points. This is representative of over-fitting. The input values have been shown in the x-axis for simplicity, and the output values have been denoted by the values in the Y-axis. The saved ANN weights and biases/thresholds are corresponding to the minimum of the validation set error.

Figure 12 illustrates the problem of over-fitting in machine learning. The blue dots represent training set data. The blue line represents the true functional relationship, while the red line shows the learned or trained function, which is shown to be under the effect of over-fitting.

It is noted that if the test set error had shown a considerable increase before the validation set error, it could be guessed that over-fitting could have happened. Thus, in Fig. 13, for instance, there are no worries about over-fitting until stopping point at epoch number 6. But surely after epoch 6, again over-fitting would happen. The minimum validation error is at epoch no. 6, but the training has continued 6 more

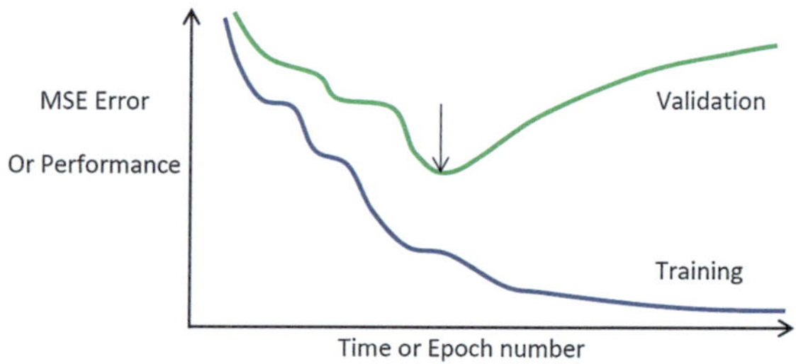

Fig. 11 The sudden increase of validation set error (*black arrow*) shows over-fitting in the graph of MSE versus time/epoch number

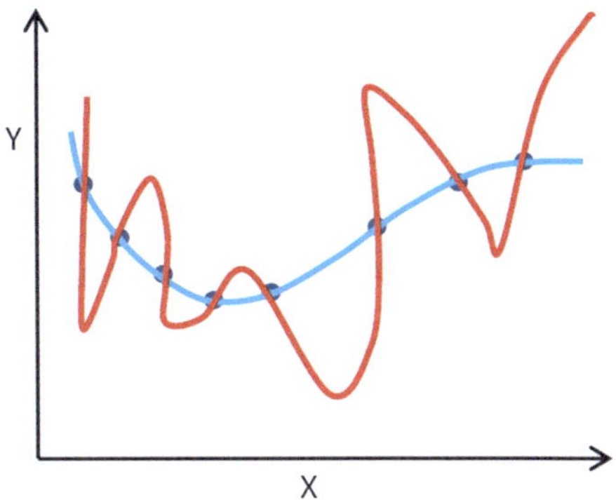

Fig. 12 Over-fitting. Too much complexity of the ANN due to too much learning (*red curve*) has caused the ANN estimated values by ANN learned function at inputs other than train values to be very different from true functional relationship or curve (*blue curve*). *x*-axis and *y*-axis are, respectively, representative of inputs and ANN outputs

epochs until epoch number 12 (the user has instructed the network to do so). Thus, it is possible to rely on the ANN weights and biases adjusted based on stopping point at epoch 6. Just to remind that as time passes by (epoch increase), the ANN is getting more and more complex.

In order to prevent over-fitting, first it is required that the network is not too complex that models even the noises. Second, before the termination of the training process, the process is commonly stopped from time to time so that the network generalization capability is checked using the test data points. If the network performed well in the test phase (if the MSE in the test phase is in the magnitude of 0.001 or a little higher), no further training is required. Since the outputs of the test data points are not utilized in the training phase, the network prediction or generalization capabilities could be analyzed by comparing the network outputs (using the input values in testing phase) with the real test output values.

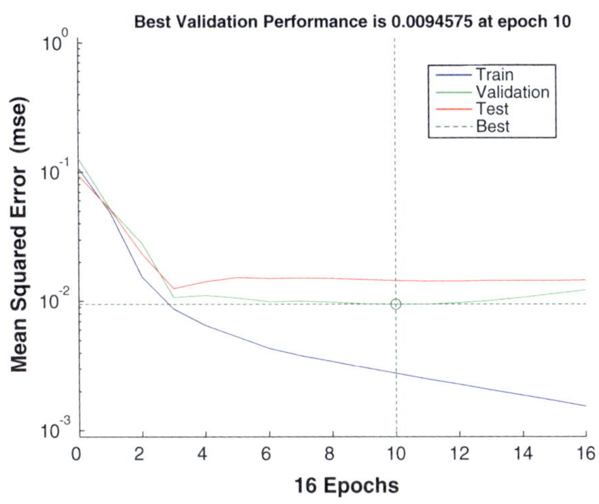

Fig. 13 At epoch no. 10, the error of the second test or validation error (MSE as its set error function) and also the first test set error increased, while training set error is still decreasing. But, as the increase of the validation set error is not too much sharp, it is no sign of over-fitting

5.3 Testing

After the training stage, we reach to the testing or calibration stage wherein the weights between the neurons are tested. Testing takes place using input–output pairs that have not been utilized in the training stage. The difference between the desired and the actual output can show if enough training has happened. It is noted that it is usually a common fault among users with not enough experience to test/calibrate the ANN using the same data points used for training.

5.3.1 Validation

Validation is the last stage in the ANN application or neural data processing by ANN. To summarize the data processing by ANN, during the training phase, a trained network is developed. Then, it is tested. If it shows good performance during testing, it is recognized as a proper network to be used with new data. During validation, the user can apply the trained network to obtain results. In this stage, new input data (with unknown outputs) is supplied to the trained ANN in order to obtain the network outputs. As the trained network has performed well in the testing stage, the outputs obtained in the validation stage are considered to be trustworthy and are taken for granted as nearest to real.

It must be noted that the validation stage is usually considered as the second test. If so, the validation and test error curves versus the number of epochs are plotted along with the training error curve. The validation and test error curves should also have the declining trend just like the training error curve before convergence. After convergence, the validation and test error curve would rise, while the training error curve would continue declining showing too much training (over-fitting). Therefore, if the training curve is declining while the validation and test ones are rising and do not show a decline before convergence, it indicates a problem with the network. To help remove the problem with the network, it is suggested to follow the steps below:

1. Re-check the data: Among the data points, there might be some wrong or faulty ones. Try to detect them and delete them from the available data points.
2. Reconsider the effective parameters: Try to consider one or more new input parameters whose effect on the output parameter might have been ignored. This requires a review of the problem and identifying the effective parameters to find its corresponding values.

According to the author's experience, the error was reduced dramatically using the above two methods.

Sometimes, the output is not dependent on some of the data, or in other words, its corresponding sensitivity is very small. This case can be found by sensitivity test analysis.

Please note that normalizing the data can also help to obtain better results from the network performance. Normalizing means arranging all the available data values in the range of 0–1. By knowing the maximum and minimum values among the data, normalizing is possible. Thus, the normalized value is found by the following relation:

$$\delta = \frac{d - d_{\min}}{d_{\max} - d_{\min}} \tag{18}$$

It is noted that some users consider the MSE error in the validation phase as the model quality. Some others may consider the error in the test phase.

6　ANN and Validation Error Statistical Parameters

As mentioned in the text, MSE is the main error function to compare the neural network models and performance and in more detail the discrepancy between the expected real values and the output by ANN. Additionally, several other statistical parameters are also utilized to report this discrepancy. Collectively, these error parameters include mean squared error (MSE), average root mean squared error (ARMSE), average percent error (APE), average absolute percent error (AAPE), Pearson correlation coefficient (R), squared Pearson correlation coefficient (R^2), standard deviation (SD), and variance (V). It is noted that variance is simply the square of standard deviation.

It is to note that R^2 can be used along with R to analyze network performance. MSE and R^2 collectively can give an indication of the performance of the network. Generally, an R^2 value greater than 0.9 indicates a very satisfactory model performance, while an R^2 value in the range 0.8–0.9 signifies a good performance, and a value less than 0.8 indicates a rather unsatisfactory model performance (Coulibaly and Baldwin 2005).

Definitions and mathematical relations of these parameters are given in the Appendix.

7　Sequential Forward Selection of Input Parameters (SFS)

As said before, one of the challenges in neural network modeling is the determination of the important effective input parameters on the output. Certainly, experience and knowledge of the problem could help in determination of the effective parameters or selection criterion. However, it is better to utilize a systematic method to determine the impact of each parameter, rank them. Using sequential forward selection (SFS), each of the input parameters is considered individually. Then, the input parameters with the highest impact on the output parameters are detected in successive stages so that ranking of the parameters could be performed on the basis of highest to lowest impact. For more clarity on the procedure, an example is given below.

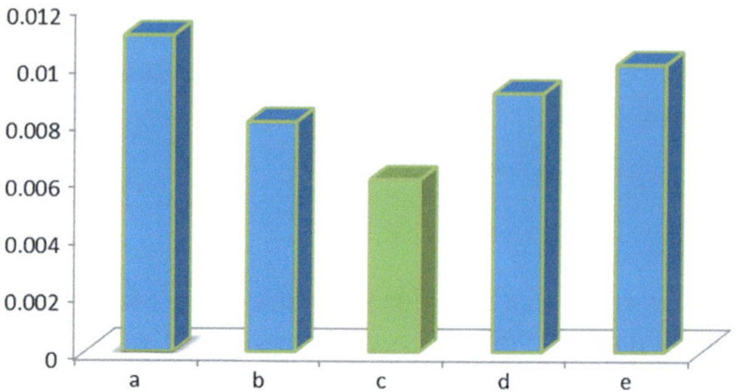

Fig. 14 1st stage of SFS. The MSE error corresponding to the 5 neural networks each just trained by a single input parameter (a to e). The input parameter with the most impact is c

Suppose 5 input parameters denoted by a to e have been characterized as important parameters of a problem and pressure is the output parameter. To utilize SFS for ranking the parameters which have the greatest impact, in the first stage, 5 neural network models are constructed each just using one single input parameter. To compare their performance, the errors of all these single input parameters are evaluated. As illustrated in Fig. 14, using single parameter c, for instance, network creation leads to the least error of all. Thus, parameter c is the most effective input parameter or with the highest impact on output parameter which is bottom hole pressure (BHP).

In the second stage, four neural networks using two input parameters are constructed such that input parameter c is surely considered (fixed parameter) and the second parameter is changed among the seven remaining parameters. In this way, four neural networks are constructed. Now, their corresponding errors are calculated. Out of these four networks, the second parameter of the network with the least error is selected as the second most effective parameter. For instance, this could be parameter no. e (Fig. 15).

At the third stage, three neural networks are constructed using three input parameters such that input parameters c and e as the first two most influential parameters are considered (fixed parameters) and the third input parameter is changed among the three remaining parameters. The above SFS process is continued until the ranking of all the input parameters is performed (Fig. 16). Collectively, 15 neural networks will have been considered to know the final ranking. After finding the input parameters with maximum to minimum effect, the errors of five neural networks corresponding

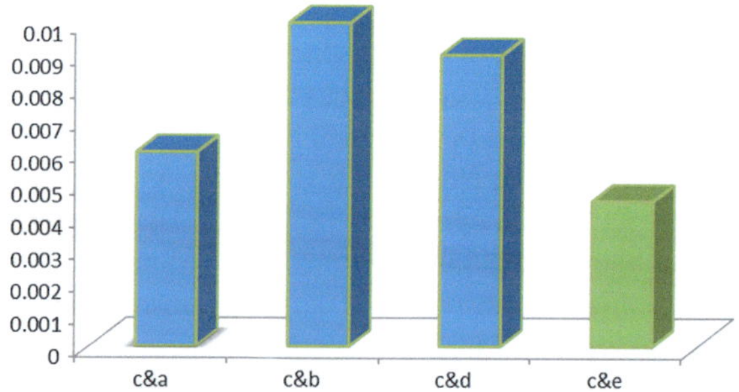

Fig. 15 The MSE error corresponding to the 4 neural networks each just trained by a single input parameter (a to e). The second most effective parameter is e

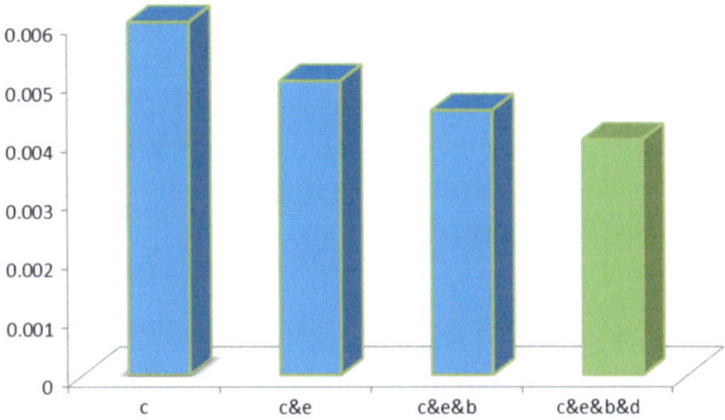

Fig. 16 Final ranking of input parameters

to each stage of SFS (first constructed with the single most effective input parameter, second constructed with two most effective input parameters, etc.) are compared with each other. Finally, Fig. 16 is yielded which shows how adding input parameters could reduce error.

8 Applied Examples of ANN in Geoscience and Petroleum Engineering

Among the many applications of ANN in geoscience and petroleum engineering, some applied examples are given as follows.

8.1 Multiphase Flow

Among really complex problems, multi-phase flow problems could be considered. One important parameter in two-phase flow in wells is the bottom hole pressure (BHP). In underbalanced drilling (UBD), the BHP should be kept at less than formation pore pressure and above wellbore collapse pressure (to prevent wellbore instability). Thus, the prediction capability of predicting BHP in UBD operations is of great importance which could leave the necessity of measuring devices bottom hole. Different approaches have their own defects. For the following reasons, the errors of the mentioned approaches are not small:

Because of the complexity of the phenomenon, indeed no analytical solutions exist for multi-phase problems. Empirical multi-phase flow correlations sometimes over-predict, make extrapolations risky, and are susceptible to uncertainties.

The use of mechanistic modeling approaches has been increasing in multi-phase flow problems in pipes; however, utilizing them in annular flow problems has not been very promising.

Taking into account the lack of analytical solutions in multi-phase problems, artificial neural network was utilized to evaluate BHP in multi-phase annular flow while UBD operations (Ashena et al. 2010). Thus, BHP was considered as the output parameter. An attempt was made to find the parameters which have an influence on the output parameter (BHP) or BHP depends on. This is obtained just by a brief study of multi-phase flow relations and also experience. Seven parameters were found as effective parameters:

- Rate of liquid or diesel injection (in gallon per minute or gpm)
- Rate of gas or nitrogen injection (in standard cubic foot per minute or scfm)
- Measured depth or MD (in meter)
- True vertical depth or TVD (in meter)
- Inclination angle from the vertical (in degrees)
- Well surface pressure (in psi)
- Well surface temperature (in degrees Celsius).

Qualitatively speaking, it must be noted that the more the value of item 1 (rate of liquid injection), the more the value of BHP. The greater the value of item 2 (rate of gas injection), the less the value of BHP. Please note that as items 3–5 (measured depth, true vertical depth, and inclination angle) collectively take the hole depth in vertical and directional wells into account, they were considered as input parameters. Normally, the neural network structure is itself responsible for recognizing the extent of the input parameter strength or importance.

The number of 163 data points with the above 7 input parameters and measured BHP (the only output parameter) was collected. About 10 data points were found to cause too much error. After their deletion, the ANN performance was enhanced.

153 collected data points should be divided for use in training and testing phases. 60% of them were allocated to training, 20% to testing-1, and 20% to validation (or testing-2) purposes. Some additional data have been used as simulation data with known measured values as well. As the number of the available collected data points was limited, only limited number of neurons and layers was tried to be utilized. Feed-forward ANN with backpropagation algorithm of learning was set as the neural network. Input data were inserted from the MATLAB workspace. All the data values were normalized (with the range between 0 and 1).

The training function was set as Levenberg–Marquardt algorithm (trainlm). It is noted that Levenberg–Marquardt algorithm makes an interpolation between the Gauss–Newton algorithm and the gradient descent method. MSE was selected as the performance or error function. The hyperbolic tangent sigmoid function (tansig) was selected as the transfer or activation function. We could have also selected the logistic sigmoid function (logsig). According to the authors' experience, it would be good to consider 2–2.5 times the number of inputs as the number of neurons in the hidden layer as one option. In this study, it was shown that the case with 3 layers (2 hidden layers) has the least error and hence is used to simulate the input data. However, normally 1 hidden layer can meet engineering needs. In this study, as the number of layers increased, the error was decreased. As can be seen from case 3–4 (Table 1), since the number of data points to train the network, 92, is not that many, increasing the number of neurons may not necessarily decrease the MSE. It is just to note that it is required to first reinitialize the weights at each time of running neural networks.

In Fig. 17, the value of MSE versus the number of epochs for one case of the above example has been given. It is shown that the value of error (MSE) of training, first testing (test), and second testing (as only named validation in Fig. 17) is decreasing with the number of epochs before convergence is reached. Upon convergence, the errors of one of the tests or both start increasing, while the training error still decreases. The criterion for MSE calculation is the error value at the convergence (MSE: 0.007 for the case given).

Please note that if the starting increase of the validation (or second test) and test (or first test) MSE curves is not too sharp at the convergence point, the neural network performance is considered to be valid. In Fig. 17, at the convergence point (epoch 3), the starting increase in the validation (second testing) MSE is not too sharp.

In Fig. 18, the values of the Pearson coefficients (R) for the training, first testing (test), and second testing (validation) have been shown. The y-axis and x-axis, respectively, show the predicted outputs by ANN and the expected real values (targets). The closer these values are to the value 1, the better the indication of convergence. In our example, these values are good enough (about 1) and thus show good convergence. This graph is an important indication of the validity of the trained neural network. Suppose only the R^2 corresponding to training was near to 1 (e.g., 0.9), and the first

Table 1 Different ANN structures used in the example and the validation error

ANN	Type	No. of hidden layers	AAPE %	APE %	SD	V	R	R^2	MSE
Case-1	FF-BB	1 with10 neurons	20.99	20.99	30.46	927.81	0.8872	0.787124	0.007
Case-2	FF-BB	1 with 20 neurons	18.55	11.08	30.12	907.21	0.9134	0.8343	0.006
Case-3	FF-BB	1 with 33 neurons	26.7	9.73	93.99	8834.1	0.8982	0.806763	0.003
Case-4	FF-BB	1 with 42 neurons	16.19	12.15	33.06	1093	0.8625	0.743906	0.005
Case-5	FF-BB	2 with 18 and 18 neurons	18.87	18.67	47.2	2227.8	0.9313	0.86732	0.003
Case-6	FF-BB	2 with 26 and 26 neurons	15.81	14.59	36.15	1306.8	0.9337	0.871796	0.004
Case-7	FF-BB	2 with 32 and 32 neurons	16.99	16.96	18.97	359.86	0.9375	0.878906	0.003

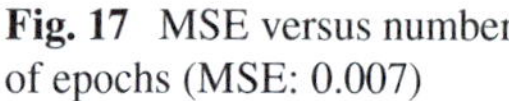

Fig. 17 MSE versus number of epochs (MSE: 0.007)

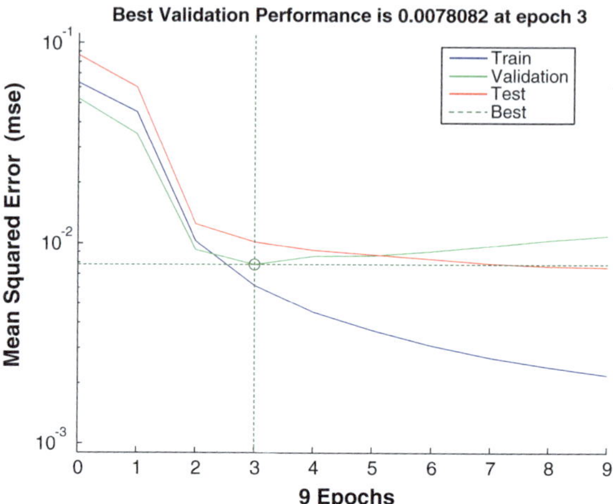

and second test values were not (e.g. 0.4). If so, the neural network performance was weak, and validity of the work was under question.

To analyze the performance of different neural network models, several statistical parameters were utilized. These parameters include average absolute percent error (AAPE), average percent error (APE), average root mean squared error (ARMSE), Pearson correlation coefficient (R), squared Pearson coefficient (R^2), standard deviation (SD), and variance (V) as shown in Table 1. In the Appendix, the required statistical parameters have been described.

The trained ANN above, which has been validated successful after two testing, could be used for simulation using new input data.

8.2 Well Hydraulics

Drilling hydraulics simulation is indeed a non-straightforward problem which is more sophisticated in complex wells (slim holes and extended reach wells). There are many unknowns in this problem. As simulation by hydraulics simulators is time consuming, they are not suitable for application in real-time drilling.

Because of the above reasons, Fruhwirth et al. (2006) utilized a number of 11 generations of a special ANN called completely connected perceptron (CCP) for prediction of pump pressure or hydraulic pressure losses.

It is noted that CCP is a more general type of perceptron which could be considered as multilayered as seen in Fig. 19. In each generation, one hidden layer was added to the CCP (first generation without any hidden layer, and the 11th generation with 10 hidden layers). Out of the available data, 50% was devoted to training. Then, 25% of the data points were allocated to each test or validation. The real data of two wells were utilized for training the ANN. The input drilling parameters considered include

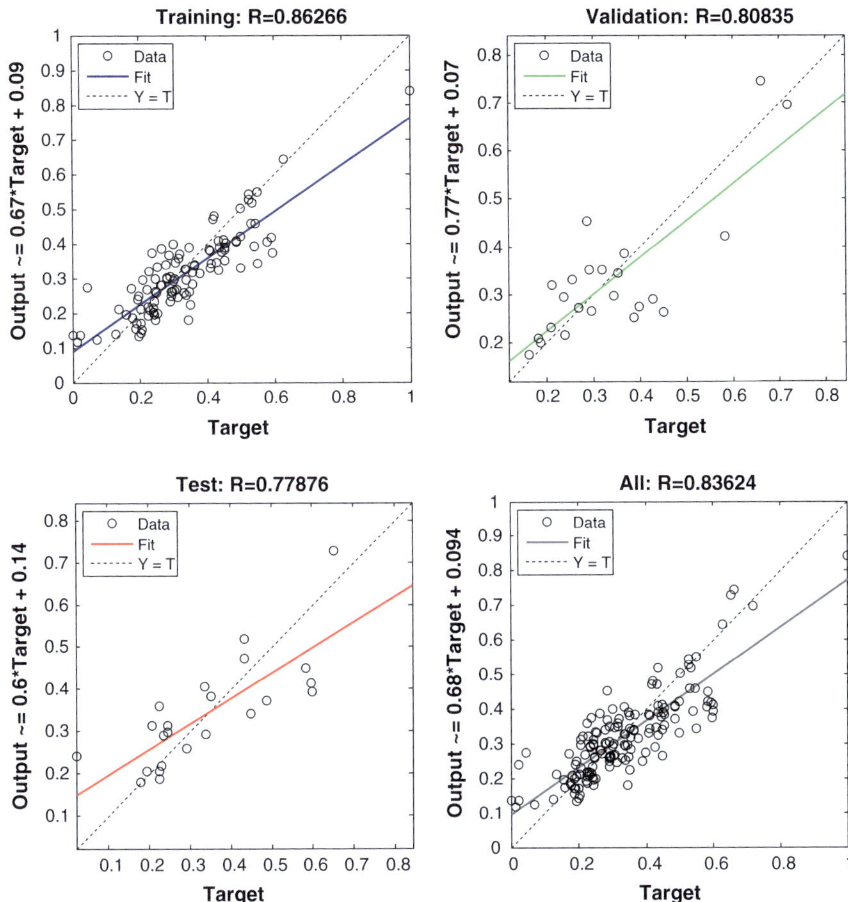

Fig. 18 The Pearson coefficients for the training, testing 1(test), and testing 2 (validation). The *y*-axis and *x*-axis, respectively, show the predicted outputs by ANN and the expected real values (target). The values are near to 1 and thus ANN has good valid performance

bit measured depth (MD in m), bit true vertical depth (TVD in m), block position (in m), rate of penetration (ROP in m/h), average mud flow rate (in m³/s), average hook load (in kg), average drill string revolutions (RPM), average weight on bit (WOB in kg), and average mud weight out of hole (in kg/m³). Average pump pressure (in bar) was considered as the output parameter.

Utilization of a completely connected perceptron (CCP) has the advantage of eliminating the need to find an optimal number of hidden layers (Fruhwirth et al. 2006). The schematics of different generations of ANN are shown in Fig. 20.

The results of modeling by ANN were promising. As seen in Fig. 21, the RMS error (in bar) is low enough, and there is not much change in the error from a generation on (the number of hidden layers does not have much effect from a number on). It

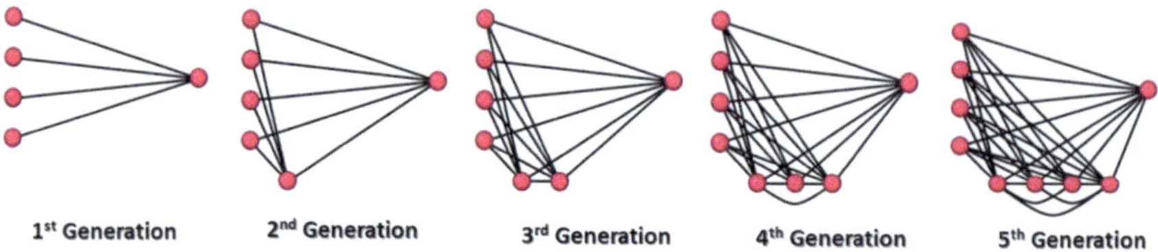

Fig. 19 Growing completely connected perceptron (cVision Manual)

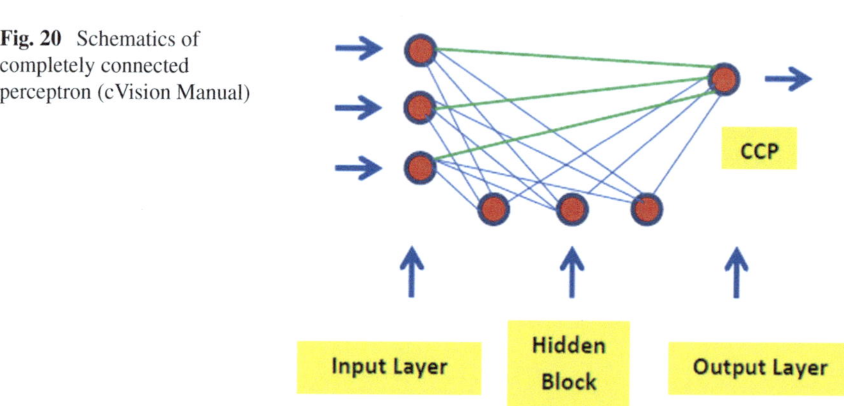

Fig. 20 Schematics of completely connected perceptron (cVision Manual)

is reminded that pump pressure is the output parameter. In Fig. 22a, the measured, calculated pump pressures by ANN (as the output) and also the corresponding error versus time have been shown for one of the wells. In Fig. 22b a good match of the data could be observed in the cross-plot of the measured (expected real data) and the calculated (or predicted) output by ANN.

Although the results obtained by the authors have enough accuracy, the authors have added some more features to the input data including MD–TVD ratio, dog leg severity (DLS), sine and cosine of well inclination, and three features corresponding to Reynolds number (N_{Re}). Reynolds number was considered because it is related to

Fig. 21 The RMS error corresponding to learning or training, test, and validation phases (Fruhwirth et al. 2006)

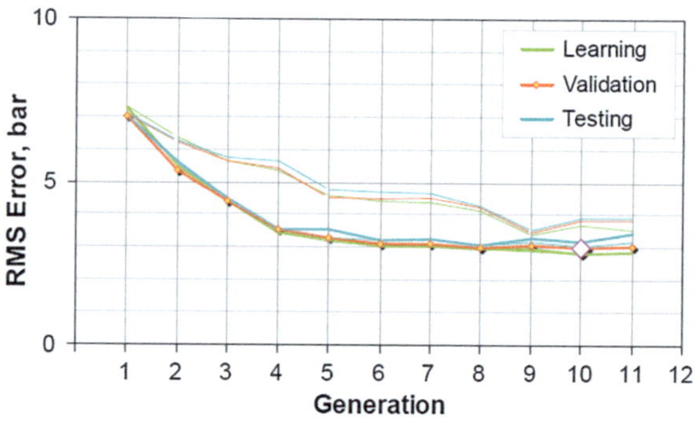

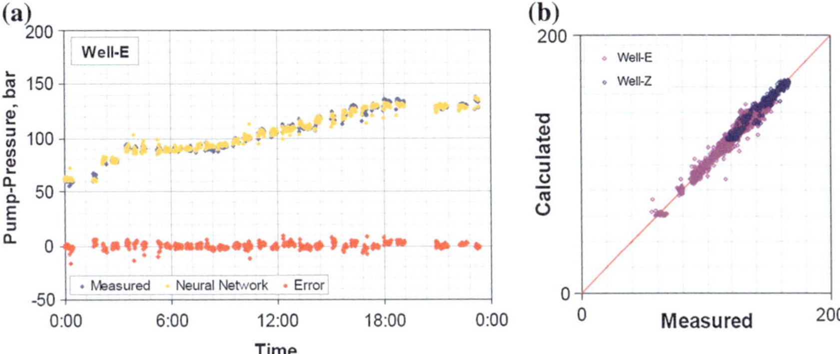

Fig. 22 **a** Measured, calculated by ANN pump pressure and error versus time and **b** cross-plot of calculated versus measured pump pressure (Fruhwirth et al. 2006)

the mud rheological properties, well geometry, and mud flow rate (which all have an effect on hydraulics).

In a further study, torque and bit-measured depth have been added as additional input parameters and it has been concluded that 95% of the theoretical predicted standpipe pressure values lie within 10 bars of the real measured data (Fig. 23). Also, in the cross-plot of predicted (y-axis) and measured pressure (x-axis) as shown Fig. 24, the squared Pearson coefficient has been found to be equal to 0.9677 (Todorov and Thonhauser 2014). Also, the simulated pressure losses clearly follow the trend of the measured standpipe pressure (Fig. 23). All the above discussions indicate the capability of ANN in handling complex hydraulics problems.

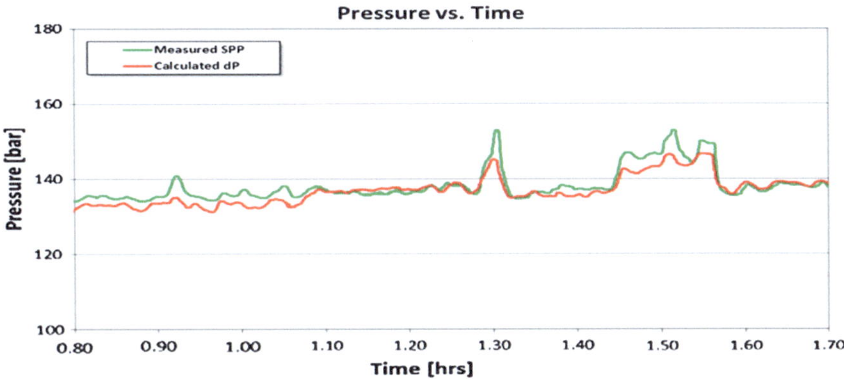

Fig. 23 Measured standpipe pressure and calculated pressure drop by ANN versus time (Todorov and Thonhauser 2014)

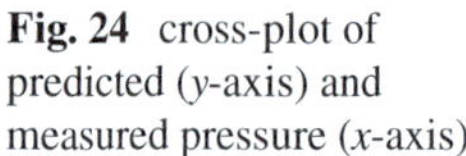

Fig. 24 cross-plot of predicted (*y*-axis) and measured pressure (*x*-axis)

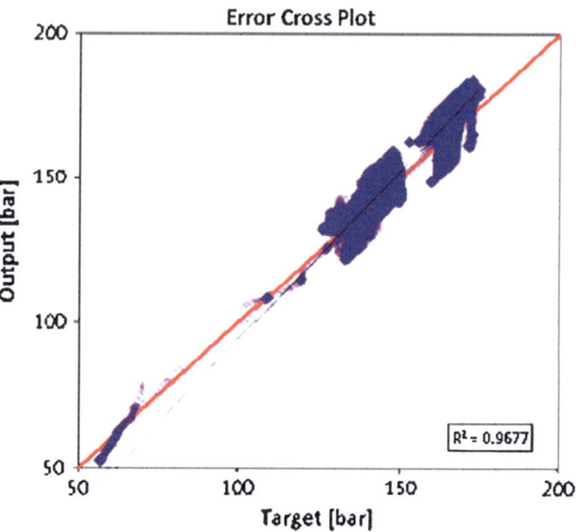

8.3 *Drilling Optimization*

Many studies have been performed on the application of ANN for drilling optimization of rate of penetration (ROP) in the literature.

In a recent work, four ROP models with, respectively 6, 9, 15, and 18 input parameters or data channels were constructed using ANN with the objective of investigating the effect of vibration parameters on ROP (Esmaeili et al. 2012). In the first two models, the vibration parameters were not considered. In the third model, formation mechanical properties were not considered though vibration parameters were taken into consideration.

In the first ROP model, only the drilling parameters (average and standard deviation of WOB, average and standard deviation of RPM of drill string, and average and standard deviation of torque) were considered as input parameters or channels.

In the second ROP model, the drilling and mechanical properties (uniaxial compressive strength UCS, Young's modulus of elasticity, and Poisson's ratio) were considered as input parameters.

In the third ROP model, the drilling and vibration parameters were considered as input parameters. The vibration parameters include standard deviation, and first and second order frequency moment of X, Y, and Z component of vibration.

In the fourth ROP model, all the drilling, mechanical properties, and vibration parameters were considered as input parameters or channels.

In Fig. 25, the ranking of input parameters (including drilling, mechanical, and vibration) has been made for the fourth model using sequential forward selection (SFS) which shows the parameters UCS, standard deviation of Z component of vibration, average WOB, etc. that have the most impact on ROP.

Eventually, Esmaeili et al. (2012) reached the results in Table 2.

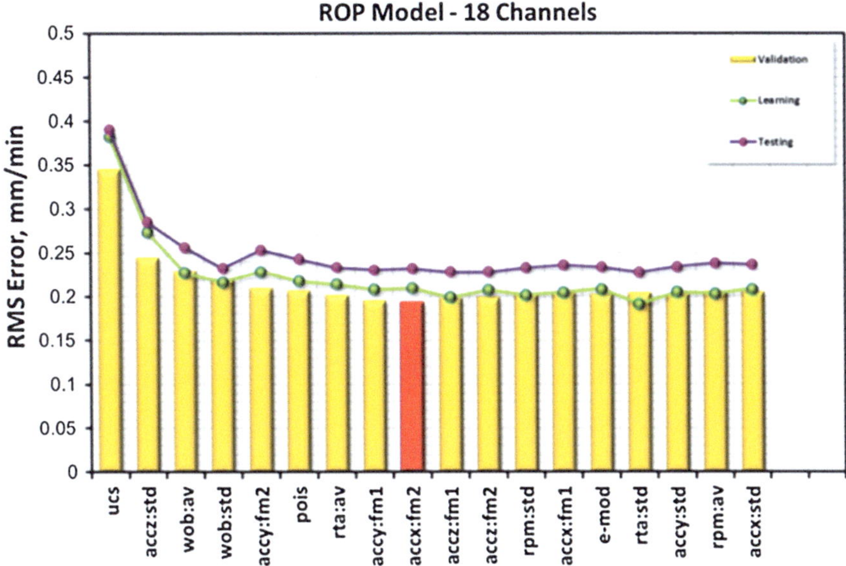

Fig. 25 Ranking of input channels (or parameters) for the fourth ROP Model using sequential forward selection (Esmaeili et al. 2012)

Table 2 Summary of ROP models and corresponding RMS errors (Esmaeili et al. 2012)

Model	No. of input channels	Number of hidden units	RMS error (mm/min)
First ROP model	6	6	0.301
Second ROP model	9	6	0.208
Third ROP model	15	6	0.231
Fourth ROP model	18	5	0.194

Most of the common ROP models (Bourgoyne, Warren, Maurer, etc.), which are based on only the drilling and mechanical properties, do not consider the vibration properties. The RMS error corresponding to the second ROP model, which looks like the mentioned models, was calculated as 0.208 mm/min. Nevertheless, in the fourth ROP model wherein the vibration parameters were considered, the RMS error was calculated as 0.194 mm/min which shows vibration parameters have important effects on ROP modeling.

In Fig. 26, the cross-plot of actual ROP values (predicted by ANN) and desired ROP values (measured or expected real values) have been shown which shows a better match in the fourth model.

Some similar drilling optimization studies using neural networks to be mentioned are as follows: Gidh et al. (2012) in prediction of bit wear, Lind and Kabirova (2014) in prediction of drilling problems, and Bataee et al. (2010).

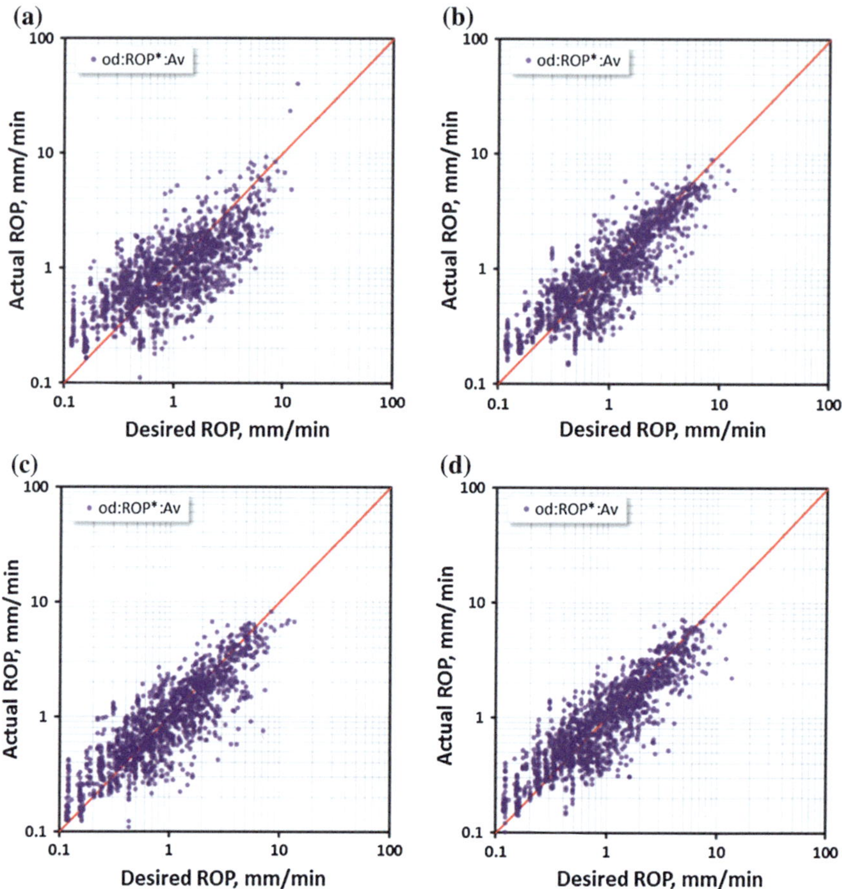

Fig. 26 Actual ROP (predicted by ANN) versus desired ROP (expected real or measured values) for all four ROP models from **a–d** (Esmaeili et al. 2012)

8.4 Permeability Prediction

Permeability is one of the important parameters which have been estimated using ANN in many applications.

In an application of ANN to predict permeability (Naeeni et al. 2010), the input parameters considered in the FF-ANN with backpropagation algorithm, include depth, CT (true conductivity), DT (sonic travel time), NPHI (neutron porosity), RHOB (bulk density), SGR (spectral gamma ray), NDSEP (neutron-density log separation), *northing of well*, *easting of well*, S_{WT} (water saturation), and FZI (flow zone indicator). The number of three hidden layers (with 13, 10, and 1 neurons) was selected for the network. In order to estimate permeability, determination of different hydraulic flow units (HFU) is the first stage because it leads to FZI values which are needed for the first stage. At the same time, FZI which is related to pore size and geometry has the following relation with RQI:

$$\mathrm{FZI} = \frac{\mathrm{RQI}}{\O_z} \tag{19}$$

It is also noted that RQI and $\O_z$ are evaluated by:

$$\mathrm{RQI}(\mu m) = 0.031\sqrt{\frac{K}{\O_e}} \tag{20}$$

$$\O_z = \frac{\O_e}{1 - \O_e} \tag{21}$$

where $\O_e$ is the effective porosity.

Thus, taking logarithms from both sides leads to:

$$\mathrm{Log}\,(\mathrm{RQI}) = \mathrm{Log}(\O_z) + \mathrm{Log}(\mathrm{FZI}) \tag{22}$$

Therefore, Log(FZI) is the intercept of the plot of RQI versus $\O_z$ in log–log scale. Thus, **first** RQI (rock quality index) values versus normalized porosity values ($\O_z$) were plotted in log–log scale using core data. **Second**, RQI values versus $\O_z$ were plotted in a log–log scale. **Third**, some straight lines are selected to intersect with line $\O_z = 1$ such that reasonable initial guesses of intercepts are obtained as mean FZI values.

Fourth, the data points are assigned to the adjacent straight lines and are considered as different HFUs (clustering technique). This stage requires considerable time. **Fifth**, the intercept or FZI of each HFU is recalculated utilizing regression equations and is compared with the guess in the fourth stage. If the difference between the recalculated FZI value and the guess value in the fourth stage is considerable, it is required to come back to the fourth stage so that the guessed FZI can be updated.

Following the above procedure, finally different rock types could be determined with good accuracy (HFU determination is done) and also permeability values could be estimated as shown in Figs. 27 and 28.

As the second step, an ANN was trained with well log and RQI data as input and permeability data as output. Eventually, it is possible to simulate the ANN using well log data in uncored wells to estimate their permeability values.

As the cross-plot of predicted versus measured values of permeability is shown in Fig. 29, the performance of the ANN in permeability prediction was promising with the Pearson coefficient of 0.85 for the validation phase.

In another study with a rather similar approach (Kharrat et al. 2009), the predicted permeability values show a good match with the core measured values. The well FZI profile versus depth has been shown on the left of Fig. 30 (with dots showing the values corresponding to core measured values). The predicted and core measured permeability values versus depth have been illustrated on the right of Fig. 30.

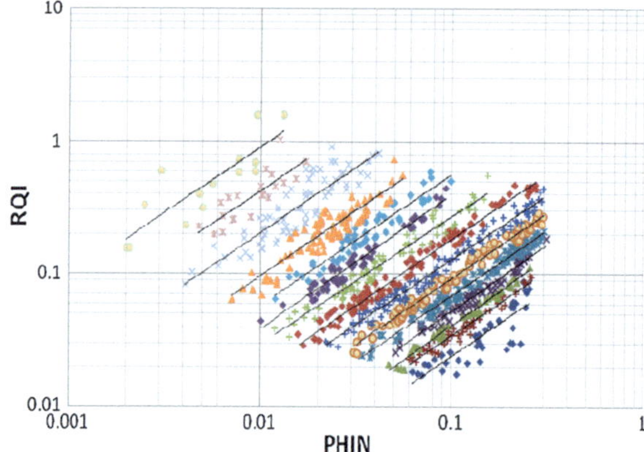

Fig. 27 The number of 15 rock types has been identified in the collected data from the reservoir (Naeeni et al. 2010)

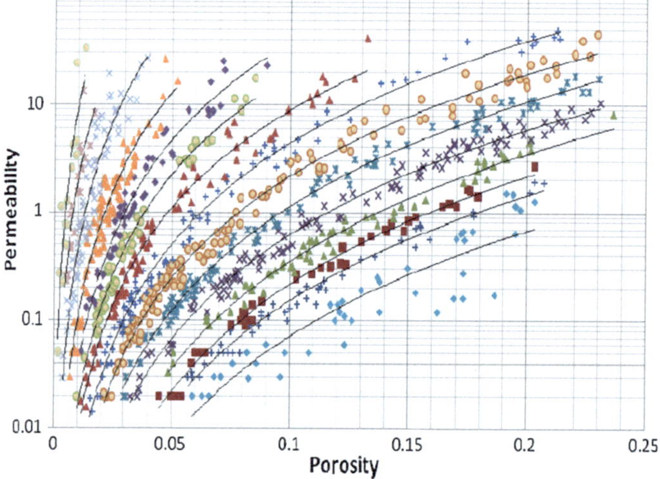

Fig. 28 For each rock type of the reservoir, one permeability curve versus porosity has been determined (Naeeni et al. 2010)

Some similar reservoir engineering studies using neural networks are as follows: Thomas and Pointe (1995) in conductive fracture identification, Lechner and Zangl (2005) in reservoir performance prediction, Adeyemi and Sulaimon (2012) in predicting wax formation, and Tang (2008) in reservoir facies classification.

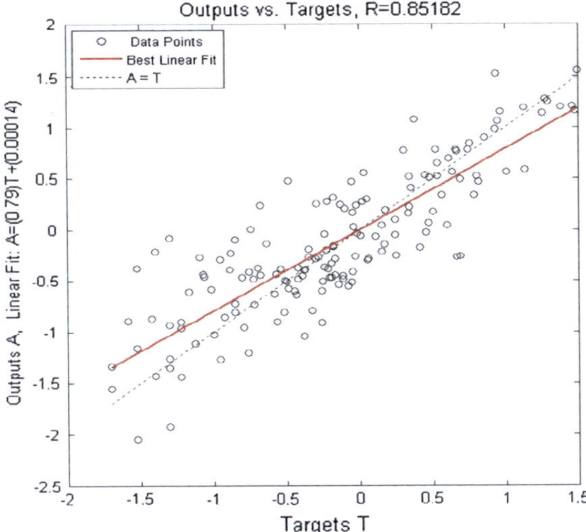

Fig. 29 Predicted permeability values or output values as *y*-axis versus core permeability values as *x*-axis (Naeeni et al. 2010)

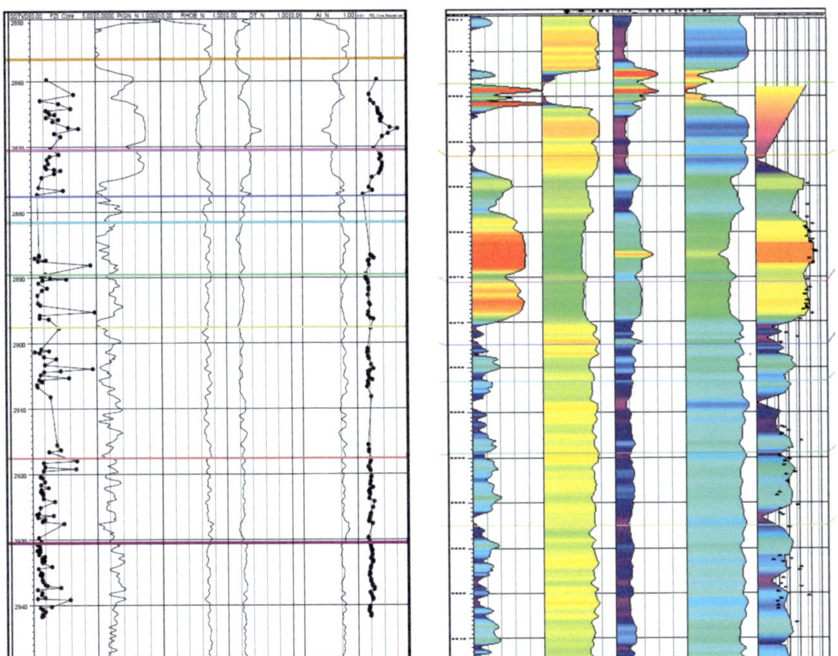

Fig. 30 FZI profile of one well based on log and core data (*left*) and its permeability predictions (Kharrat et al. 2009)

9 Conclusion

Artificial neural networks (ANNs) have been shown to be an effective tool to solve complex problems with no analytical solutions. In this chapter, the following topics have been covered: artificial neural networks basics (neurons, activation function, ANN structure), feed-forward ANN, backpropagation and learning, perceptrons and backpropagation, multilayer ANNs and backpropagation algorithm, data processing by ANN (training, over-fitting, testing, validation), ANN and statistical parameters, and some applied examples of ANN in geoscience and petroleum engineering.

Appendix A: Appendix: Important Statistical Parameters

The corresponding relations of a few important statistical parameters to compare performance and accuracy of different neural network models are given as follows:

1. Average percent relative error (APE):

   This error is defined as the relative deviation from the measured data.

$$APE = \frac{1}{n} \sum_{i=1}^{n} E_i \tag{23}$$

$$E_i = \left[\frac{P_m - P_e}{P_m} \right]_i \quad i = 1, 2, 3, \ldots, n, \tag{24}$$

2. Average absolute percent relative error (AAPE):

   This error gives an idea of absolute relative deviation of estimated outputs from the measured or expected output data.

$$AAPE = \frac{1}{n} \sum_{i=1}^{n} |e_i| \tag{25}$$

$$e_i = [P_m - P_e]_i \tag{26}$$

3. Mean squared error (MSE):

   This error is corresponding to the expected value of the squared error loss.

$$MSE = \frac{1}{n} \sum_{i=1}^{n} e_i^2 \tag{27}$$

4. Average root-mean-square error (ARMSE):

This error is an indeed measure of scatter or lack of accuracy of the estimated data.

$$\mathrm{ARMSE} = \sqrt{\frac{1}{n}\sum_{i=1}^{n} e_i^2}$$ (28)

5. Standard deviation (SD):

This error shows the dispersion of the values from the average value or mean.

$$\mathrm{SD} = \left[\frac{n\sum_{i=1}^{n} E_i^2 - \left(\sum_{i=1}^{n} E_i\right)^2}{n^2}\right]^{\frac{1}{2}}$$ (29)

6. Variance or V, σ^2:

This error is the square of the standard deviation.

$$\sigma^2 = \frac{\sum (X - M)^2}{N}$$ (30)

7. Correlation coefficient or Pearson coefficient (R):

It represents the degree of success in the reduction of the standard deviation (SD). It is normally used as a measure of the extent of the linear dependence between two variables. The nearer R is to 1, the better the convergence and ANN performance is.

$$R = \frac{\sum_{i=1}^{n} \left[\left(P_{m,i} - P_{m,\mathrm{av}}\right) \times \left(P_{e,i} - P_{e,\mathrm{av}}\right)\right]}{\sqrt{\sum_{i=1}^{n} \left[\left(P_{m,i} - P_{m,\mathrm{av}}\right)^2\right] \times \sum_{i=1}^{n} \left(P_{e,i} - P_{e,\mathrm{av}}\right)^2}}$$ (31)

$$P_{\mathrm{av}} = \frac{1}{n}\sum_{i=1}^{n} P_i$$ (32)

8. Squared Pearson coefficient: R^2.

References

Adeyemi BJ & Sulaimon AA (2012) Predicting wax formation using artificial neural network. In: SPE-163026-MS, Nigeria annual international conference and exhibition, Lagos, Nigeria

Ashena R, Moghadasi J, Ghalambor A, Bataee M, Ashena R, Feghhi, A (2010) Neural networks in BHCP prediction performed much better than mechanistic models. In: SPE 130095, international oil and gas conference and exhibition, Beijing, China

Bataee M, Edalatkhah S, Ashena R (2010) Comparison between bit optimization using artificial neural network and other methods base on log analysis applied in Shadegan oil field. In: SPE 132222-MS, international oil and gas conference and exhibition, Beijing, China

Bertsekas DP, Tsitsiklis JN (1996) Neuro-dynamic programming. Athena Scientific, Belmont, MA. ISBN 1-886529-10-8

Cacciola M, Calcagno S, Lagana F, Megali G, Pellicano D (2009) Advanced integration of neural networks for characterizing voids in welded strips. In: 19th international conference, Cyprus

Coulibaly P, Baldwin CK (2005) Nonstationary hydrological time series forecasting using nonlinear dynamic methods. J Hydrol 307:164–174

CVision Software Manual, NGS-Neuro Genetic Solutions GmbH

Darken C, Moody J (1992) Towards faster stochastic gradient search. In: Moody JE, Hanson SJ and Lippmann RP (eds)

Esmaeili A, Elahifar B, Fruhwirth R, Thonhauser G (2012) ROP modelling using neural network and drill string vibration data. In: SPE 163330, Kuwait international petroleum conference and exhibition

Fruhwirth RK, Thonhauser G, Mathis W (2006) Hybrid simulation using neural networks to predict drilling hydraulics in real time. In: SPE 103217, SPE annual technical conference and exhibition in San Antonia, Texas, USA

Gidh YK, Purwanto A, Ibrahim H (2012) Artificial neural network drilling parameter optimization system improves ROP by predicting/managing bit wear. In: SPE 149801-MS, SPE intelligent energy, Utrecht, The Netherlands

Kharrat R, Mahdavi R, Bagherpour M, Hejri S (2009) Rock type and permeability prediction of a heterogenous carbonate reservoir using artificial neural networks based on flow zone index approach. In: SPE 120166, SPE middle east oil and gas show and conference, Bahrain

Lechner JP, Zangl G (2005) Treating uncertainties in reservoir performance prediction with neural networks. In: SPE-94357-MS, SPE Europec/EAGE annual conference, 13–16 June, Madrid, Spain

Lind YB, Kabirova AR (2014) Artificial neural networks in drilling troubles prediction. In: SPE 171274-MS, SPE Russian oil and gas exploration and production technical conference and exhibition, Moscow, Russia

Mohaghegh S (2000) Virtual intelligence application in petroleum engineering: part I-artificial neural networks. J Pet Technol 52:64–72

Naeeni MN, Zargari H, Ashena R, Kharrat R (2010) Permeability prediction of uncored intervals using IMLR method and artificial neural networks: a case study of Bangestan field, Iran. In: SPE 140682, 34th annual SPE international conference and exhibition, Nigeria

Tang H (2008) Improved carbonate reservoir facies classification using artificial neural network method. In: PETSOC-2008-122, Canadian international petroleum conference, Calgary, Alberta

Thomas AL, Pointe PR (1995) Conductive fracture identification using neural networks. In: ARMA-95-0627, the 35th U.S. symposium on rock mechanics (USRMS), Reno, Nevada

Todorov D, Thonhauser G (2014) Hydraulics monitoring and well control event detection using model based analysis. In: SPE 24803, offshore technology conference Asia, Kuala lumpur, Malasia

On Support Vector Regression to Predict Poisson's Ratio and Young's Modulus of Reservoir Rock

A. F. Al-Anazi and I. D. Gates

Abstract Accurate prediction of rock elastic properties is essential for wellbore stability analysis, hydraulic fracturing design, sand production prediction and management, and other geomechanical applications. The two most common required material properties are Poisson's ratio and Young's modulus. These elastic properties are often reliably determined from laboratory tests by using cores extracted from wells under simulated reservoir conditions. Unfortunately, most wells have limited core data. On the other hand, wells typically have log data. By using suitable regression models, the log data can be used to extend knowledge of core-based elastic properties to the entire field. Artificial neural networks (ANNs) have proven to be successful in many reservoir characterization problems. Although nonlinear problems can be well resolved by ANN-based models, extensive numerical experiments (training) must be done to optimize the network structure. In addition, generated regression models from ANNs may not perfectly generalize to unseen input data. Recently, support vector machines (SVMs) have proven successful in several real-world applications for their potential to generalize and converge to a global optimal solution. SVM models are based on the structural risk minimization principle that minimizes the generalization error by striking a balance between empirical training error and learning machine capacity. This has proven superior in several applications to the empirical risk minimization (ERM) principle adopted by ANNs that aims to reduce the training error only. Here, support vector regression (SVR) to predict Poisson's ratio and Young's modulus is described. The method uses a fuzzy-based ranking algorithm to select the most significant input variables and filter out dependency. The learning and predictive capabilities of the SVR method is compared to that of a backpropagation neural network (BPNN). The results demonstrate that SVR has similar or superior learning and prediction capabilities to that of the BPNN. Parameter sensitivity analysis was performed to investigate the effect of the SVM regularization parameter, the regression tube radius, and the type of kernel function used. The result shows that the capability of the SVM approximation depends strongly on these parameters.

A. F. Al-Anazi · I. D. Gates (✉)
Department of Chemical and Petroleum Engineering, Schulich School of Engineering, University of Calgary, Calgary, Canada
e-mail: ian.gates@ucalgary.ca

© The Author(s), under exclusive license to Springer Nature Switzerland AG 2024   155
C. Cranganu (ed.), *Artificial Intelligent Approaches in Petroleum Geosciences*,
https://doi.org/10.1007/978-3-031-52715-9_5

Keywords Poisson's ratio · Young's modulus · Artificial neural networks · Support vector machines · Log data · Core data

Nomenclature

AAE   Absolute average error
b Bias term
BPNN   Backpropagation neural networks
C Regularization parameter
E Young's modulus, psi
f An unknown function
g Overburden stress, psi
G Shear modulus, psi
h Vapnik–Chervonenkis dimension
K Bulk modulus, psi
L Lagrangian equation for a dual programming problem or Loss function
MAE   Maximum absolute error
r Correlation coefficient
RBF   Radial basis function
RMSE   Root mean square error
R_{emp} Empirical risk
R Structural risk
P Pressure, psi
SVMs   Support vector machines
SVR   Support vector regression
x Input variable
y Output variable
$\hat{y}$ Estimated output value
v Velocity
w Weight vector

Greek Symbols

α, α^{*} Lagrangian multiplier to be determined
ε Error accuracy/strain
φ Porosity, fraction
φ Mapping function from input space into a high-dimensional feature space
η, η^{*} Lagrangian multipliers
κ, ϑ Sigmoid function parameters
μ Poisson's ratio, dimensionless
ρ Density, g/cc

σ	Stress/variance
σ^2	Standard deviation
ξ, ξ^*	Slack variables

Subscripts and Superscripts

a	Axial
b	Bulk
h	Horizontal
k, l	Indices
N	Number of samples
n	Input space dimension
p	Pore/compressional
r	Radial
s	Shear

1 Introduction

Elastic properties of reservoir rock are important geomechanical data required in hydraulic fracture design, wellbore stability analysis, reservoir dilation, surface heave (especially for high-pressure processes such as cyclic steam stimulation), and sand production anticipation as in cold heavy oil production. Rock properties including Poisson's ratio, μ, shear modulus, G, Young's modulus, E, and bulk modulus, K, can be determined under static test conditions by using triaxial stress cells or under dynamic conditions by measuring compressional and shear velocities and density of reservoir core samples measured by acoustic or sonic logging tools (Gatens et al. 1990; Barree et al. 2009; Khaksar et al. 2009).

Core-measured elastic properties of reservoir rock obtained from detailed laboratory analysis are often considered to be the most direct and accurate method (Ameen et al. 2009). However, due to the costs of obtaining and handling core samples, most often, a limited number of samples are analyzed and often correlations are established between core and log data to estimate elastic properties of other reservoir rocks where core data are not available. Acoustic log measurements provide compressive and shear wave velocities which can be combined with density log and elasticity theory to estimate elastic properties of reservoir rock (Gatens et al. 1990; Abdulraheem et al. 2009). In complex reservoirs, acoustic theory may be insufficient to describe its actual behavior, thus limiting the accuracy of the derived rock correlations (Barree et al. 2009; Khaksar et al. 2009). Therefore, an integrated knowledge of the logging tool responses and understanding of underlying geology in addition to the use of advanced statistical techniques are necessary to determine an interpretation model that can effectively map the dependency between well log data and elastic rock properties. Due to several different factors that control log responses and the construction of the mapping function between log data and the elastic properties, this can be a challenging task since the mapping can be nonlinear. To deal with this nonlinearity, artificial neural networks (ANNs), fuzzy logic (FL), and functional networks (FN) approaches to predict Poisson's ratio and Young's modulus have been used in the literature with varying degrees of success (Widarsono et al. 2001; Abdulraheem et al. 2009).

A further complication is that the data available to train artificial intelligence methods is scarce. The smaller the training data set, the greater the risk of poor estimation of rock properties. This is especially the case if data are missing from specific intervals along a well. Recently, support vector machines (SVMs) have been recognized as an efficient and accurate tool with strong learning and prediction capabilities (Al-Anazi and Gates 2010a, b, c, d). The SVM learning machine approach is novel and is based on the principle of structural risk minimization (SRM), which aims to minimize an upper bound of the generalization error. On the other hand, ANNs follow the principle of empirical risk minimization (ERM) which attempts to minimize the training error. The SRM principle is based on bounding the generalization error by minimizing the sum of the training error and a confidence interval term depending on the Vapnik–Chervonenkis (VC) dimension (Vapnik and Chervonenkis 1974; Vapnik 1982, 1995). To accomplish this, in support vector regression (SVR), a regularization term is used to determine the trade-off between the training error and the VC confidence term. Consequently, the SVM approach provides a promising tool to generalize to unseen data. To capture the nonlinear behavior of the mapping, kernel functions are used to project the input space into a higher dimensional feature space where a linear regression hyperplane is devised (Kecman 2005).

The accuracy and robustness of the SVM approach have been robustly demonstrated in many real-world applications; for example, face recognition, object detection, hand writing recognition, text detection, speech recognition and prediction, porosity and permeability determination from log data, and lithology classification (Li et al. 2000; Lu et al. 2001; Choisy and Belaid 2001; Gao et al. 2001; Kim et al. 2001; Ma et al. 2001; Van Gestel et al. 2001; Al-Anazi and Gates 2010a, b, c, d). In this study, the potential of SVR to establish an interpretation model that relates core and log data to elastic rock properties is evaluated from data originating from a well in a hydrocarbon reservoir. Two separate interpretation models were constructed: one for Poisson's ratio and the other for Young's modulus. The first model for Poisson's ratio was developed by using the density and compressive and shear wave velocities, whereas the second one for Young's modulus was devised by using variables chosen from a fuzzy selection scheme. For the well used in this study, the available core-derived input variables are porosity, minimum horizontal stress, pore pressure, and overburden stress, whereas the available well log data include bulk density and compressional and shear velocities.

In this research, the performance of SVM was compared with that of a back-propagation neural network (BPNN) to evaluate its potential to predict elastic rock properties under scarce data conditions. For SVM nonlinear approximation, a radial basis function (RBF) kernel function is used. During the machine learning stage of the SVM, the kernel function parameter, insensitivity tube parameter, ε, and penalty parameter were selected through grid search and pattern search schemes. To avoid overfitting the data, a ten-fold cross-validation was used to select the optimal parameter to control the trade-off between the bias and variance of the model. Error analysis was done by examining the correlation coefficient, r, root mean square error (RMSE), absolute average error (AAE), and maximum absolute error (MAE) between target and predicted values. The study also investigated the impact of the penalty parameter, insensitivity tube radius, and the type of kernel function on the SVM approximation capability.

2 Backpropagation Neural Network (BPNN)

Backpropagation multilayer perceptron neural networks (BPNN) have been extensively used to interpret rock physical and elastic properties in hydrocarbon reservoirs (Rogers et al. 1995; Huang et al. 1996, 2001; Fung et al. 1997; Helle and Ursin 2001; Helle and Bhatt 2002; Abdulraheem et al. 2009). BPNN can approximate any continuous nonlinear function over a compact interval to any desirable accuracy depending on the number of hidden layers. They use activation functions in the processing neurons to facilitate the modeling of nonlinear mapping functions (Suykens et al. 2002). During learning, in BPNNs, the input patterns are propagated forward through hidden layers toward the output, while error is backpropagated toward the input layer. Here, a conjugate gradient algorithm is used to train the BPNN by minimizing the square of the residuals between the target and training data. One well-known issue with such training algorithms is that they may become trapped in local minima since it is sensitive to the starting weight values (Hastie et al. 2001). Here, a Nguyen–Widrow algorithm was used to select the initial range of weight values, and the conjugate gradient algorithm was then used to optimize the weights. Optimization is done several times with different random weight values to ensure convergence to the global optimal solution. One shortcoming of BPNNs is its susceptibility to overfit training data which results in poor generalization capability to interpret new input data. In this work, three layers (input, hidden, and output) were used with neurons being automatically optimized. A ten-fold cross-validation technique was used to stop training (DTREG v9.1 2009).

3 Support Vector Regression

SVMs are learning algorithms originally developed to solve classification problems. By using Vapnik's ε-insensitive loss function, SVMs can be used to solve nonlinear regression problems by using kernel functions (Vapnik 1995). Given a data set $(y_k, \mathbf{x}_k)$

of dimension N where y_k is the output value at input variable values expressed in the vector $\mathbf{x}_k$, the relationship between the input and output variables can be expressed by the following linear regression function:

$$f(\mathbf{x}_k) = \mathbf{w}^T \mathbf{x}_k + b \tag{1}$$

where $\mathbf{w}$ is a set of weights and b is an offset or bias. The empirical risk (to be minimized) can be expressed by

$$R_{\text{emp}} = \frac{1}{N} \sum_{k=1}^{N} v\left(y_k - \mathbf{w}^T \mathbf{x}_k - b\right) \tag{2}$$

where $v(\cdot)$ is Vapnik's ε-insensitive loss function defined by

$$v(y - f(x)) = \begin{cases} 0 & \text{if } |y - f(x)| \le \varepsilon \\ |y - f(x)| - \varepsilon & \text{otherwise} \end{cases} \tag{3}$$

The optimization problem in the primal space that has to be solved to achieve an optimal linear function in terms of the weights, $\mathbf{w}$, and bias, b, is given by

$$\min \ J_P(\mathbf{w}) = \frac{1}{2} \mathbf{w}^T \mathbf{w} \tag{4}$$

$$\text{subject to} \begin{cases} y_k - \mathbf{w}^T \mathbf{x}_k - b \le \varepsilon, \ k = 1, \ldots, \\ N \mathbf{w}^T \mathbf{x}_k + b - y_k \le \varepsilon, \ k = 1, \ldots, N \end{cases}.$$

The value of ε in Vapnik's loss function, v, characterizes the radius of an approximation tube which in turn controls the accuracy of the model. Slack variables, ξ_k, ξ_k^* for $k = 1, \ldots, N$, are introduced to the optimization problem given by Eq. 4:

$$\min \ J_P\left(\mathbf{w}, \xi, \xi^*\right) = \frac{1}{2} \mathbf{w}^T \mathbf{w} + C \sum_{k=1}^{N} \left(\xi_K + \xi_K^*\right) \tag{5}$$

$$\text{subject to} \begin{cases} y_k - \mathbf{w}^T \mathbf{x}_k - b \le \varepsilon + \xi_k, \ k = 1, \ldots, N \\ \mathbf{w}^T \mathbf{x}_k + b - y_k \le \varepsilon + \xi_k^*, \ k = 1, \ldots, N \\ \xi_k, \xi_k^* \ge 0, \qquad\qquad\quad k = 1, \ldots, N \end{cases}.$$

The regularization (penalty) constant, C, is positive and determines how large the deviation from the desired accuracy is tolerated. The Lagrangian form of the problem is expressed by

$$L\left(\mathbf{w}, b, \xi, \xi^*; \alpha, \alpha^*, \eta, \eta^*\right) = \frac{1}{2} \mathbf{w}^T \mathbf{w} + C \sum_{k=1}^{N} \left(\xi_k + \xi_k^*\right)$$

$$-\sum_{k=1}^{N} \alpha_k \left(\varepsilon + \xi_k - y_k + \mathbf{w}^T \mathbf{x}_k + b\right)$$

$$-\sum_{k=1}^{N} \alpha_k^* \left(\varepsilon + \xi_k^* + y_k - \mathbf{w}^T \mathbf{x}_k - b\right)$$

$$-\sum_{k=1}^{N} \left(\eta_k \xi_k + \eta_k^* \xi_k^*\right) \tag{6}$$

where $\alpha_k, \alpha_k^*, \eta_k, \eta_k^*$ are positive Lagrange multipliers. The solution is obtained by solving a saddle point problem: The Lagrangian, L, must be minimized with respect to $\mathbf{w}, b, \xi, \xi^*$ and maximized with respect to $\alpha, \alpha^*, \eta, \eta^*$. The following conditions are satisfied at the saddle point:

$$\begin{cases} \dfrac{\partial L}{\partial \mathbf{w}} = 0 \rightarrow \mathbf{w} = \sum_{k=1}^{N} \left(\alpha_k - \alpha_k^*\right)\mathbf{x}_k \\[2ex] \dfrac{\partial L}{\partial b} = 0 \rightarrow \sum_{k=1}^{N} \left(\alpha_k - \alpha_k^*\right)\mathbf{x}_k = 0 \\[2ex] \dfrac{\partial L}{\partial \xi_k} = 0 \rightarrow C - \alpha_k - \eta_k = 0 \\[2ex] \dfrac{\partial L}{\partial \xi_k^*} = 0 \rightarrow C - \alpha_k^* - \eta_k^* = 0 \end{cases} \tag{7}$$

The primal optimization problem can be re-formulated as a dual problem as follows:

$$\max J_D\left(\alpha, \alpha^*\right) = -\frac{1}{2} \sum_{k,l=1}^{N} \left(\alpha_k - \alpha_k^*\right)\left(\alpha_l - \alpha_l^*\right) x_k^T x_l$$

$$-\varepsilon \sum_{k=1}^{N} \left(\alpha_k + \alpha_k^*\right) + \sum_{k=1}^{N} y_k\left(\alpha_k - \alpha_k^*\right) \tag{8}$$

subject to $\begin{cases} \sum_{k=1}^{N} \left(\alpha_k - \alpha_k^*\right) \\ \alpha_k, \alpha_k^* \in [0, c] \end{cases}$.

By solving Eq. 8, the optimal Lagrange multiplier pairs can be found and the linear regression is then given by

$$f(\mathbf{x}_k) = \sum_{k=1}^{N} \left(\alpha_k - \alpha_k^*\right)\mathbf{x}_k^T \mathbf{x}_k + b \tag{9}$$

with

Table 1 Common kernel function and corresponding mathematical expression

Kernel function	Mathematical expression
Linear	$k(\mathbf{x}_i, \mathbf{x}) = \langle \mathbf{x}_i, \mathbf{x} \rangle$
Gaussian radial basis function	$k(\mathbf{x}_i, \mathbf{x}) = e^{-\frac{\|\mathbf{x}_i - \mathbf{x}\|^2}{2\sigma^2}}$
Sigmoid	$k(\mathbf{x}_i, \mathbf{x}) = \tanh(\kappa \langle \mathbf{x}_i, \mathbf{x} \rangle + \theta)$

$$\mathbf{w} = \sum_{k=1}^{N} \left(\alpha_k - \alpha_k^* \right) \mathbf{x}_k \tag{10}$$

The training points with nonzero α_k values allow the calculation of the bias term, b (Kecman 2005). Generalization of the SVM approach to nonlinear regression estimation is accomplished by using kernel functions. Typical kernel fusssnctions that are used are linear ones, Gaussian RBF, and polynomial and sigmoid functions as listed in Table 1. In the primal weight space, the regression model is given by

$$f(\mathbf{x}) = \mathbf{w}^T \varphi(\mathbf{x}) + b \tag{11}$$

with given training data $\{\mathbf{x}_k, y_k\}_{k=1}^{N}$ and $\varphi(\cdot) : R^n \rightarrow R^{n_h}$ is a kernel mapping function which projects the input space to a higher dimensional feature space. The primal problem is then formulated as follows:

$$min \; J_P\left(\mathbf{w}, \xi, \xi^*\right) = \frac{1}{2}\mathbf{w}^T\mathbf{w} + C \sum_{k=1}^{N} \left(\xi_k + \xi_k^* \right) \tag{12}$$

$$\text{subject to} \begin{cases} y_k - \mathbf{w}^T \varphi(\mathbf{x}_k) - b \leq \varepsilon + \xi_k, & k = 1, \ldots, k \\ \mathbf{w}^T \varphi(\mathbf{x}_k) + b - y_k \leq \varepsilon + \xi_k^*, & k = 1, \ldots, k \\ \xi_k, \xi_k^* \geq 0, & k = 1, \ldots, N \end{cases}$$

The dual problem is then formulated as follows:

$$\max J_D\left(\alpha, \alpha^*\right) = -\frac{1}{2} \sum_{k,l=1}^{N} \left(\alpha_k - \alpha_k^* \right)\left(\alpha_l - \alpha_l^* \right) K\left(\mathbf{x}_k, \mathbf{x}_l \right)$$

$$-\varepsilon \sum_{k=1}^{N} \left(\alpha_k + \alpha_k^* \right) + \sum_{k=1}^{N} y_k \left(\alpha_k - \alpha_k^* \right) \tag{13}$$

$$\text{subject to} \begin{cases} \sum_{k=1}^{N} \left(\alpha_k - \alpha_k^* \right) = 0 \\ \alpha_k, \alpha_k^* \in [0, c] \end{cases}$$

By setting $K(x_k, x_l) = \varphi(x_k)^T \varphi(x_l)$ for $k, l = 1, \ldots, N$, the explicit calculation of the kernel function is avoided. The nonlinear regression representation of the dual problem is given by

$$f(\mathbf{x}) = \sum_{k=1}^{N} \left(\alpha_k - \alpha_k^*\right) K(\mathbf{x}, \mathbf{x}_k) + b \tag{14}$$

where α_k, α_k^* are the solution of Eq. 13 and the bias term b is calculated as an average value over the support vectors corresponding to the training data set. The solution to Eq. 13 is unique and is a global minimum so long as the kernel function is positive definite (Suykens et al. 2002).

4   Bounds on the Generalization Error

The VC-theory underlying the SVM formulation characterizes the generalization error instead of the training (empirical) error. Given a set of functions $f(\mathbf{x}, \boldsymbol{\theta})$ characterized by different adjustable parameter vector $\boldsymbol{\theta}$ with a training data set $\{(\mathbf{x}_k, y_k)\}_{k=1}^{N}$ where $\mathbf{x}_k \in R^{N \times n}$ and $y_k \in R^n$. The empirical error is defined as follows:

$$R_{\text{emp}}(\boldsymbol{\theta}) = \frac{1}{N} \sum_{k=1}^{N} (y_k - f(\mathbf{x}, \boldsymbol{\theta}))^2 \tag{15}$$

whereas the generalization error is defined by

$$R(\boldsymbol{\theta}) = \int (y_k - f(\mathbf{x}, \boldsymbol{\theta}))^2 p(\mathbf{x}, y) dx dy \tag{16}$$

measures the error over all patterns that are extracted from an underlying probability distribution $p(\mathbf{x}, y)$ which is typically unknown in practical applications. However, from Eqs. 15 and 16, the upper bound on the generalization error is given as follows:

$$R(\boldsymbol{\theta}) \leq R_{\text{emp}}(\boldsymbol{\theta}) + \left(\frac{1}{1 - c\sqrt{\frac{h(\ln(aN/h)+1) - \ln(\eta)}{N}}} \right)_+ \tag{17}$$

where h is the VC dimension of the set of approximating functions and the notation $(x)_+$ indicates that $(x)_+ = x$ if $x > 0$ and 0 otherwise. The upper bound given by Eq. 17 holds for probability $1 - \eta$. . This is the probability (or level of confidence) to approximate functions at which the generalization bound holds (Schölkopf and Smola 2002; Kecman 2005). The confidence term (second one in Eq. 17) also depends on the VC dimension which in turn characterizes the capacity of the set of approximating functions which in turn reflects model complexity (Vapnik 1998; Schölkopf and

Smola 2002; Suykens et al. 2002). In this research, $a = c = 1$ (Cherkassky and Shao 2001).

5 SVR Parameter and Model Selection

To determine the optimal set of parameters (C, ε, and kernel function shape factors, e.g., the variance of the Gaussian RBF) for the SVR model, iterative grid and pattern searches are used. The grid search aims to use values extracted from a specified range controlled by geometric steps, whereas pattern search is based on the idea that a range is specified and that the search starts at the center of the range and takes trial steps in each direction for each parameter. If the parameters at the new point enhance the fit of the regression function, the search center moves to the new point and the process is repeated. Otherwise, the step size is reduced, and the search is resumed. This iterative process is stopped after the step size drops below a pre-defined tolerance. Grid search is computationally demanding since the model must be evaluated at many points within the grid for each parameter. This limitation may be exaggerated if cross-validation has been adopted as a model selection technique. Pattern search requires far fewer evaluations of the model than that of grid search. However, pattern search can potentially converge to a local instead of a global optimum. Here, both search methods are used to overcome the shortcomings of each method: The optimization process starts with grid search seeking to locate a region close to the global optimal point. Next, pattern search is done over a narrow range that surrounds the best point located by the grid search.

6 Elastic Properties Prediction Methodology

6.1 Well Log and Core Data Description

Figure 1 displays a subset of the core and log data versus depth used in this study. The input variables used to model Poisson's ratio, μ, and Young's modulus, E, at each depth, $\mathbf{x}_k$, are core-derived porosity ϕ, minimum horizontal stress σ_h, pore pressure P_p, and overburden stress g, and log data including bulk density ρ_b, compressional wave velocity v_p, and shear wave velocity v_s. Elastic rock properties were extracted from samples tested in a laboratory-based triaxial pressure test cell. The input data set consists of 601 multidimensional data points spanning over 300 ft. of wellbore length. To evaluate the methods, the input data are randomly separated into training subsets consisting of 10, 20, 30, 40, 50, and 60% of the total data set. The complements of the training data subsets are the testing subsets.

Acoustic logging tools measure the characteristic propagation speed of the P (compression) and S (shear) waves which are related to the elastic properties of the

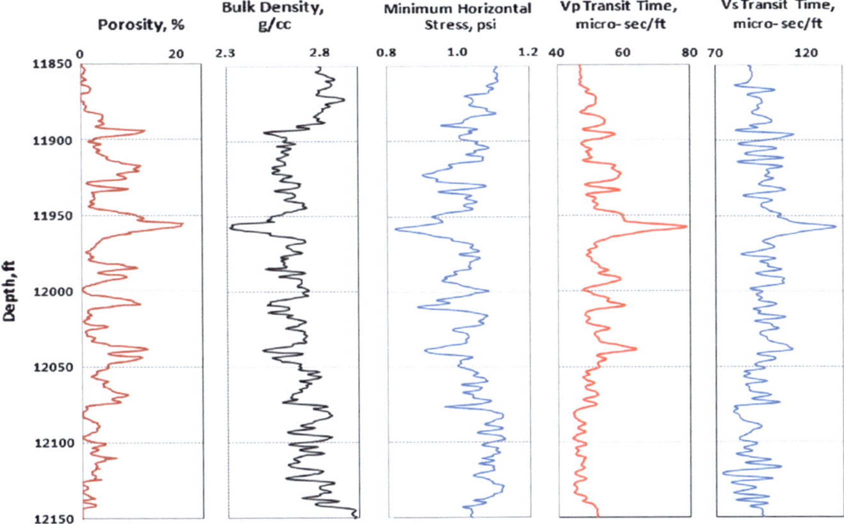

Fig. 1 Subset of raw log data

formation (Serra 1984). For core samples, elastic rock properties are determined from the stress–strain relationship that results from changes of stress with changes of the core strain (Montmayeur and Graves 1985, 1986):

$$E = \frac{\mathrm{d}\sigma_a}{\mathrm{d}\varepsilon_r}$$

where σ_a is the axial stress applied to the core sample and ε_r is the radial strain, and

$$\mu = \frac{\mathrm{d}\varepsilon_r}{\mathrm{d}\varepsilon_a}$$

where μ is Poisson's ratio and ε_a is the axial strain.

7 Modeling of Poisson's Ratio, *M*

Feature Selection

Here, a model is developed that uses core-derived porosity ϕ, minimum horizontal stress σ_h, pore pressure P_p, and overburden stress g, and log data including bulk density ρ_b, compressional wave velocity v_p, and shear wave velocity v_s. Following Al-Anazi and Gates (2009), a Two-stage fuzzy ranking algorithm is used to identify information-rich core and log measurements and filter out data dependencies. The result of fuzzy ranking analysis, listed in Table 3, shows that the significant input variables, in order of importance, are σ_h, v_s, v_p, ρ_b, P_p, and ϕ. Therefore, these variables are the best correlators of Poisson's ratio and will be used in SVM model construction.

Poisson's Ratio Training and Prediction

The BPNN and SVR methods were compared by using different data training fractions (10, 20, 30, 40, 50, and 60%) of the total available data to examine the impact of data scarcity on the predictive capabilities of the generated models. For example, a 10% data training fraction means that 10% of the total data available is selected randomly and used to train the model. The remaining 90% of the data are used to test the correlation. For the SVR model, the Gaussian RBF kernel function was used.

Table 4 lists the error measures for the BPNN model results at different data training fractions when it is used to reproduce the training data set values. Learning performance is measured by correlation coefficient (r) and error statistics including RMSE, average absolute error (AAE), and MAE as defined in Table 2. The results reveal that the correlation coefficient is low and that as the data training fraction is enlarged, the capability of the BPNN to reproduce the training data set does not improve; in other words, the errors do not diminish as the training fraction grows. Table 5 lists the results of the SVR method. Even with training fraction as low as 10%, the correlation coefficient is essentially equal to 1 and the errors are much smaller than that of the BPNN. In other words, the SVR method fully reproduces the training data set. These results demonstrate that for this data set, the SVR has superior learning capability to that of the BPNN.

Table 6 lists the error measures for the BPNN model results at different training fractions when it is used to predict the testing data set. The results show that the method does not provide a good prediction of the testing data set. The SVR results for the predicted testing data set are presented in Table 7. The results demonstrate an excellent prediction performance by the SVM revealing that the trained correlation models have captured the underlying relationships and have the potential to generalize accurately to new data. It can also be observed that consistent prediction capabilities are maintained over all data training fractions.

Table 2 Error measures used for accuracy assessment

Accuracy measure	Mathematical expression
Correlation coefficient, r	$\dfrac{\sum_{i=1}^{l}\left(y_i-\bar{y}_i\right)\left(\hat{y}_i-\bar{\hat{y}}_i\right)}{\sqrt{\sum_{i=1}^{l}\left(y_i-\bar{y}_i\right)^2 \sum_{i=1}^{N_P}\left(\hat{y}_i-\bar{\hat{y}}_i\right)^2}}$
Root mean square error, RMSE	$\sqrt{\dfrac{1}{l}\sum_{i=1}^{l}\left(y_i-\hat{y}_i\right)^2}$
Average absolute error, AAE	$\dfrac{1}{l}\sum_{i=1}^{l}\left|y_i-\hat{y}_i\right|$
Maximum absolute error, MAE	$\max\ \left|y_i-\hat{y}_i\right|,\quad i=1,\ldots,l$

Table 3 Two-stage fuzzy ranking analysis for Poisson's ratio

Fuzzy curves ranking for Poisson's ratio data

Input x^j	P_{c^i}	P_{c^i} / P_{c^R}	$P_{y_c^i}$	$P_{v_c^i}$
σ_h	0.6494	0.6512	0.6812	0.0489
v_s	0.8643	0.8667	0.8776	0.0154
v_p	0.8748	0.8773	0.8859	0.0127
ϕ	0.9194	0.9220	0.9279	0.0092
ρ_b	0.9540	0.9567	0.9723	0.0192
P_p	0.9740	0.9767	0.9794	0.0056
g	0.9740	0.9767	0.9794	0.0056

Second fuzzy surfaces ranking performance for Poisson's ratio data

Input x^j	$P_{s^{1,j}}$	$P_{s^{1,j}} / P_{s^{1,R}}$	$P_{s^{1,j}} / P_{c^1}$	$P_{y_s^{1,j}}$	$P_{v_s^{1,j}}$
v_s	0.3530	0.5475	0.5436	0.4110	0.1642
ρ_b	0.5759	0.8931	0.8869	0.6169	0.0711
P_p	0.5997	0.9300	0.9235	0.6496	0.0831
g	0.5997	0.9300	0.9235	0.6496	0.0831
ϕ	0.6282	0.9742	0.9674	0.6685	0.0641
v_p	0.6340	0.9832	0.9763	0.6763	0.0667

Third fuzzy surfaces ranking performance for Poisson's ratio data

Input x^j	$P_{s^{2,j}}$	$P_{s^{2,j}} / P_{s^{2,R}}$	$P_{s^{2,j}} / P_{c^2}$	$P_{y_s^{2,j}}$	$P_{v_s^{2,j}}$
v_p	0.6325	0.7365	0.7319	0.6665	0.0538
ϕ	0.7135	0.8308	0.8256	0.7468	0.0467
g	0.7878	0.9173	0.9116	0.8048	0.0215
P_p	0.7878	0.9173	0.9116	0.8048	0.0215
ρ_b	0.8085	0.9414	0.9355	0.8436	0.0434

Fourth fuzzy surfaces ranking performance for Poisson's ratio data

Input x^j	$P_{s^{3,j}}$	$P_{s^{3,j}} / P_{s^{3,R}}$	$P_{s^{3,j}} / P_{c^3}$	$P_{y_s^{3,j}}$	$P_{v_s^{3,j}}$
ρ_b	0.8380	0.9621	0.9578	0.8580	0.0239
P_p	0.8489	0.9746	0.9703	0.8708	0.0258
g	0.8489	0.9746	0.9703	0.8708	0.0258
ϕ	0.8495	0.9753	0.9710	0.8665	0.0200

(continued)

Table 3 (continued)

Fourth fuzzy surfaces ranking performance for Poisson's ratio data

Input x^j	$P_{s^{5,j}}$	$P_{s^{5,j}} / P_{s^{5,R}}$	$P_{s^{5,j}} / P_{c^5}$	$P_{y_s^{5,j}}$	$P_{v_s^{5,j}}$
P_p	0.8789	0.9294	0.9212	0.9190	0.0457
g	0.8789	0.9294	0.9212	0.9190	0.0457
ϕ	0.8794	0.9299	0.9218	0.9000	0.0235

Fifth fuzzy surfaces ranking performance for Poisson's ratio data

Input x^j	$P_{s^{6,j}}$	$P_{s^{6,j}} / P_{s^{6,R}}$	$P_{s^{6,j}} / P_{c^6}$	$P_{y_s^{6,j}}$	$P_{v_s^{6,j}}$
ϕ	0.8461	0.8785	0.8687	0.8768	0.0363

Table 4 Comparison of partition-based training error performance of BPNN using minimum horizontal stress, P and S wave velocities, RHOB, pore pressure, and porosity for prediction of Poisson's ratio

	10%	20%	30%	40%	50%	60%
r	0.6375	0.5394	0.5269	0.5940	0.5202	0.5628
RMSE	0.0291	0.0278	0.0289	0.0266	0.0295	0.0261
AAE	0.0239	0.0227	0.0224	0.0212	0.0229	0.0203
MAE	0.0852	0.0847	0.0984	0.0852	0.1043	0.1055

Table 5 Comparison of partition-based training error performance of SVR using minimum horizontal stress, P and S wave velocities, RHOB, pore pressure, and porosity for prediction of Poisson's ratio

	10%	20%	30%	40%	50%	60%
r	1.000	1.000	1.000	1.000	1.000	1.000
RMSE	0.0003	0.0003	0.0002	0.0003	0.0002	0.0002
AAE	0.0003	0.0002	0.0002	0.0002	0.0002	0.0002
MAE	0.0006	0.0006	0.0008	0.0012	0.0006	0.0006

Table 6 Comparison of partition-based testing error performance of BPNN using minimum horizontal stress, P and S wave velocities, RHOB, pore pressure, and porosity for prediction of Poisson's ratio

	10%	20%	30%	40%	50%	60%
r	0.5247	0.5330	0.5554	0.5005	0.5519	0.4997
RMSE	0.0277	0.0216	0.0271	0.0286	0.0256	0.0294
AAE	0.0214	0.0216	0.0213	0.0220	0.0204	0.0231
MAE	0.1062	0.1004	0.1050	0.1061	0.1014	0.1014

Table 7 Comparison of partition-based testing error performance of SVR using minimum horizontal stress, P and S wave velocities, RHOB, pore pressure, and porosity for prediction of Poisson's ratio

	10%	20%	30%	40%	50%	60%
r	0.9992	0.9998	1.000	0.9999	1.000	1.000
RMSE	0.0014	0.0007	0.0003	0.0004	0.0002	0.0003
AAE	0.0007	0.0004	0.0002	0.0003	0.0002	0.0002
MAE	0.0090	0.0051	0.0021	0.0030	0.0008	0.0013

8 Modeling of Young's Modulus, *E*

Feature Selection

Similar to the analysis on Poisson's ratio described above, a two-stage fuzzy ranking analysis was done to identify the most important variables to predict Young's modulus. The results, listed in Table 8, reveal that the ranking of input data, in order of importance, is v_s, ρ_b, v_p, P_p, ϕ, g, and σ_h. As expected, the ranking of the variables makes it clear that the acoustic and density logs are most important with respect to Young's modulus.

Young's Modulus Training and Prediction

As above, Gaussian RBFs was the kernel function used in the SVM nonlinear regression. Tables 9 and 10 list the results of the learning capabilities (the capability of the methods to reproduce the training data set) of the BPNN and SVR, respectively, at different training data fractions. For the BPNN, the results show that the correlation coefficient is high for all training data fractions. However, the RMSE and AAE do not exhibit a strong reducing trend as the training data fraction is enlarged and the MAE, in fact, shows a growth trend as the training data fraction increases. For the SVR, the correlation coefficients are high and are similar to that of the BPNN, but there is a decreasing trend of the RMSE, AAE, and MAE as the size of the training data fraction grows. Beyond 30% training data set, all of the error measures of the SVR are lower than that of the BPNN. The results suggest that the SVM approach has a higher learning capability than that of the BPNN for the data set used here.

Tables 11 and 12 list the correlation coefficient and error measures for the BPNN and SVR methods, respectively, at different training data fraction to predict the testing data subset. The results reveal excellent prediction performance can be observed by both techniques in terms of correlation coefficient. However, the analysis of the error performance indicates that the errors of the trained SVR models decline faster versus the size of the training data fraction than that of the BPNN and that the errors are lower than that of the BPNN when the training data fraction is 40% or higher.

Table 8 Two-stage fuzzy ranking analysis for Young's modulus

Fuzzy curves ranking for Young's modulus data

Input x^j	P_{c^i}	$P_{c^i}\big/P_{cR}$	$P_{y_c^i}$	$P_{v_c^i}$
v_s	0.2484	0.2507	0.3054	0.2295
v_p	0.4010	0.4048	0.4620	0.1521
ρ_b	0.4676	0.4720	0.5206	0.1135
ϕ	0.4868	0.4913	0.5468	0.1234
σ_h	0.5026	0.5073	0.5589	0.1120
P_p	0.5848	0.5903	0.6181	0.0571
g	0.5848	0.5903	0.6181	0.0571

Second fuzzy surfaces ranking performance for Young's modulus data

Input x^j	$P_{s^{1,j}}$	$P_{s^{1,j}}\big/P_{s^{1,R}}$	$P_{s^{1,j}}\big/P_{c^1}$	$P_{y_s^{1,j}}$	$P_{v_s^{1,j}}$
ρ_b	0.1344	0.5432	0.5412	0.1823	0.3559
v_p	0.1447	0.5847	0.5826	0.1967	0.3596
g	0.1497	0.6051	0.6029	0.1993	0.3307
P_p	0.1497	0.6051	0.6029	0.1993	0.3307
σ_h	0.1514	0.6117	0.6095	0.2065	0.3642
ϕ	0.1567	0.6332	0.6309	0.2122	0.3540

Third fuzzy surfaces ranking performance for Young's modulus data

Input x^j	$P_{s^{3,j}}$	$P_{s^{3,j}}\big/P_{s^{3,R}}$	$P_{s^{3,j}}\big/P_{c^3}$	$P_{y_s^{3,j}}$	$P_{v_s^{3,j}}$
v_p	0.2415	0.5325	0.5166	0.3052	0.2636
P_p	0.3136	0.6912	0.6706	0.3856	0.2298
g	0.3136	0.6912	0.6706	0.3856	0.2298
σ_h	0.3141	0.6924	0.6717	0.3842	0.2231
ϕ	0.3551	0.7827	0.7594	0.4259	0.1995

Fourth fuzzy surfaces ranking performance for Young's modulus data

Input x^j	$P_{s^{2,j}}$	$P_{s^{2,j}}\big/P_{s^{2,R}}$	$P_{s^{2,j}}\big/P_{c^2}$	$P_{y_s^{2,j}}$	$P_{v_s^{2,j}}$
P_p	0.2770	0.6983	0.6909	0.3477	0.2551
g	0.2770	0.6983	0.6909	0.3477	0.2551
ϕ	0.3054	0.7698	0.7616	0.3808	0.2469
σ_h	0.3436	0.8661	0.8569	0.4229	0.2307

Fourth fuzzy surfaces ranking performance for Young's modulus data

Input x^j	$P_{s^{6,j}}$	$P_{s^{6,j}}\big/P_{s^{6,R}}$	$P_{s^{6,j}}\big/P_{c^6}$	$P_{y_s^{6,j}}$	$P_{v_s^{6,j}}$
ϕ	0.2909	0.5072	0.4975	0.3645	0.2531
σ_h	0.3599	0.6276	0.6155	0.4370	0.2140
g	0.5175	0.9023	0.8850	0.5717	0.1046

(continued)

Table 8 (continued)

Fourth fuzzy surfaces ranking performance for Young's modulus data

Input x^j	$P_{s^4,j}$	$\dfrac{P_{s^4,j}}{P_{s^4,R}}$	$\dfrac{P_{s^4,j}}{P_{c^4}}$	$P_{y_s^{4,j}}$	
g	0.2909	0.6115	0.5977	0.3645	0.2531
σ_h	0.3595	0.7556	0.7385	0.4371	0.2160

Fifth fuzzy surfaces ranking performance for Young's modulus data

Input x^j	$P_{s^7,j}$	$P_{s^7,j}\big/P_{s^7,R}$	$P_{s^7,j}\big/P_{c^7}$	$P_{y_s^{7,j}}$	$P_{v_s^{7,j}}$
σ_h	0.3599	0.6276	0.6155	0.4370	0.2140

Table 9 Comparison of partition-based training error performance of BPNN using minimum horizontal stress, P and S wave velocities, pore pressure, overburden stress, and porosity for prediction of Young's modulus

	10%	20%	30%	40%	50%	60%
r	0.9995	0.9995	0.9996	0.9995	0.9995	0.9995
RMSE	0.2336	0.2415	0.2635	0.2705	0.2864	0.2594
AAE	0.1903	0.1806	0.1990	0.2032	0.2181	0.1792
MAE	0.5223	0.9198	1.2780	1.2645	1.4517	2.0796

Table 10 Comparison of partition-based training error performance of SVR using minimum horizontal stress, P and S wave velocities, pore pressure, overburden stress, and porosity for prediction of Young's modulus

	10%	20%	30%	40%	50%	60%
r	0.9944	0.9993	0.9999	0.9999	0.9999	1.0000
RMSE	0.9389	0.3098	0.1619	0.1390	0.0954	0.0759
AAE	0.5270	0.1584	0.0845	0.0764	0.0460	0.0419
MAE	4.6358	1.7137	0.9692	1.0244	1.1114	0.4942

Table 11 Comparison of partition-based testing error performance of BPNN using minimum horizontal stress, P and S wave velocities, pore pressure, overburden stress, and porosity for prediction of Young's modulus

	10%	20%	30%	40%	50%	60%
r	0.9981	0.9989	0.9997	0.9987	0.9994	0.9994
RMSE	0.5426	0.4209	0.249	0.4658	0.2922	0.3159
AAE	0.3421	0.2731	0.1964	0.2609	0.2337	0.2002
MAE	4.0785	3.2514	1.3903	3.4581	1.5281	2.0031

Table 12 Comparison of partition-based testing error performance of SVR using minimum horizontal stress, P and S wave velocities, pore pressure, overburden stress, and porosity for prediction of Young's modulus

	10%	20%	30%	40%	50%	60%
r	0.9905	0.9986	0.9996	0.9994	0.9998	0.9999
RMSE	1.5067	0.5016	0.2405	0.3153	0.1547	0.1445
AAE	0.922	0.2747	0.1337	0.1378	0.0764	0.0786
MAE	7.5939	3.596	1.9554	2.2864	1.4729	1.0244

9 SVM Parameter Sensitivity Analysis

The generalization capability of SVMs depends on the regularization parameter, C, (see Eq. 5) that controls the trade-off between the training error (empirical error) and the VC dimension (complexity) of the regression model. If the value of C is very small, the training error is the primary error that is minimized, whereas if its value is very large, the estimate on the prediction error dominates and the training error plays a lesser role to construct the SVR model. Additional inputs of the SVR model include the cost function parameter (the regression tube radius ε of the ε-insensitivity cost function) and the type of kernel function used and its associated input parameters (for example, the variance of the Gaussian RBF kernel function). In the above analysis, these parameters were chosen by using the cross-validation method. Here, the impact of these parameters on the SVR model is investigated. Here, 10% data training fraction subset (with 90% remaining for testing) is used.

10 Impact of C on SVM Regression Performance

The effect of the value of C on the SVR training regression model for Poisson's ratio and Young's modulus were investigated with RBF kernel function with both approximation tube radius ε and the variance of the kernel function σ kept constant at values selected in the above analysis. Tables 13 and 14 list the training performance as examined by correlation coefficient and error statistics to predict Poisson's ratio and Young's modulus. The results indicate that the constructed regression model fits the data perfectly as the values of the regularization parameter, C, increase. The selection of higher values of C, however, does not always improve the performance of the SVR especially if the data itself does not describe the underlying function that relates the input vector to the output scalar as would be the case with data polluted with severe noise. In this case, the SVR would model the noise rather than the true relationship between the input and output data.

Table 13 Comparison of partition-based training error performance of SVM over different regularization constant, C values for prediction of Poisson's ratio ($\sigma = 0.1877$, $\varepsilon = 9 \times 10^{-5}$)

	1×10^{-4}	1×10^{-3}	1×10^{-2}	1×10^{-1}	1	1×10^{1}	1×10^{2}
r	0.6849	0.6849	0.7297	0.9865	0.9994	0.9999	1.0000
RMSE	0.0356	0.0351	0.0312	0.0092	0.0014	0.0004	0.0003
AAE	0.0287	0.0283	0.0250	0.0067	0.0009	0.0003	0.0003
MAE	0.0964	0.0959	0.0895	0.0292	0.0067	0.0010	0.0006

Table 14 Comparison of partition-based training error performance of SVM over different regularization constant, C values for prediction of Young's modulus ($\sigma = 0.1208$, $\varepsilon = 4 \times 10^{-5}$)

	1×10^{3}	1×10^{4}	1×10^{5}	1×10^{6}	1×10^{7}	1×10^{8}	1×10^{9}
r	0.0000	0.7908	0.8246	0.9331	0.9944	0.9995	1.0000
RMSE	7.5444	7.4695	6.8751	3.8890	0.9389	0.2379	0.0432
AAE	5.8183	5.7562	5.2528	2.8527	0.5270	0.1476	0.0262
MAE	25.0131	24.7374	21.9113	12.1491	4.6358	0.9137	0.2444

11 Impact of ε on SVM Regression Performance

The radius of the regression tube, ε, within which the regression function must lay, is a measure of the error tolerance of the predictive capability of the regression model. If a predicted value lies within the tube radius, that is, its absolute value is less than ε and the loss (error) is set equal to zero. For a predicted value lying outside the ε-tube, the loss (error) equals the difference between the predicted value and the radius of the tube ε. Here, to investigate the effect of the regression tube radius on the capability of the SVR to generalize to unseen data, the regularization parameter, C, and the variance, σ, of the RBF kernel function were fixed at the values determined above. The results listed in Tables 15 and 16 reveal that the constructed training model of Poisson's ratio and Young's modulus fits perfectly the data as the values of the radius of the regression tube decrease as indicated by the correlation coefficient and error statistics. On the contrary, as the size of the insensitivity tube increases, the accuracy of the prediction model drops. The input data for Poisson's ratio and Young's modulus data are quite clustered, and there are few outliers. As a result, when the radius of the regression tube enlarges, a fewer number of training points is used, thus the number of support vectors drops, and thus, the quality of the SVR model degrades. A further increase in the radius causes the SVR model to overshoot the test data. The results reveal that an increase in the radius of the insensitivity tube has a smoothing effect on modeling, whereas a decrease in the size may lead to overfitting the data.

Table 15 Comparison of partition-based training error performance of SVM over different approximation tube radius, ε values for prediction of Poisson's ratio ($\sigma = 0.1877$, $C = 45.66$)

	1×10^{-4}	1×10^{-3}	1×10^{-2}	1×10^{-1}
r	1.0000	0.9997	0.9907	0.000
RMSE	0.0003	0.0009	0.0060	0.0378
AAE	0.0003	0.0007	0.0051	0.0319
MAE	0.0005	0.0014	0.0104	0.0776

Table 16 Comparison of partition-based training error performance of SVM over different approximation tube radius, ε values for prediction of young's modulus ($\sigma = 0.1208$, $C = 1 \times 10^7$)

	1×10^{3}	1×10^{4}	1×10^{5}	1×10^{6}	1×10^{7}
r	0.9944	0.9945	0.9934	0.9155	0.0000
RMSE	0.9389	0.9228	0.9868	3.8853	7.9617
AAE	0.5270	0.5303	0.7088	3.2992	6.0465
MAE	4.6358	4.5669	4.2223	8.4958	22.5317

12 Impact of Kernel Function on SVM Regression Performance

The kernel function is an integral part of the SVM formulation because it is necessary to solve nonlinear regression problems. The kernel function maps the nonlinear input space into a high-dimensional feature space where a linear SVM formulation can be applied. As such, the kernel function is assumed to have the capability to provide or approximately provide the nonlinear mapping. Thus, the choice of the kernel function will depend on the nature of the regression problem being solved. Here, three kernel functions are investigated including the linear, sigmoid, and Gaussian RBF functions. The prediction capabilities of SVR approximation models based on these kernels to predict Poisson's ratio and Young's modulus are compared are shown in Figs. 2 and 3, respectively. The corresponding correlation coefficients and error statistics are listed in Tables 17 and 18. The results reveal that all three kernel functions exhibit reasonably good capabilities to approximate the core-measured Poisson's ratio values. The best results are achieved with the RBF kernel function. Similarly, all three kernel functions performed well to predict the core-measured Young's modulus, but the RBF function performs slightly better than the linear and sigmoid kernel functions.

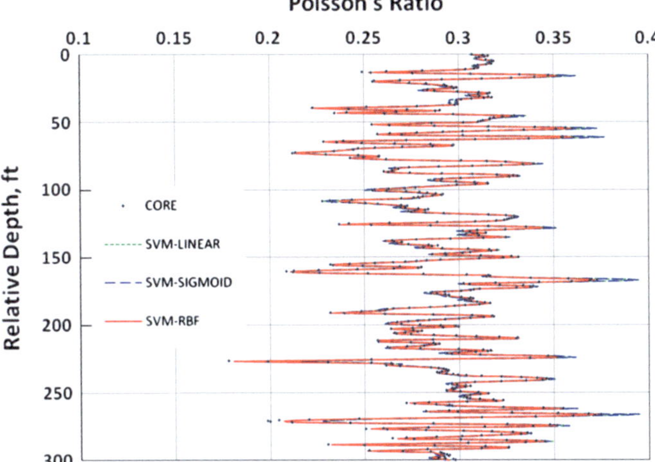

Fig. 2 Comparison of SVM-LINEAR, SVM-SIGMOID, and SVM-RBF predictions of Poisson's ratio using 10% of the data for training and 90% for testing. The dots are data obtained from triaxial tests of core samples

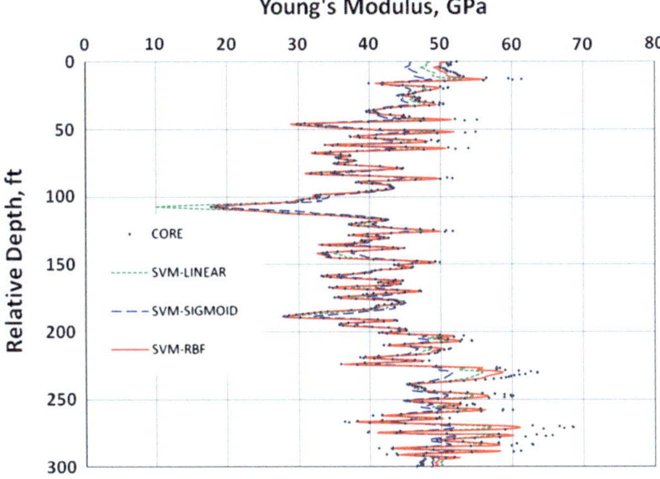

Fig. 3 Comparison of SVM-LINEAR, SVM-SIGMOID, and SVM-RBF predictions of young's modulus using 10% of the data for training and 90% for testing. The dots are data obtained from triaxial tests of core samples

Table 17 Comparison of correlation coefficient, RMSE, AAE, and MAE of the errors between SVM-LINEAR, SVM-SIGMOID, and SVM-RBF predictions of Poisson's ratio using 10% of the data for training and 90% for testing

	Linear	Sigmoid	RBF
r	0.9935	0.9937	1.0000
RMSE	0.0040	0.0040	0.0003
AAE	0.0029	0.0029	0.0003
MAE	0.0193	0.0184	0.0006

Table 18 Comparison of correlation coefficient, RMSE, AAE, and MAE of the errors between SVM-LINEAR, SVM-SIGMOID, and SVM-RBF predictions of Young's modulus using 10% of the data for training and 90% for testing

	Linear	Sigmoid	RBF
r	0.9612	0.9203	0.9905
RMSE	2.8075	4.5299	1.5067
AAE	1.9659	3.3711	0.9219
MAE	11.6000	16.8300	7.6000

13 Conclusions

SVMs integrated with a fuzzy-based curve and surface input variable ranking analysis have demonstrated their potential applicability to develop interpretation models of Poisson's ratio and Young's modulus under limited core data conditions. The main conclusions are as follows:

1. The SVM model is easier to construct than that of an artificial neural network model.
2. The SVMs formulation facilitate a unique global solution compared to BPNN which often suffers multiple local minima.
3. The SVMs formulation offers a natural means to deal with sparse data given that the number of support vectors used is equal to the number of training data points.
4. The SVMs show high learning capability for both Poisson's ratio and Young's modulus under the presence of limited core data.
5. The results show that SVR yields a better model to predict Poisson's ratio than a backpropagation neural networks model.
6. The results demonstrate that the prediction errors for Young's modulus obtained from the SVR decrease faster as the training data size grows than that of the backpropagation neural network.
7. Linear, sigmoid, and Gaussian RBF kernel functions show high prediction generalizability for Poisson's ratio and Young's modulus. The RBF function exhibits slightly better performance than the other two kernel functions.

References

Abdulraheem A, Ahmed M, Vantala A, Parvez T (2009) Prediction of rock mechanical parameters for hydrocarbon reservoirs using different artificial intelligence techniques. Paper SPE 126094 presented at the SPE Saudi Arabia section technical symposium, Alkhobar, Saudi Arabia, pp 9–11

Al-Anazi A, Gates ID (2009) Fuzzy logic data-driven permeability prediction for heterogeneous reservoirs. Paper SPE 121159 presented at the 2009 SPE EUROPEC/EAGE annual conference and exhibition, Amsterdam, The Netherlands, pp 8–11

Al-Anazi A, Gates ID (2010b) Support vector regression for permeability prediction in a heterogeneous reservoir: a comparative study. SPE Res Eval Eng 13(3):485–495. SPE-126339-PA. https://doi.org/10.2118/126339-PA

Al-Anazi A, Gates ID (2010a) On the capability of support vector machines to classify lithology from well logs. Nat Resour Res 19(2):125–139. https://doi.org/10.1007/s11053-010-9118-9

Al-Anazi A, Gates ID (2010c) A support vector machine algorithm to classify lithofacies and model permeability in heterogeneous reservoirs. Eng Geol 114:267–277. https://doi.org/10.1016/j.enggeo.2010.05.005

Al-Anazi A, Gates ID (2010d) Support vector regression for porosity prediction in a heterogeneous reservoir: a comparative study. Comput Geosci 36(12):1494–1503

Ameen MS, Smart Brian GD, Mc J, Hammilton SS, Naji Nassir A (2009) Predicting rock mechanical properties of carbonates from wireline logs (a case study: Arab-D reservoir, Ghawar field, Saudi Arabia). Mar Petrol Geol 26(4):430–444

Barree RD, Gilbert JV, Conway MW (2009) Stress and rock property profiling for unconventional reservoir stimulation. Paper SPE 118703 presented at the SPE hydraulic fracturing technology conference, The Woodlands, Texas, 19–21 Jan 2009

Cherkassky V, Shao X (2001) Signal estimation and denoising using VC-theory. Neural Netw 14:37–52

Choisy C, Belaid A (2001) Handwriting recognition using local methods for normalization and global methods for recognition. In: Proceedings of sixth international conference on document analysis and recognition, pp 23–27

DTREG (2009) Predictive modeling software user's manual, version 9.1 (http://www.dtreg.com)

Fung C, Wong K, Eren H (1997) Modular artificial neural network for prediction of petrophysical properties from well log data. IEEE Trans Instrum Measure 46(6):1295–1299

Gao D, Zhou J, Xin L (2001) SVM-based detection of moving vehicles for automatic traffic monitoring. IEEE Intell Transp Syst 745–749

Gatens JM III, Harrison CW III, Lancaster DE, Guldry FK (1990) In-situ stress tests and acoustic logs determine mechanical properties and stress profiles in the devonian shales. SPE Formation Eval 5(3):248–254

Hastie T, Tibshirani R, Friedman J (2001) The elements of statistical learning: data mining, inference, and prediction. Springer, New York

Helle H, Bhatt A (2002) Fluid saturation from well logs using committee neural networks. Petrol Geosci 8:109–118

Helle H, Ursin B (2001) Porosity and permeability prediction from wireline logs using artificial neural networks: a North Sea case study. Geophys Prospect 49:431–444

Huang Z, Shimeld J, Williamson M, Katsube J (1996) Permeability prediction with artificial neural network modelling in the venture gas field, offshore eastern Canada. Geophysics 61:422–436

Huang Y, Gedeon TD, Wong PM (2001) An integrated neural-fuzzy-genetic-algorithm using hyper-surface membership functions to predict permeability in petroleum reservoirs. J Eng Appl Artif Intell 14:15–21

Kecman V (2005) Support vector machines—an introduction. In: Wang L (ed) Support vector machines: theory and applications, Chap. 1. Springer, Berlin, pp 1–47

Khaksar A, Taylor PG, Fang Z, Kayes T, Salazar A, Rahman K (2009) Rock strength from core and logs, where we stand and ways to go. Paper SPE 121972 presented at the EUROPEC/EAGE conference and exhibition, Amsterdam, The Netherlands, 8–11 June 2009

Kim K, Jung K, Park S, Kim HJ (2001) Support vector machine-based text detection in digital video. Pattern Recognit 34:527–529

Li Z, Weida Z, Licheng J (2000) Radar target recognition based on support vector machine. In: Proceedings of 5th international conference on signal processing, vol 3, pp 1453–1456

Lu J, Plataniotis K, Ventesanopoulos A (2001) Face recognition using feature optimization and v-support vector machine. IEEE Neural Netw Signal Processing 11:373–382

Ma C, Randolph M, Drish J (2001) A support vector machines-based rejection technique for speech recognition. Proc IEEE Int Conf Acoust, Speech, Signal Process 1:381–384

Montmayeur H, Graves RM (1985) Prediction of static elastic/mechanical properties of consolidated and unconsolidated sands from acoustic measurements: basic measurements. Paper SPE 14159 presented at the 60th SPE annual technical conference and exhibition, Las Vegas, NV, 22–25 Sept 1985

Montmayeur H, Graves RM (1986) Prediction of static elastic/mechanical properties of consolidated and unconsolidated sands from acoustic measurements: correlations. Paper SPE 15644 presented at the 61st SPE annual technical conference and exhibition, Orleans, LA, 5–8 Oct 1986

Rogers SJ, Chen HC, Kopaska-Merkel DC, Fang JH (1995) Predicting permeability from porosity using artificial neural networks. AAPG Bull 79:1786–1797

Schölkopf B, Smola AJ (2002) Learning with kernels: support vector machines, regularization, optimization, and beyond. MIT Press, Cambridge, MA

Serra O (1984) Fundamentals of well-log interpretation. Elsevier, Amsterdam

Suykens JAK, Van Gestel T, Brabanter J, De Moor B, Vandewalle J (2002) Least squares support vector machines. World Scientific, Singapore

Van Gestel T, Suykens J, Baestaens D, Lambrechts A, Lanckriet G, Vandaele B, De Moor B, Vandewalle J (2001) Financial time series prediction using least squares support vector machines within the evidence framework. IEEE Trans Neural Netw 12(4):809–821

Vapnik VN (1982) Estimation of dependences based on empirical data. Springer, Berlin

Vapnik V (1995) The nature of statistical learning theory. Springer, New York

Vapnik V (1998) Statistical learning theory. Wiley, New York

Vapnik V, Chervonenkis A (1974) Theory of pattern recognition [in Russian]. Nauka, Moscow (German trans: Wapnik W, Tscherwonenkis A., Theorie der Zeichenerkennung, Akademie, Berlin, 1979)

Widarsono B, Wong PM, Saptono F (2001) Estimation of rock dynamic elastic property profiles through a combination of soft computing, acoustic velocity modeling and laboratory dynamic test on core samples. Paper SPE 68712 presented at the SPE Asia Pacific oil and gas conference and exhibition, Jakarta, Indonesia, 17–19 Apr 2001

Use of Active Learning Method to Determine the Presence and Estimate the Magnitude of Abnormally Pressured Fluid Zones: A Case Study from the Anadarko Basin, Oklahoma

Constantin Cranganu and Fouad Bahrpeyma

Abstract We discuss the active learning method (ALM) as an artificial intelligent approach for predicting a missing log (DT or sonic log) when only two other logs (GR and REID) are present. Applying the ALM approach involves three steps: (1) supervised training of the model, using available GR, REID, and DT logs; (2) confirmation and validation of the model by blind-testing the results in a well containing both the predictors (GR, REID) and the target (DT) values; and (3) applying the predicted model to wells containing the predictor data and obtaining the synthetic (simulated) DT values. Our modeling approach indicates that the performance of the algorithm is satisfactory, while the time performance is significant. The quality of our simulation procedure was assessed by three parameters, namely mean square error (MSE), mean relative error (MRE), and Pearson product–momentum correlation coefficient (R). The values obtained for these three quality-control parameters appear congruent, with the exception of MRE, regardless of the training set used (*reduced* vs. *complete*). ALM performance was measured also by the time required to attain the desirable outcomes: five depth levels of investigation took a little more than one minute of computing time during which MSE dropped significantly. We performed twice the regression analysis: with and without normalization of input data sets (training well and validation well) using the procedure indicated by previous works. The results show minimum differences in quality assessment parameters (MSE, MRE, and R), suggesting that data normalization is not a necessary step in all regression algorithms. We employed both the measured and simulated sonic logs DT to predict the presence and estimate the depth intervals where an overpressured fluid zone may develop in the Anadarko Basin, Oklahoma. Based on our interpretation of the sonic log trends, we inferred that overpressure regions are developing between ~1250 and 2500 m depth and the overpressured intervals have thicknesses varying between ~700 and

C. Cranganu (✉)
Department of Earth and Environmental Sciences, Brooklyn College, The City University of New York, 2900 Bedford Avenue, New York City, NY 11210, USA
e-mail: cranganu@brooklyn.cuny.edu

F. Bahrpeyma
Faculty of Electrical and Computer Engineering, Shahid Beheshti University G.C, Tehran, Iran

© The Author(s), under exclusive license to Springer Nature Switzerland AG 2024
C. Cranganu (ed.), *Artificial Intelligent Approaches in Petroleum Geosciences*,
https://doi.org/10.1007/978-3-031-52715-9_6

1000 m. These results match very well our previous results reported in the Anadarko Basin, using the same wells, but different artificial intelligent approaches.

Keywords Active learning method · Well logs · Sonic log · Abnormally pressured zones · Anadarko Basin

1 Introduction

In petroleum geosciences, it is necessary most of the time to characterize pore-fluid pressures, rock lithologies, and a large number of petrophysical properties of reservoir rocks (porosity, permeability, water/oil saturation, etc.). Those parameters are best determined by in situ and core measurements. When such measurements are not available, due to financial, technical, or other impediments, the next best available way of obtaining those data is using a suite of geophysical and petrophysical logs.

Common recorded logs, such as gamma ray (GR), dual induction, deep resistivity (REID), density porosity, photoelectric absorption factor (PEF), or self-potential (SP), provide information about a specific physical characteristic of the rocks penetrated by and surrounding the borehole (e.g., natural radioactivity content, electric resistivity, density, mineralogy, etc.). Among various logs recorded by the petroleum industry, the compressional acoustic or sonic log (DT) has been used many times to predict rock porosity (so-called "acoustic porosity", Ellis and Singer (2007)), to evaluate petrophysical properties or to characterize areas containing pore fluids with abnormal pressure (overpressurized) (Hearst et al. 2000; Cranganu 2007; Cranganu and Bautu 2010; Cranganu and Breaban 2013).

Despite its proven value in estimating rock porosity, or studying and estimating the presence of abnormally pressured pore fluids in sedimentary basins containing oil and gas reservoirs, the sonic log is not always a part of a commonly recorded well logs. The reasons include lacking a full suite of logs, lacking of data due to incomplete logging, instrument failure (damage or faulty), or poor-quality recordings (Bahrpeyma et al. 2013; Cranganu and Bautu 2010; Cranganu and Breaban 2013). As a result, some wells may lack the sonic log entirely or the log is only partially recorded.

To overcome such situations, we propose using a soft computing method—active learning method (ALM)—to synthesize missing sonic (DT) logs when only other common logs (e.g., GR and REID) are present. Thus, we will be able to more effectively map porosity variations, to detect changes in pore-fluid pressure, and to increase the density of control data in wells without recorded DT.

As hinted above, a major use of DT logs is determining the presence and estimating the amplitude of areas with abnormal pore-fluid pressures via effective rock stress estimations. In other words, the sonic log can be used as a predictor of pore-fluid pressures because it responds to changes in porosity or compaction produced when abnormal pore-fluid pressures are present in a sedimentary basin. A case study from the Anadarko Basin, Oklahoma, will be presented to illustrate this use of DT log.

2   Active Learning Method—Background

ALM was originally introduced by Shouraki and Honda (1999) as a soft computing tool for modeling unknown multiple inputs-single output (M.I.S.O.) systems. ALM performs simulation based on the intelligent information-handling processes existing inside the human brain. The engine of ALM is the ink drop spread (IDS) (Saeed Bagheri and Honda 1997) operator, which acts as a fuzzification operator. IDS works based on a non-exact processing procedure without suffering from complex formulas (Murakami and Honda 2005a, b, c).

ALM has proved its ability for estimating missing logs in hydrocarbon reservoirs (Bahrpeyma et al. 2013), modeling chlorophyll and pigment retrieval (Shahraini et al. 2005), modeling geophysical variables (Shahraiyni et al. 2007), and controlling design problems (Behnia et al. 2011; Ghorbani et al. 2010; Sakurayi et al. 2003).

In ALM, the behavior of the output is depicted (projected) onto a two-dimensional (2D) surface with respect to each input parameter in order to reduce the complexity of the modeling procedure for M.I.S.O. systems. In fact, the projected 2D planes are single input–single output (S.I.S.O.) candidate models for the given output parameter. The fuzzy interpolation is then responsible for aggregating the candidate S.I.S.O. models into a unified model according to the degree of relativity the candidate models exhibit for describing the behavior of the output (Fig. 1).

As shown in Fig. 1, ALM breaks an M.I.S.O. system into some S.I.S.O. sub-systems in order to cope with less computational complexities when trying to extract the relationships. The idea is originated from the fact that handling the projected

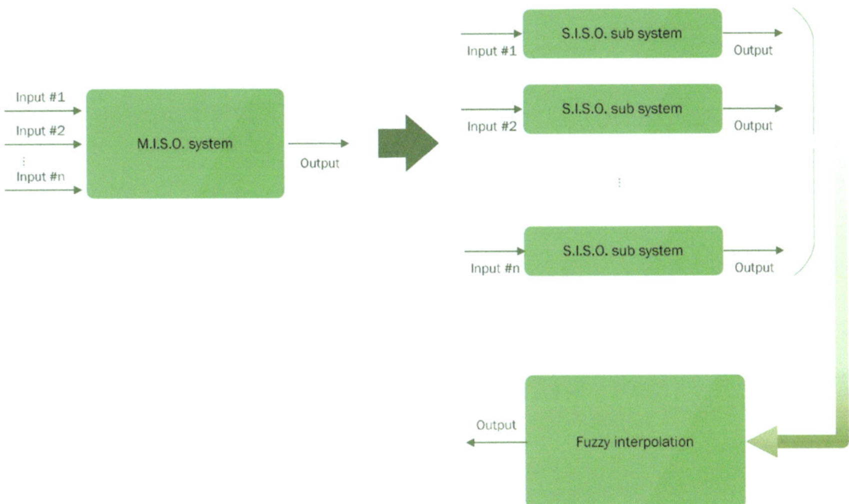

Fig. 1 Handling complexity by breaking the M.I.S.O. system into some S.I.S.O. sub-systems and aggregating them into a unified model inside ALM

Input–Output 2D surfaces reduces the complexity imposed by the extraction of multi-dimensional relations existing between the dimensions inside an M.I.S.O. system. Therefore, complexity is meaningfully reduced. In the end, the fuzzy interpolation revives the impact of each input parameter on the output parameter.

The flowchart of ALM is illustrated in Fig. 2.

At the 2D surface of each S.I.S.O. sub-system, the general behavior of the output with respect to the input is called the narrow path (NP). Figure 3 illustrates the process of NP extraction through IDS and COG (Bahrpeyma et al. 2013; Saeed Bagheri and Honda 1997) operators in an operation board.

The NP is a continuous and function-like path representing the output as the function of the input in a 2D space. The function-like path, as a model of an S.I.S.O. system, should be extracted from a continuous path to provide generalization ability

Fig. 2 The flowchart of ALM

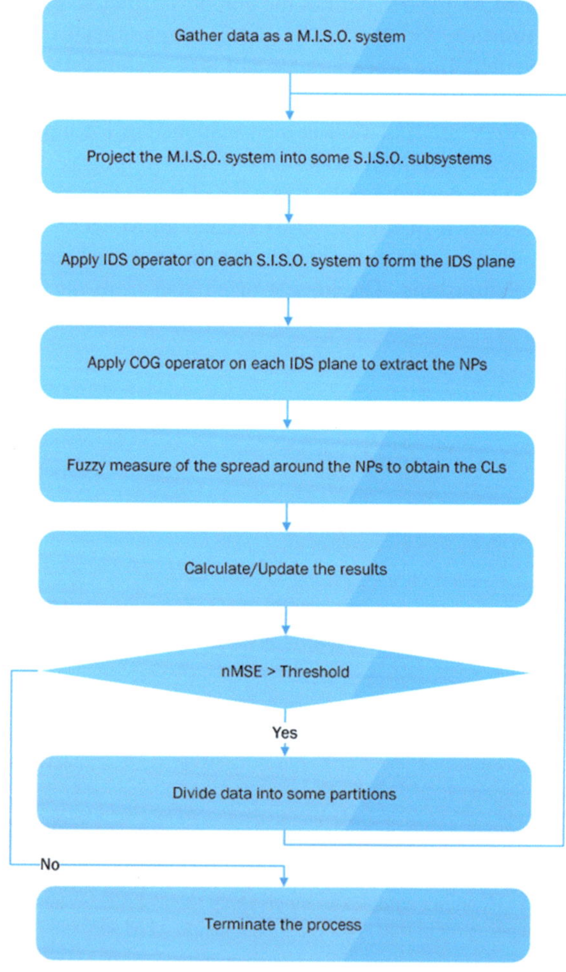

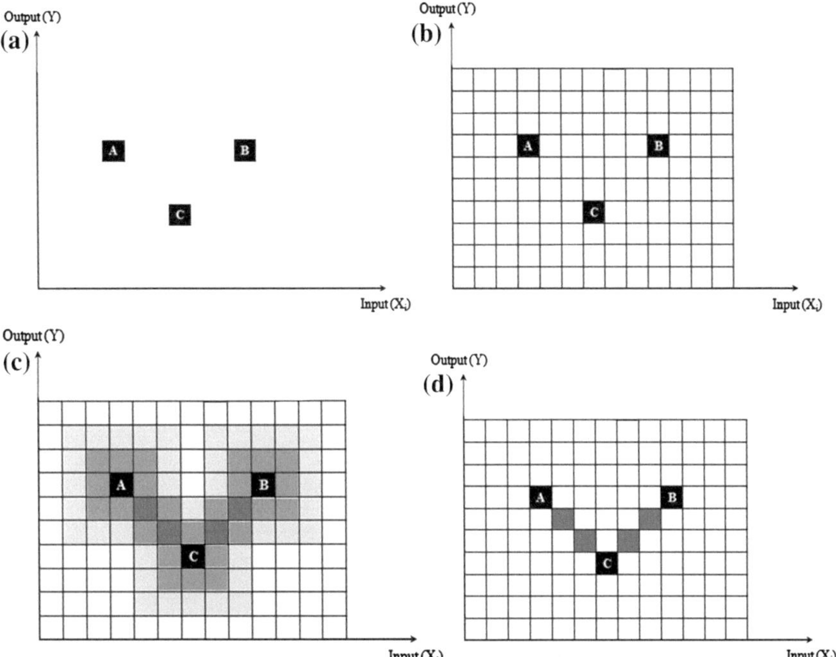

Fig. 3 Extracting the NPs by applying IDS and COG operators. **a** Projection of the M.I.S.O. system onto a 2D surface. **b** Mapping the observations onto an operation board. **c** Applying IDS operator on observed data points to form IDS planes. **d** Applying COG operator on IDS planes to extract the NPs

for unseen data. However, the pure S.I.S.O. sub-system consists of some discrete points in a 2D surface (Fig. 3a and b) which display no aspect of continuity.

The IDS operator is used inside ALM as a fuzzification method to extract fuzzy continuity from a set of discrete points inside a 2D surface. Thus, inside the 2D surface of a S.I.S.O. sub-system, for each projected data point, the IDS operator propagates membership values to the neighborhood in an attenuated pattern. After application of IDS operator on each data point in the 2D surface, the outcome is a fuzzy continuous area that is called the IDS plane (Fig. 3c). The fuzzy continuous area is then converted to a narrow function-like path by the application of the COG operator. The outcome is a function that represents the general behavior of the output with respect to the input, i.e., the NP (Fig. 3d).

To provide an environment for implementing the IDS operator, Bahrpeyma et al. (2013) used a $d\%$ margined operation board (Fig. 3), insuring that membership propagation can be simulated through the neighborhood of rooms inside a 2D board. The domain of ith S.I.S.O. sub-system $(X_i - Y)$ is defined as

$$
\begin{cases}
D_{X_i} : x \,|\, min(X_i) < x < max(X_i) \\
D_Y : y \,|\, min(Y) < y < max(Y)
\end{cases}
\tag{1}
$$

The domain of $d\%$ margined domain is considered as

$$\begin{cases} D_{\widehat{X_i}} : x \mid min(X_i) - \text{margin}_{X_i}^{d\%} < x < max(X_i) + \text{margin}_{X_i}^{d\%} \\ D_{\widehat{Y}} : y \mid min(Y) - \text{margin}_{Y}^{d\%} < y < max(Y) + \text{margin}_{Y}^{d\%} \end{cases} \tag{2}$$

where $\text{margin}_X^{d\%}$ is calculated by

$$\text{margin}_X^{d\%} = [max(X_i) - min(X_i)] \times \frac{d}{100} \tag{3}$$

Finally, the size of each unit $\left[U_{\widehat{X_i}}^M, U_{\widehat{Y}}^M \right]$ in ith S.I.S.O. sub-system for an $M \times M$ board is

$$\begin{cases} U_{\widehat{X_i}}^M = \dfrac{max\left(\widehat{X_i}\right) - min\left(\widehat{X_i}\right)}{M} \\ U_{\widehat{Y}}^M = \dfrac{max\left(\widehat{Y}\right) - min\left(\widehat{Y}\right)}{M} \end{cases} \tag{4}$$

Propagation of fuzzy membership values is attenuated by distance to provide more membership values for the closer neighbors. Regarding a linear function for attenuation, a room inside the board receives μ from a projected point:

$$\mu = R - \sqrt{u^2 + v^2} + 1; \quad -R \le u, v \le R,$$
$$\Delta d(x_s + u, y_s + v) \Rightarrow \begin{cases} \mu; & \text{if } \mu > 0 \\ 0; & \text{otherwise} \end{cases} \tag{5}$$

where R is the radius of the IDS operator, (x_s, y_s) are the coordinates of the propagator, (u, v) is the distance between the receiver and:

$$\Delta d(x_s + u, y_s + v) = \sqrt{u^2 + v^2} \tag{6}$$

After forming the IDS planes, to extract the NP $\psi(x)$, the COG operator is applied on the IDS plane:

$$\psi(x) = \frac{\sum_{j \in Y(x)} \mu_j Y_j}{\sum_{j \in Y(x)} Y_j} \tag{7}$$

where Y is the output axis of the board.

After extracting NPs, the NPs of multiple S.I.S.O. systems are unified into a single model through a fuzzy interpolation. The fuzzy interpolation is a weighted averaging mechanism in which each S.I.S.O. system participates in the aggregation with a weight called the confidence level (CL). For each S.I.S.O. system, the spread of data is measured around the corresponding NP:

$$\text{Spread}_k = \frac{1}{n} \sum_{i=1}^{n} \left(\psi_k(x_i^k) - y_i \right)^2 \tag{8}$$

where x_i^k and y_i are the coordinates of ith data point and ψ_k is the projection of the ith point on kth NP.

The interpolation mechanism for an m inputs, single output system is implemented by

$$y_{\text{final}} = \frac{\sum_{k=1}^{m} w_k y_k}{\sum_{k=1}^{m} w_k}, \tag{9}$$

where w_k (CL) is the weight of kth candidate model:

$$w_k = \frac{1}{\text{Spread}_k} \tag{10}$$

An example of the NPs for a three inputs-single output system in presented in Fig. 4.

For the example in Fig. 4, the output of the unseen sample $x_1 = [x_1^{\ 1}, x_1^{\ 2}, x_1^{\ 3}]$ is calculated by

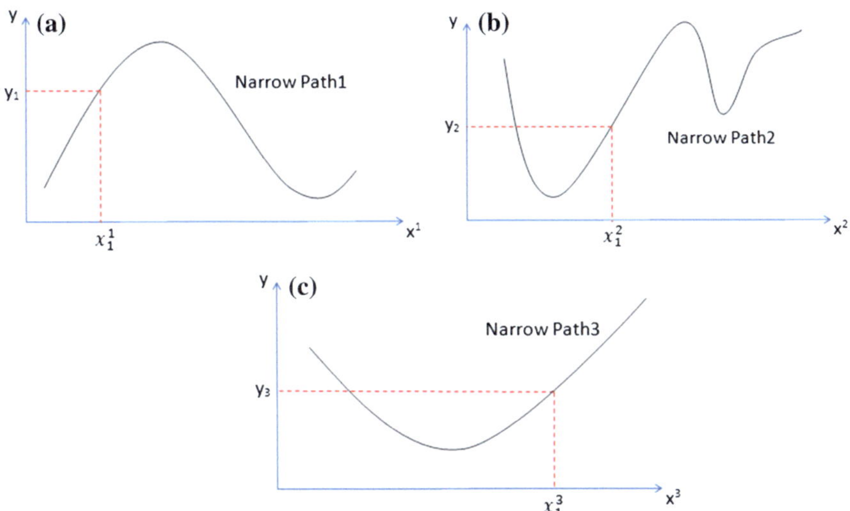

Fig. 4 The narrow paths of a rule for a 3 inputs-single output system

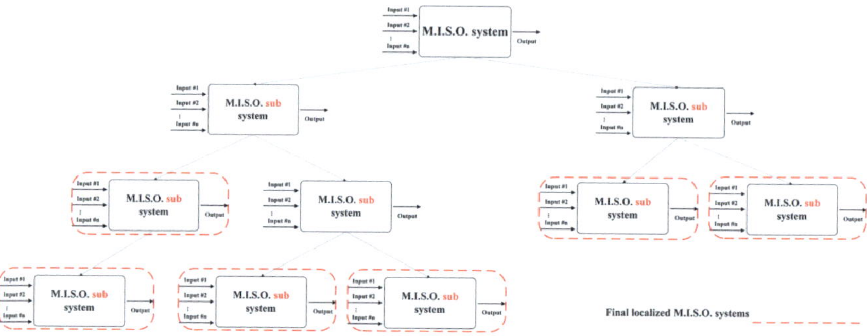

Fig. 5 Partitioning the M.I.S.O. domain into some sub-domains and modeling based on localized M.I.S.O. sub-models

$$\text{Out}\left(x_1^1, x_1^2, x_1^3\right) = \frac{C_1 \times \psi_1\left(x_1^1\right) + C_2 \times \psi_2\left(x_1^2\right) + C_3 \times \psi_3\left(x_1^3\right)}{C_1 + C_2 + C_3} \tag{11}$$

In the end, ALM recursively localizes the entire modeling process in order to improve the accuracy of the algorithm. The localization process, which is performed by recursively partitioning the domain of the M.I.S.O. system into some sub-domains, is performed to algorithmically model the local behaviors of the system that is damped by the general IO behavior extraction procedure. Localization can help including more local behaviors and improving the accuracy of the entire process. Therefore, after localization, multiple localized M.I.S.O. systems are processed by information-handling procedure of ALM (Fig. 5).

The recursive partitioning procedure inside ALM is similar to the training process of a decision tree. Thus, to avoid overfitting, early stopping (which is one of the well-known solutions to overfitting) is performed by ALM. Early stopping inside ALM includes measuring the overall generalization error through cross-validation. Thus, if the generalization error, which is measured via normalized nMSE, is less than the threshold, partitioning will not happen. nMSE is calculated by

$$\text{nMSE} = \frac{\sum_{i=1}^{N}\left(Y_i - T_i\right)^2}{\sum_{i=1}^{N}\left(Y_i - \overline{Y}\right)^2} \tag{12}$$

Figure 6 depicts the decision process of partitioning for *Threshold* $= 0.1$; partitioning stops at the terminating node.

In this paper, if the decision implies to carry on partitioning in the current stat (node), the data are divided into two partitions from the midpoint of the input axis inside the S.I.S.O. system (2D surface), which has gained the greatest weight in the interpolation mechanism. Greater weight/confidence level implies on that the underlying input parameter has more impact on determination of the behavior of the output. Therefore, dividing the data based on this S.I.S.O. system helps preserving the greatest CL, while the CLs of the other S.I.S.O systems may be improved in

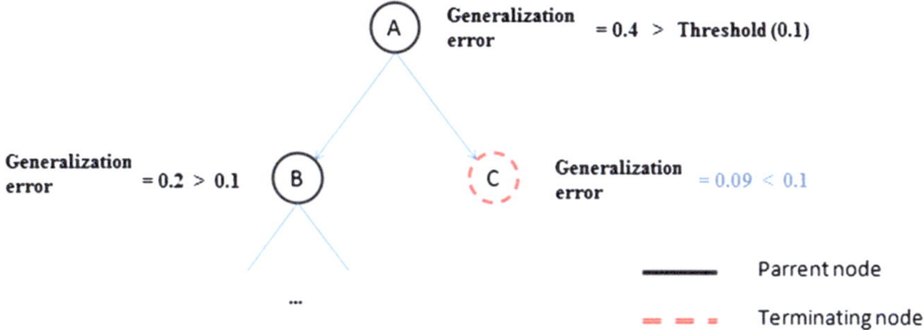

Fig. 6 Choosing from inputs to divide the data at each node in ALM tree based on the spread

aspect of accuracy. This approach is a heuristic, although the heuristic search is a guided search and does not guarantee to present the optimal solution; it can cause to reach to satisfactory solutions (Abbas et al. 2002).

3 Data Sets and Features

In order to use ALM to estimate missing sonic logs in the Anadarko Basin, Oklahoma, we employed a 3-step approach: supervised training to create a simulation model, verification and confirmation of the model, and application of the model to new wells. More details about this approach can be found in our previous work (Cranganu 2007; Cranganu and Bautu 2010; Cranganu and Breaban 2013).

The data sets are represented by GR,[1] deep resistivity[2] (REID), sonic[3] (DT), and caliper[4] (CAL) logs obtained from four wells drilled in the Anadarko Basin, Oklahoma. The CAL log is not participating directly in the modeling process, because it is not related unequivocally to a physical property. Rather, CAL log serves as an indicator of the borehole environmental "health": it points out to the presence of such drilling problems as caverns, mud cake, wash out zones, etc. The use of CAL log and the selection of most suitable depth intervals for simulation (*reduced data set* versus *complete data set*) are fully described in Cranganu (2005, 2007), Cranganu and Bautu (2010), and Cranganu and Breaban (2013).

---

[1] Natural gamma radiation, in uAPI.

[2] Electric resistivity variation, in Ωm.

[3] Propagation time of seismic waves in and around a borehole, in μs/ft.

[4] Measurement of the borehole diameter variations, in in.

4 Evaluation Measures

The quality of the training and verification models was assessed by using the following parameters:

The MSE was computed as the average overall squared deviations of the prediction from the real values:

$$\mathrm{MSE}(f) = \frac{1}{n} \sum_{i=1}^{n} \left(f\left(x^{(i)}\right) - y^{(i)} \right)^2 \tag{13}$$

The mean relative error (MRE) was computed as the average deviation reported to the real value:

$$\mathrm{MRE}(f) = \frac{1}{n} \sum_{i=1}^{n} \frac{f\left(x^{(i)}\right) - y^{(i)}}{y^{(i)}} \tag{14}$$

The *Pearson product–moment correlation coefficient, R,* is also used throughout this paper to measure the linear correlation between the measured and predicted values.

5 Training, Validation, and Application

We used a 3-step approach to perform our ALM simulation:

- *Training*: In this step, we computed the regression function based on a training set consisting of selected values of GR, REID, and DT from well T&T 1–10 (aka the *training well*). Because in ALM, the learned regression function depends on the values of input parameters and the procedural steps described in Sect. 2, we had to run several configurations of the learning pattern. The best model was considered the one that achieved the most accurate predictions (equivalent to the lowest MSE) (Fig. 7).
- *Validation*: In this step, the ALM algorithm created during training phase was evaluated on a data set from a *validation well*. This last data set consisted from a suite of GR, REID, and DT log values, but only GR and REID were used to simulate the DT values. Then, the simulated values were compared to the recorded ones. This step is meant to verify and validate the algorithm ability to simulate a DT log. The *validation well* was Whittenberg 3–29 (Fig. 8).
- *Application*: Once the ALM algorithm was verified and validated, it was applied to two *application wells* (Ledbetter 1–18, Fig. 9, and Smith 1–13, Fig. 10) where only GR and REID were recorded.

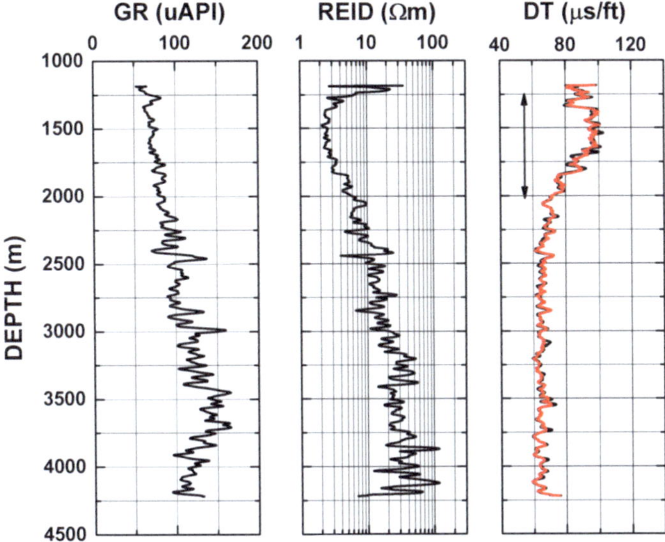

Fig. 7 Training results using T&T 1–10 well data set. The *red DT curve* is the simulated one. The *arrow* indicates the location and depth extension of the overpressure

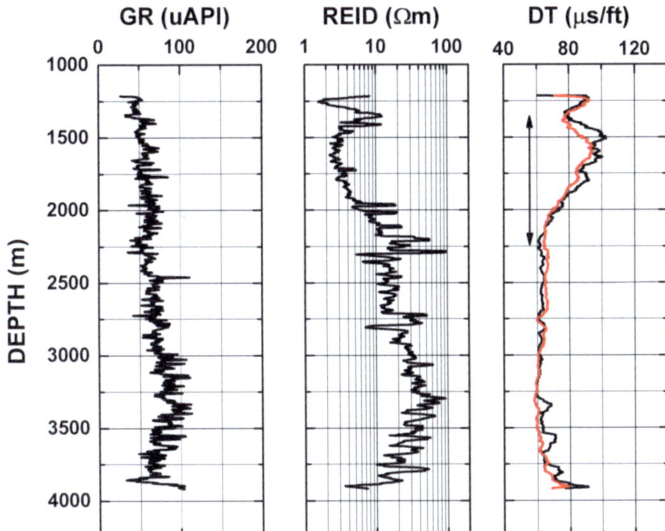

Fig. 8 Verification and validation results using Whittenberg 3–29 well data set. The *red DT curve* is the simulated one. The *arrow* indicates the location and depth extension of the overpressure

 C. Cranganu and F. Bahrpeyma

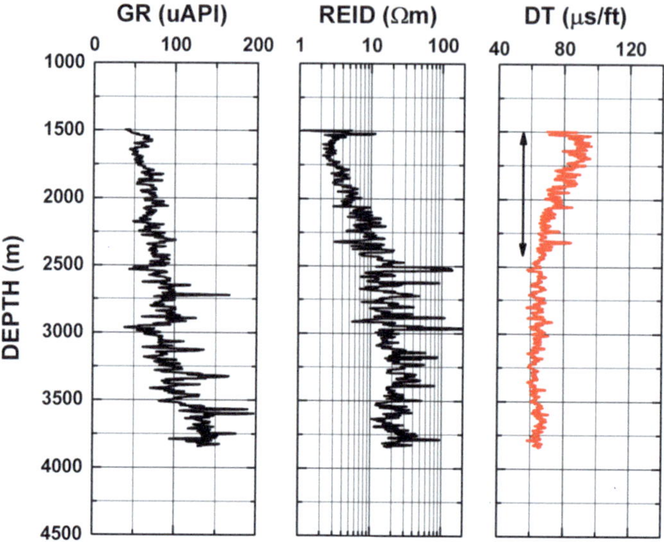

Fig. 9 Predicted DT values for application well Smith 1–13. The *arrow* indicates the location and depth extension of the overpressure

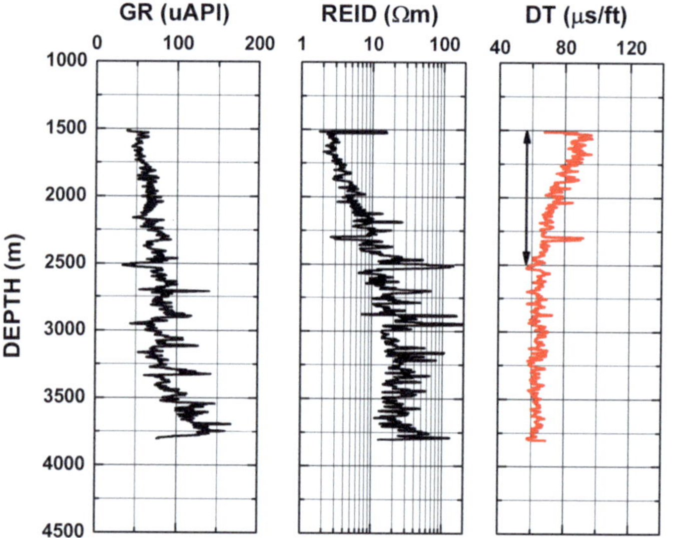

Fig. 10 Predicted DT values for application well Ledbetter 1–18. The *arrow* indicates the location and depth extension of the overpressure

Table 1 The impact of normalization procedure on the prediction accuracy of the trained model (DT = f(GR, REID)) measured by MSE, MRE, and R on the training and validation sets

	T&T 1–10			Whittenberg 3–29		
	MSE	MRE	R	MSE	MRE	R
Without normalization	45.1622	0.0000886	0.8900	90.9971	−0.01251	0.8032
With normalization	46.2572	0.0000980	0.8890	91.6786	0.00049	0.8020

6 Data Normalization

When doing data analysis, normalization is a popular pre-processing step. One main goal of data normalization is speeding up of the convergence of the gradient descent algorithm, which represents the basis of many machine learning jobs.

When using support vector regression, Cranganu and Breaban (2013) normalized all their input data sets and found that normalization yielded significant impacts on the accuracy of predictions and a better compression of the training model.

However, data normalization is not a necessary step in all regression algorithms. For example, while performing symbolic regression with genetic programming, (Cranganu and Bautu 2010) found that data normalization does not have a significant influence on the final results of regression analysis.

In this paper, we performed twice the regression analysis: with and without normalization of input data sets using the procedure indicated by Cranganu and Breaban (2013). The results are presented in Table 1.

7 Results

ALM took 67.09 s to perform five-level localization through the domain of the problem (Fig. 11). At the end, results of validation in the Whittenberg well including MSE = 90.99, MRE = − 0.01251 and R = 0.8032, indicate that the performance of the algorithm is satisfactory while the time performance is significant. Table 2 illustrates the results of the ALM process for training, and validation tests.

Table 3 and Fig. 12 illustrate the process of ALM for reaching to the desired performance in the localized sub-models at different localization depth levels.

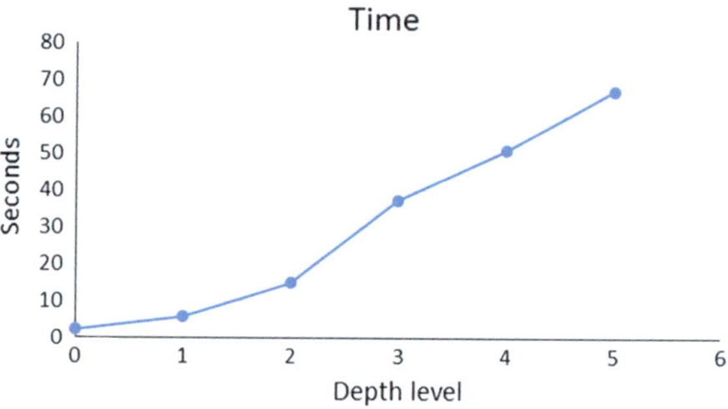

Fig. 11 Time necessary for reaching the depth levels of the localization process

Table 2 Results obtained when the regression function (DT = f(GR, REID)) was trained by ALM on the T&T 1–10 well and validated on Whittenberg 3–29 well

	MSE	MRE	R
Training set: T&T 1–10 reduced	44.8996	0.0095	0.9032
Test set: T&T 1–10 complete	45.1622	0.0000886	0.8900
Validation set: Whittenberg 3–29	90.9971	− 0.01251	0.8032

Table 3 The performance of ALM for time and accuracy in the process of localization in different depth levels

Depth level	Time	MSE
0	2.3784	207.534
1	5.8774	114.947
2	14.9953	97.838
3	37.337	52.005
4	51.0026	46.885
5	67.0907 +	43.345

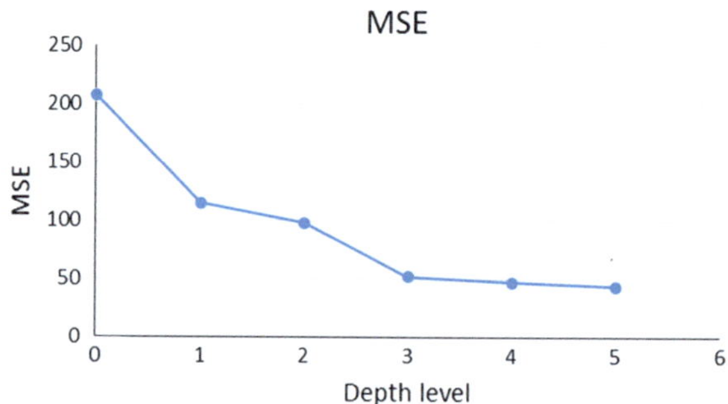

Fig. 12 MSE in the process of localization in different depth levels

8   Estimating the Presence and Depth Extension of Overpressured Zones in the Anadarko Basin, Based on Sonic Logs

As detailed in our previous work (Cranganu 2007; Cranganu and Bautu 2010; Cranganu and Breaban 2013) and references therein, many sedimentary basins may experience pore-fluid pressures that differ from the hydrostatic ("normal") pressure: *overpressure* (pore-fluid pressure exceeds the hydrostatic pressure) or *underpressure* (pore-fluid pressure is lower than hydrostatic pressure).

The existence of overpressured fluid zones, their depth locations, along with estimated magnitude of overpressure, represents important decision factors when drilling through those zones. The main danger is that unknowingly penetrating an overpressured zone can create potential hazards, such as blowouts.

Along with other information (mud weight, repeat formation testing, drill-stem testing), the sonic log (DT) proved to be a reliable predictor of overpressure zone because it responds to changes in porosity or compaction generated by abnormal pore-fluid pressure. It seems that DT log is preferred in oil industry because it is accurate and sensitive to changes in rock stress subject to overpressured pore fluids, especially in sequences dominated by shales (Asquith and Gibson 2004; Hearst et al. 2000). Determining trends in DT log variations with depth is a conventional method to predict the presence and estimate the depth intervals of overpressured zones (Cranganu and Bautu 2010; Cranganu and Breaban 2013). Because the acoustic transit time (recorded by DT log) in sedimentary rock is dependent on the effective rock stress, an increase in DT values is indicating a lower effective rock stress and, consequently, a zone with overpressure.

In Figs. 7, 8, 9, and 10, we predicted the existence of overpressured pore-fluid zones in four wells drilled in the Anadarko Basin, Oklahoma. The vertical arrows in those figures suggest that the overpressure regime is developing between ~1250 and 2500 m depth and the overpressured intervals have thicknesses varying between ~700 and 1000 m. These results match very well with previous results reported in the Anadarko Basin, using the same wells, but different artificial intelligent approaches (Cranganu 2007; Cranganu and Bautu 2010; Cranganu and Breaban 2013).

9   Conclusions

This paper discusses the Active Learning Method as a tool for predicting a missing log (DT or sonic log) when only two other logs (GR and REID) are present. Applying the ALM approach involves three steps: (1) supervised training of the model, using available GR, REID, and DT logs; (2) confirmation and validation of the model by blind-testing the results in a well containing both the predictors (GR, REID) and the target (DT) values; and (3) applying the predicted model to wells containing the predictor data and obtaining the synthetic (simulated) DT values.

Our modeling approach indicates that the performance of the algorithm is satisfactory, while the time performance is significant. The quality of our simulation procedure was assessed by three parameters, namely MSE, MRE, and Pearson product–momentum correlation coefficient (R).

The values obtained for these three quality-control parameters appear congruent, with the exception of MRE), regardless of the training set used (*reduced* vs. *complete*): MSE values are 44.8996 and 45.1622, respectively, while R values are 0.9032 and 0.8900, respectively. MRE values are more different: 0.0095 and 0.0000886, respectively, and we suppose that the difference is attributable to way MRE formula works.

ALM performance was measured also by the time required to attain the desirable outcomes: five depth levels of investigation took a little more than one minute of computing time during which MSE dropped from 207.534 to 43.345.

We performed twice the regression analysis: with and without normalization of input data sets (training well and validation well) using the procedure indicated by Cranganu and Breaban (2013). The results show minimum differences in quality assessment parameters (MSE, MRE, and R), suggesting that data normalization is not a necessary step in all regression algorithms. For example, while performing symbolic regression with genetic algorithm programing, Cranganu and Bautu (2010) found that data normalization does not have a significant influence on the final results of regression analysis.

We employed both the measured and simulated sonic logs DT to predict the presence and estimate the depth intervals where overpressured fluid zone may develop in the Anadarko Basin, Oklahoma. Based on our interpretation of the sonic log trends, we inferred that overpressure regions are developing between ~1250 and 2500 m depth and the overpressured intervals have thicknesses varying between ~700 and 1000 m. These results match very well our previous results reported in the Anadarko Basin, using the same wells, but different artificial intelligent approaches (Cranganu 2007; Cranganu and Bautu 2010; Cranganu and Breaban 2013).

ALM has the following advantages:

1. ALM diminishes the heavy load of computational complexities, which naturally exist in approximations methods like ANN, multiple regression, and TSK-FIS. The computational complexities are always due to complex formulas.
2. ALM has fast computational speed, which introduces itself as a proper suit for real-time applications.
3. Because ALM devotes confidence levels to the input parameters for aggregation, the method can identify and exclude the unimportant inputs parameters from the modeling process. Therefore, extra parameters cannot mislead the algorithm to non-optimal partitioning and inference solutions.

ALM has the following disadvantages:

1. Since the sub-domains are declared for NP extraction, ALM is limited to the domain of the training data. Therefore, any test data outside of the training boundaries can be denied by the algorithm and ALM cannot provide the estimation.

2. The radius of the IDS operator should match the nature of the training data, such that the continuity of the IDS plane is guaranteed. Therefore, constant radius for IDS cannot provide the best optimality.
3. ALM uses a heuristic for partitioning, and heuristics cannot guarantee that the best solution will be found.

References

Abbas HA, Sarker RA, Newton CS (2002) Data mining: a heuristic approach. Idea Group Publishing, Hershey

Asquith GB, Gibson CR (2004) Basic well log analysis: AAPG methods in exploration series, vol 9. American Association of Petroleum Geologists, Tulsa, Okla, p 244

Bahrpeyma F, Golchin B, Cranganu C (2013) Fast fuzzy modeling method to estimate missing logs in hydrocarbon reservoirs. J Pet Sci Eng 112:310–321

Behnia H, Arab-Khaburi D, Ejlali A (2011) Enhanced direct torque control for doubly fed induction machine by active learning method. In: Proceedings of 46th international universities' power engineering conference, UPEC 2011, Soest-Germany, VDE VERLAG GMBH, Berlin, Offenbach, pp 1–6

Cranganu C (2005) Using artificial neural networks to predict abnormal pressures in the Anadarko Basin, Oklahoma. J Balkan Soc 8:343–348

Cranganu C (2007) Using artificial neural networks to predict abnormal pressures in the Anadarko Basin, Oklahoma. Pure Appl Geophys 164:2067–2081

Cranganu C, Bautu E (2010) Using gene expression programming to estimate sonic log distributions based on the natural gamma ray and deep resistivity logs: a case study from the Anadarko Basin, Oklahoma. J Pet Sci Eng 70:243–255

Cranganu C, Breaban M (2013) Using support vector regression to estimate sonic log distributions: a case study from the Anadarko Basin, Oklahoma. J Pet Sci Eng 103:1–13

Ellis DV, Singer JM (2007) Well logging for earth scientists. Springer, The Netherlands, p 692

Ghorbani MJ, Akhbari M, Mokhtari H (2010) Direct torque control of induction motor by active learning method. In: Proceedings of 1st power electronic and drive systems and technologies conference. IEEE, Tehran, Iran, pp 267–272

Hearst JR, Nelson PH, Paillet FL (2000) Well logging for physical properties: a handbook for geophysicists, geologists, and engineers: Chichester, vol viii. Wiley, New York, p 483

Murakami M, Honda N (2005a) Classification performance of the IDS method based on the two-spiral benchmark. IEEE Int Conf Syst Man Cybern 3797–3803

Murakami M, Honda N (2005b) A Comparative study of the IDS method and feedforward neural networks. In: International joint conference on neural networks, Montreal, Canada, pp 1776–1781

Murakami M, Honda N (2005c) A study on the modeling ability of the IDS method: a soft computing technique using pattern-based information processing. Int J Approx Reason 45:470–487

Saeed Bagheri S, Honda N (1997) A new method for establishment and saving fuzzy membership functions. In: Proceedings of 13th fuzzy symposium, Toyama, Japan, pp 91–94

Sakurayi Y, Honda N, Nishino J (2003) Acquisition of control knowledge of nonholonomic system by active learning method. IEEE Int Conf Syst Man Cybern 2400–2405

Shahraini HT, Bagheri S, Fell F, Abrishamchi A, Tajrishy M, Fischer J (2005) Investigating the ability of active learning method for chlorophyll and pigment retrieval in case I waters using seaWiFS wavelengths. In: International conference on advanced remote sensing, Riyadh, Kingdom of Saudi Arabia

Shahraiyni TH, Schaale M, Fell F, Fischer J, Preusker R, Vatandoust M, Shouraki BS, Tajrishy M, Khodaparast H, Tavakoli A (2007) Application of the active learning method for the estimation of geophysical variables in the Caspian Sea from satellite ocean colour observations. Int J Remote Sens 28:4677–4684

Shouraki SB, Honda N (1999) Recursive fuzzy modeling based on fuzzy interpolation. J Adv Comput Intell 3:114–125

Active Learning Method for Estimating Missing Logs in Hydrocarbon Reservoirs

Fouad Bahrpeyma, Constantin Cranganu, and Behrouz Zamani Dadaneh

Abstract Characterization and estimation of physical properties are two of the most important key activities for successful exploration and exploitation in the petroleum industry. Pore-fluid pressures as well as estimating permeability, porosity, or fluid saturation are some of the important examples of such activities. Due to various problems occurring during the measurements, e.g., incomplete logging, inappropriate data storage, or measurement errors, missing data may be observed in recorded well logs. This unfortunate situation can be overcome by using soft computing approximation tools that will estimate the missing or incomplete data. Active learning method (ALM) is such a soft computing tool based on a recursive fuzzy modeling process meant to model multi-dimensional approximation problems. ALM breaks a multiple-input single-output system into some single-input single-output sub-systems and estimates the output by an interpolation. The heart of ALM is a fuzzy measuring of the spread. In this paper, ALM is used to estimate missing logs in hydrocarbon reservoirs. The regression and normalized mean squared error (MSE) for estimating density log using ALM were equal to 0.9 and 0.042, respectively. The results, including errors and regression coefficients, proved that ALM was successful in processing the density estimation. ALM is illustrated by an example of a petroleum field in the NW Persian Gulf.

Keywords Active learning method (ALM) · Ink drop spread (IDS) · Hydrocarbon reservoir · Well logs

F. Bahrpeyma (✉) · B. Z. Dadaneh
Faculty of Electrical and Computer Engineering, Shahid Beheshti University G.C, Tehran, Iran
e-mail: f.bahrpeyma@mail.sbu.ac.ir

C. Cranganu
Department of Earth and Environmental Sciences, Brooklyn College of the City University of New York, Brooklyn, NY, USA
e-mail: cranganu@brooklyn.cuny.edu

    197
C. Cranganu (ed.), *Artificial Intelligent Approaches in Petroleum Geosciences*,
https://doi.org/10.1007/978-3-031-52715-9_7

1 Introduction

In the petroleum industry, petrophysical and geophysical logs are considered as important tools for reservoir evaluation, and they are frequently used to obtain reservoir parameters. Loss of data is a common characteristic of the logging procedures which is commonly due to incomplete logging and other failures which are addressed in Bahrpeyma et al. (2013), Cranganu and Bautu (2010), Cranganu and Breaban (2013), El-Sebakhy et al. (2012), Rezaee et al. (2008), and Saramet et al. (2008).

Soft computing tools have recently become very useful in the field of the petroleum industry because of their capability to estimate petrophysical and geophysical parameters of oil wells (Cranganu and Bautu 2010; Cranganu and Breaban 2013).

Soft computing is a set of computational techniques for finding solutions to problems for which there is no way to solve them analytically, or problems which could be solved theoretically but practically impossible, due to the necessity of huge resources and/or enormous time required for computation. For these problems, methods inspired by nature are sometimes efficient and effective. Although the solutions obtained by these methods are not always identical to analytical solutions, a near optimal solution is sometimes enough in most practical purposes. The main goal of soft computing tools is to support systems which need dealing with huge amounts of data storage, complexity, and uncertainties. Time performance is another factor which demands soft computing tools to improve their performance wherever time is an important factor.

Active learning method (ALM) (Saeed Bagheri and Honda 1999) is a fuzzy-based soft computing tool inspired from the ability of the human brain to model the systems it faces. ALM as a macro-level brain imitation has been inspired by some behavioral specifications of brain and human active learning ability. The process of ALM focuses on the remarkable human ability to effortlessly deal with intricate information.

Taheri-Shahraiyni et al. (2011) used ALM for simulating runoff in the Karoon basin. Taheri-Shahraiyni et al. (2007), showed how ALM can be used for the estimation of geophysical variables in the Caspian Sea employing satellite ocean color observations.

Bahrpeyma et al. (2013), used two ALM operators to model physical properties of hydrocarbon reservoirs. Although they used the IDS method for modeling purposes, they just used one step modeling procedure. This paper presents the complete implementation of the ALM method to accurately synthesize missing log data. In fact, this paper is an improvement on Bahrpeyma et al. (2013) to provide more accurate and reliable modeling procedure by recursively localizing the modeling process inside the problem domain.

2 Methodology

Active learning method (ALM) is a powerful soft computing tool inspired by the way the human brain deals with complexity when solving problems (Saeed Bagheri and Honda 1997a). The human brain can solve modeling problems without dealing with a huge amount of computational complexities. Therefore, simulating the way the human brain models a system enables us to achieve a modeling technique with low complexity.

ALM uses different types of devoting membership functions called ink drop spread (IDS), which acts as a powerful engine. IDS method (Saeed Bagheri and Honda 1997a) is analogous to fuzzy logic since it is algorithmically modeled on the processes of the information handling in the human brain (Afrakoti et al. 2013; Sagha et al. 2008). Considering that conventional fuzzy logic uses linguistic and logical processing, the IDS method employs intuitive image information. This concept is derived from the hypothesis that humans interpret information as images rather than in numerical or logical forms (Murakami and Honda 2004). The main idea behind ALM is to break a multiple-input single-output (M.I.S.O) system into some single–input–single–output (S.I.S.O) sub-systems and aggregate the behavior of sub-systems to obtain the final output. This idea resembles the brain activity which stores the *behavior of data* instead of their *exact values.* Each S.I.S.O sub-system is expressed as a data plane (called the IDS plane), resulting from the projection of the gathered data on each input–output plane (Murakami and Honda 2005; Saeed Bagheri and Honda 1997b).

Two types of information can be extracted from an IDS plane. One is the behavior of the output with respect to each input variable. The other is the confidence level for each input variable which is proportional to the reciprocal of variance of data around the behavior of the output with respect to each input. Narrow paths (NP) are estimated by applying IDS on data points and Center of Gravity (COG) on data planes. IDS and COG are the first two main operators of ALM (Saeed Bagheri and Honda 1999).

The flowchart of the modified ALM presented in this paper is illustrated in Fig. 1.

The algorithm includes four steps:

Step 1: Projecting data on each X_i–Y plane

To observe the behavior of the target variable (as output of the system), the given data which is a M.I.S.O system is projected into some two-dimensional (2D) input–output systems. In fact, an M.I.S.O. system is broken into some S.I.S.O sub-systems (Fig. 2).

The benefits of projection are as follows: (a) observing and analyzing the behavior of output with respect to each input; (b) recognizing the most effective inputs, and (c) easing the processes of identification and computation.

The major problems of this projection are as follows: (a) loosing the relationship between inputs, and (b) calculating the final output.

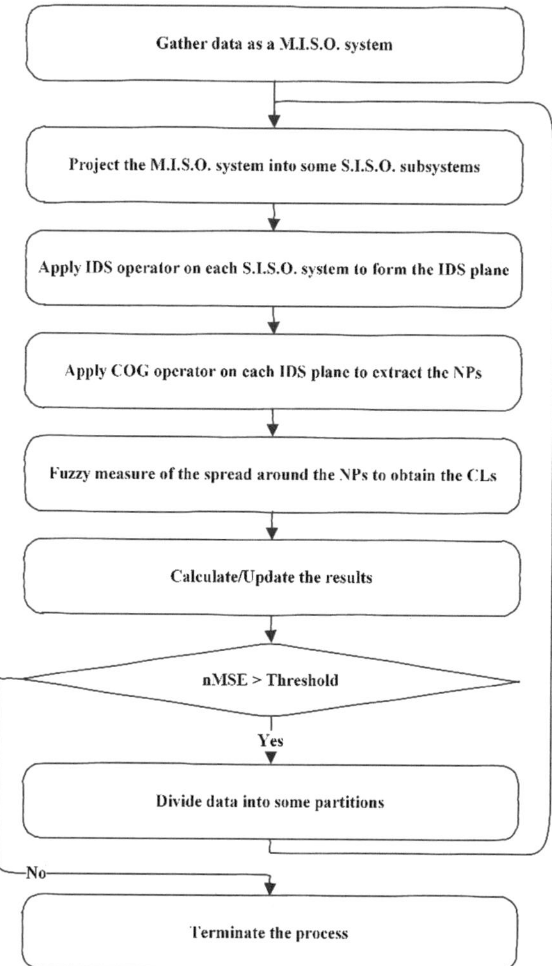

Fig. 1 The flowchart of the proposed ALM

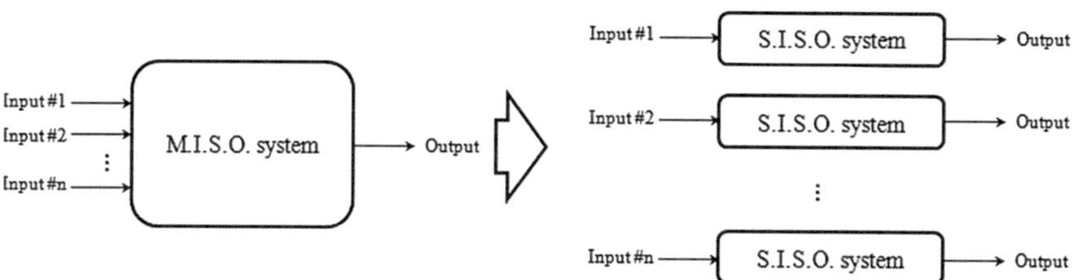

Fig. 2 Projecting an M.I.S.O system into some S.I.S.O sub-systems

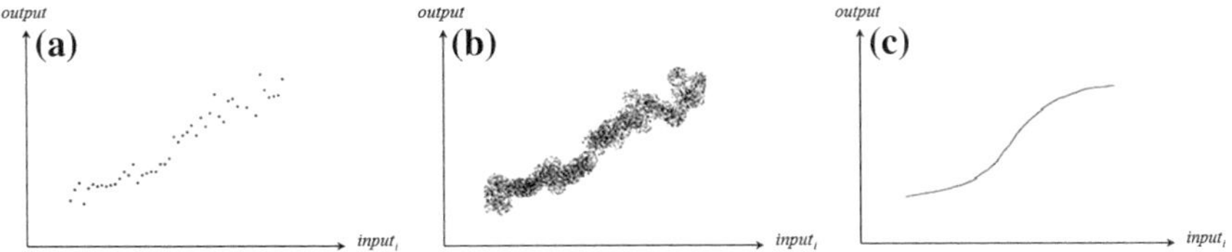

Fig. 3 Extracting an NP. **a** Discrete observations. **b** Thickening operation. **c** Thinning operation

Step 2: Finding the NPs for each 2D plane

To observe the behavior of the output with respect to each input, ALM extracts a function from each projected data samples. The function, called the NP, is a curve which shows just a unique value for each value in the domain of the input. The goal of extracting the NP function (as illustrated in Fig. 2) is to achieve a single model of the output based on each input parameter, separately:

$$y_i^{\text{out}} = \psi_i(x_i) \tag{1}$$

where i, $1 \leq i \leq n$ specifies the input index for n system inputs, y_i^{out} is the model of the output of the model constructed based on ith input, ψ_i is the NP of the ith input, and x_i is a value in the domain of ith input.

Extraction of the NP in ALM is a two-stage process: *Thickening* and *Thinning*. First, a thickening operator is applied on data samples in order to make a continuous path (fuzzification). The process of thickening is a fuzzification operation on the 2D plane of an S.I.S.O system. While observed data samples on input$_i$–output plane are set of discrete points (Fig. 3a), the fuzzification operator thickens the 2D planes such as to extract continuity (Fig. 3b) as a requirement of the generalization ability of a model. The fuzzification operation in ALM is performed through the IDS operator (Bahrpeyma et al. 2013; Saeed Bagheri and Honda 1997a).

The implementation of propagating membership function for IDS operator with a radius R is performed through Eq. 2. The amount of propagated darkness μ is calculated by (Murakami and Honda 2006):

$$\begin{aligned} \tau &= R - \Delta d + 1, \quad -R \leq u, \ v \leq R; \\ \Delta d(x_s + u, y_s + v) &= \sqrt{u^2 + v^2}; \\ \mu &= \begin{cases} \tau, & \text{if } \tau > 0 \\ 0, & \text{otherwise,} \end{cases} \end{aligned} \tag{2}$$

where (x_s, y_s) are the coordinates of the propagator, which is an observed point in the training dataset, u and v are the distances of x and y axes from the propagation point, and Δd is the Euclidean distance between the membership receiver and the propagator.

Second, to extract a function-like curve ψ, the continuous area obtained by IDS operation on the IDS plane should be thinned. The thinning operation in ALM (Fig. 3c) is performed as a defuzzification operation on the IDS plane by COG operator (Bahrpeyma et al. 2013; Saeed Bagheri and Honda 1997a) to extract the NP, i.e., ψ (Eq. 3):

$$\psi(x) = \frac{\sum_{j \in Y(x)} \mu_j Y_j}{\sum_{j \in Y(x)} Y_j} \tag{3}$$

Step 3: Calculating the spread of each X_i–Y plane

By extracting the NPs from the IDS planes, there are multiple S.I.S.O models for the given M.I.S.O. model which express the output separately. ALM uses a criterion to specify a confidence level for each model and uses the confidence level as a fuzzy weight to calculate the model output. In ALM, the confidence level of input$_i$–output plane is the reciprocal value of the spread of data samples around the extracted NP of that plane. The confidence level indicates how much the NP is proper to show the behavior of the output with respect to the underlying input. As the spread increases, the variation of the data samples around the NP increases, too. The spread of ith S.I.S.O. plane is calculated by Eq. (4):

$$v_i = \frac{1}{n} \sum_{k=1}^{n} (\psi_i(x_k) - y_k)^2 \tag{4}$$

where n is the number of data points, and (x_k, y_k) is the coordinate of the kth data sample of ith S.I.S.O. plane.

Thus, the confidence level is calculated by Eq. (5):

$$\theta_i = \frac{1}{v_i} \tag{5}$$

Step 4: Evaluation of the model and dividing the data samples into several partitions

The spread of data calculated by Eq. 3 is a measure to evaluate the accuracy of the model. When the spread is big, it means that the model is not reliable enough. ALM uses the division of data samples in order to reduce the spread of data and extract more accurately localized NPs. The evaluation of the models can be performed by the means of measuring the generalization error which can be implemented through a cross-validation operation (Jack 1983; Mori and Stephen 1991). To measure the generalization error, the output is calculated by an interpolation mechanism:

$$\text{output}_k = \frac{\sum_{i=1}^{n} \theta_i \times \psi_i(x_k)}{\sum_{i=1}^{n} \theta_i} \tag{6}$$

The generalization error is calculated by the FVU relation (Murakami and Honda 2006):

$$\text{FVU} = \frac{\sum_{k=1}^{K} (\hat{y}(x_k) - y_k)^2}{\sum_{k=1}^{K} (y_k - \overline{y})^2} \tag{7}$$

where $\hat{y}$ is the output of the model, K is the number of test points and $\overline{y}$ is

$$\overline{y} = \frac{1}{K} \sum_{k=1}^{K} y_k \tag{8}$$

If FVU satisfies the error threshold of ALM (FVU $\leq$ Err$_{\text{threshold}}$), the division is not happening because the model is satisfactory; otherwise, ALM divides data samples into partitions in order to gain more accurately localized sub-models.

According to the algorithm of ALM (Fig. 1), dividing the data samples into partitions is performed in this step. An approach is to divide data based on a point in one of the projected S.I.S.O planes. A heuristic to choose a proper plane is dividing the plane with the smallest spread, i.e., choosing the plane that has a higher degree of confidence. The main reason to choose this heuristic is that at least it holds the plane with the greatest degree of confidence unchanged while changing the others (Taheri-Shahriyani et al. 2006). Choosing a proper place to divide the chosen IDS plane requires another heuristic. Takagi and Sugeno (1985) used the midpoint to divide a plane. The heuristic search is a guided search, and it does not guarantee an optimal solution. However, it can often find satisfactory solutions (Takagi and Sugeno 1985).

Figure 4 illustrates how ALM divides data into two parts at each stage:

In fact, ALM decides on dividing data samples based on a tree structure which is implied on the recursive nature of ALM. Figure 5 illustrates the decision tree in which its nodes are labeled by their generalization error.

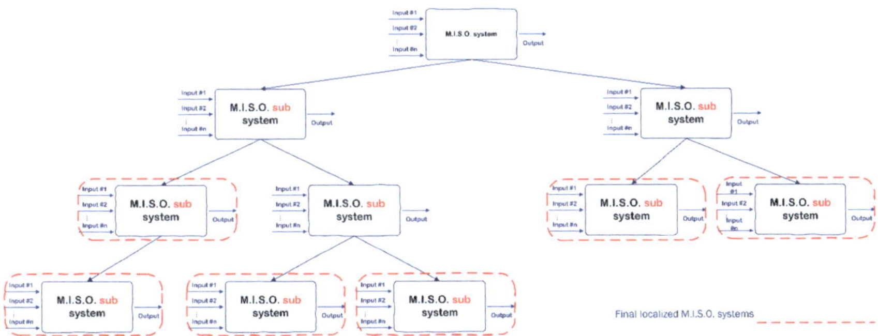

Fig. 4 Localized M.I.S.O sub-systems resulted by recursive partitioning

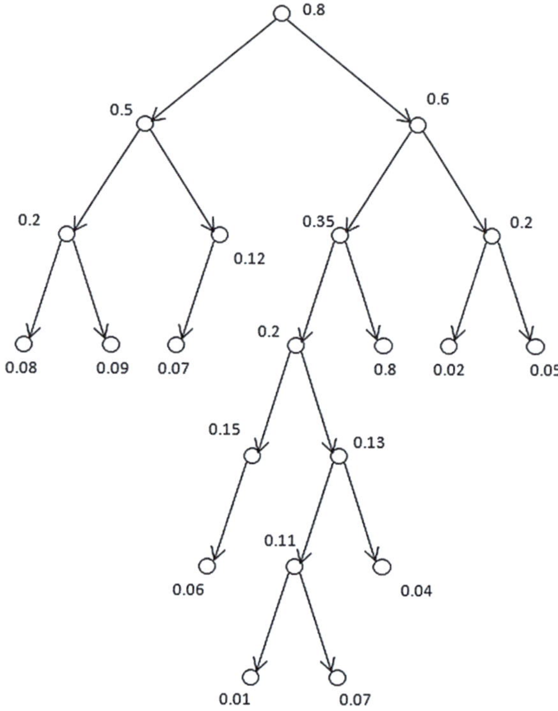

Fig. 5 An example of an ALM decision tree in which the nodes are labeled by generalization error

Each node in the decision tree of ALM correspondents to a sub-model constructed by a partition of data samples. Figure 6 is an example decision-making about the division in ALM for reaching to a state with a value of 0.1 for generalization error.

In the example of Fig. 6, the node labeled by "terminating node" corresponds to a state satisfying the threshold of generalization error for stop division. At each state which corresponds to a node in the ALM decision tree, the generalization error of the node is compared with the threshold so as to decide on continuing or stop dividing.

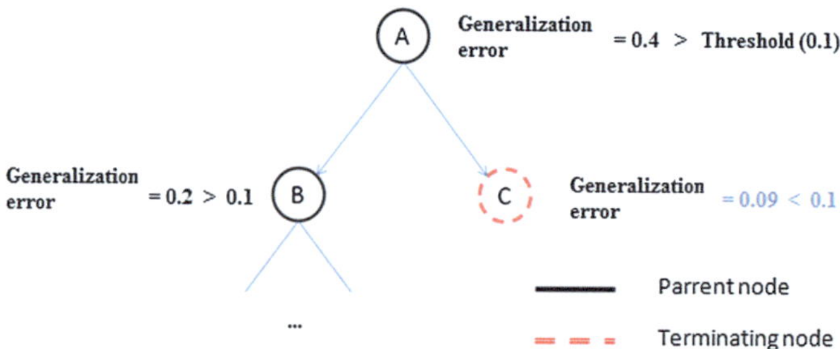

Fig. 6 An example of decision-making about division in ALM at each state

3 Experimental Results

In this paper, the experimental results are provided through modeling the log density based on photoelectric log, neutron log, and sonic log.

3.1 A Brief Overview of Petrophysical Logs

- **Density log (RHOB)**: The bulk density of the formation is measured by recording the number of returning gamma rays affected by Compton scattering; this is proportional to the electron density of the formation (Rider 2002). RHOB is used to estimate porosity (so-called *density porosity*), identify mineral lithology (mainly evaporite minerals), detect gas-bearing zones, determine hydrocarbon density, evaluate source rocks, and recognize fractures.
- **Photoelectric log (PEF)**: The photoelectric (Litho-Density) log is a continuous record of Pe (the cross-section index of effective photoelectric absorption) related to the formation composition (Rider 2002). Lithology and mineral identification is the major usage of this log.
- **Neutron log (NPHI)**: The hydrogen density of the formation is measured via the neutron log. Neutron log is used as a hydrocarbon indicator for estimation of porosity and lithology identification.
- **Sonic log (DT)**: Sonic log is the log which is the result of sending acoustic signal through the rocks. DT is mainly used for lithology identification.

The dataset of this study is the collection of logs from a petroleum field in Southern Iran (NW Persian Gulf), also used in (Bahrpeyma et al. 2013). The dataset contains 4500 records which is separated into training and test sets in which 500 records are randomly selected as the test set, and the remaining are used as the training set.

3.2 Modeling RHOB

Based on the methodology presented in Sect. 2, the collected data have to be expressed as an M.I.S.O system. ALM projects an M.I.S.O system into some S.I.S.O sub-systems. In this paper, the goal is to estimate RHOB. Therefore, inside each recursion, first the M.I.S.O system {*DT, NPHI, PEF → RHOB*} is expressed as three S.I.S.O sub-systems: {*DT-RHOB*}, {*NPHI-RHOB*}, and {*PEF-RHOB*}.

The stage of fuzzification and extracting the NPs begins by applying IDS operator on the data for each IDS plane which is shown in Fig. 7.

The process of estimation using ALM needs a numbers of iterations which are the results of dividing planes/sub-planes. At each stage, the recursive algorithm of ALM divides data into two parts, and the process is performed on data for each part. Then, the algorithm estimates the output for the test data. The results of matching

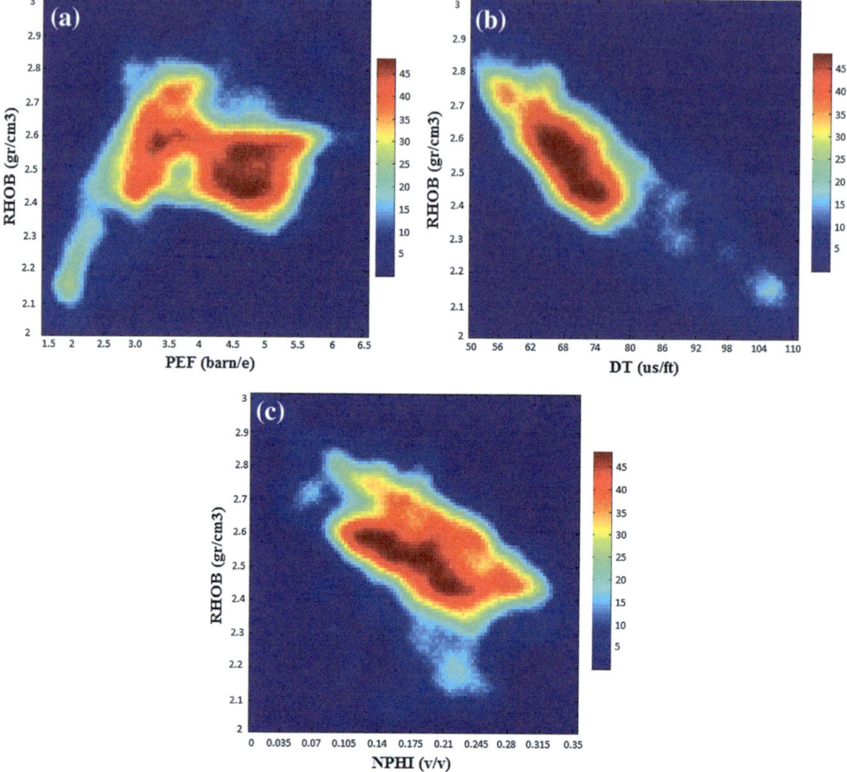

Fig. 7 Applying IDS operator as the thickening operator on data: **a** IDS plane of PEF-RHOB, **b** IDS plane of DT-RHOB, **c** IDS plane of NPHI-RHOB

measured and estimated data for 4 arbitrary exploration levels (3rd, 6th, 9th, and 12th levels) are shown in Fig. 8.

In this paper, the performance of the method is measured via two popular performance measures: normalized mean squared error (nMSE) and regression (R). R is the reliability which ranges from -1 to 1 to show how much linear relationship exists between the measured and the simulated values of the output. R is calculated by

$$R = \frac{\frac{1}{N} \sum_{i=1}^{N} \left(Y_i - \overline{T}\right)\left(T_i - \overline{Y}\right)}{\sqrt{\frac{1}{N} \sum_{i=1}^{N} \left(Y_i - \overline{T}\right)^2} \times \sqrt{\frac{1}{N} \sum_{i=1}^{N} \left(T_i - \overline{Y}\right)^2}} \tag{9}$$

And nMSE is defined as

$$nMSE = \frac{\sum_{i=1}^{N} \left(Y_i - T_i\right)^2}{\sum_{i=1}^{N} \left(Y_i - \overline{Y}\right)^2} \tag{10}$$

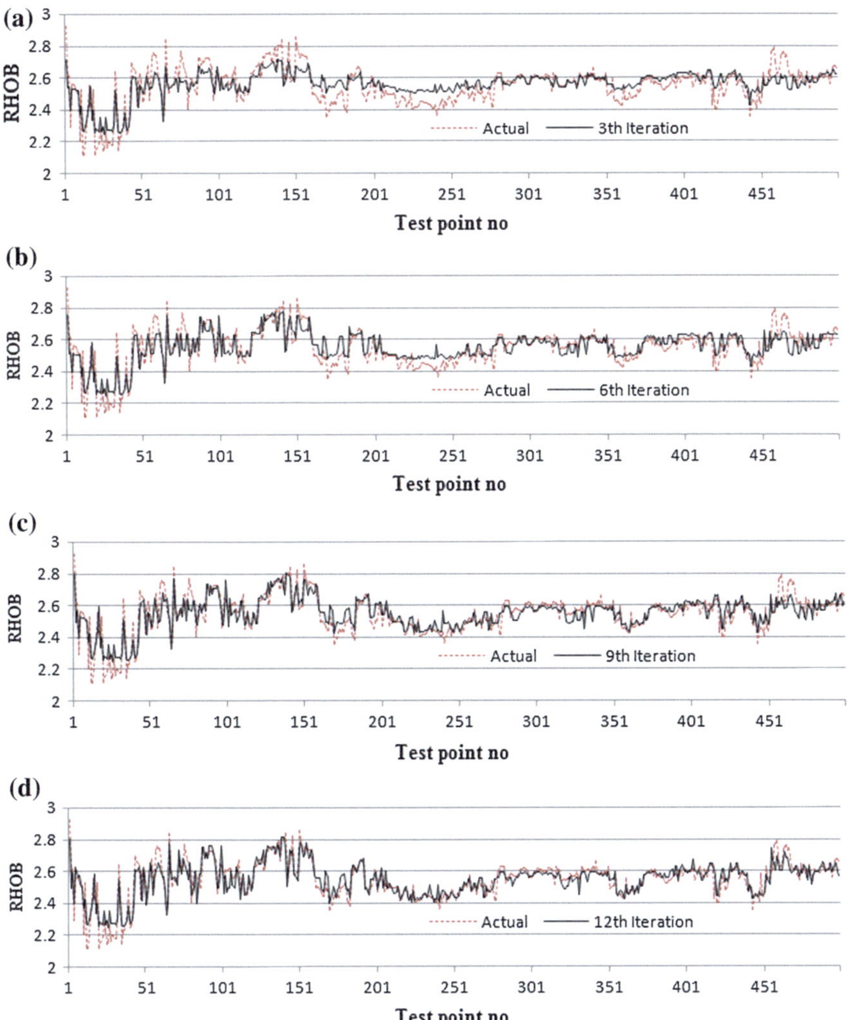

Fig. 8 Comparison between measured and estimated outputs. **a** 3rd level of exploration; **b** 6th level of exploration; **c** 9th level of exploration; **d** 12th level of exploration

R for four depth levels (3rd, 6th, 9th, and 12th levels) is shown in Fig. 9.

Figure 9 shows that ALM tries to model data more accurately when divides data at each stage and converges better when models data locally on the divided data samples.

The process of modeling depends on the accuracy of the modeling which needs to be illustrated to show how much ALM was successful on estimation. Figure 10a illustrates the sequences of resulted R between measured and estimated data. Figure 10b

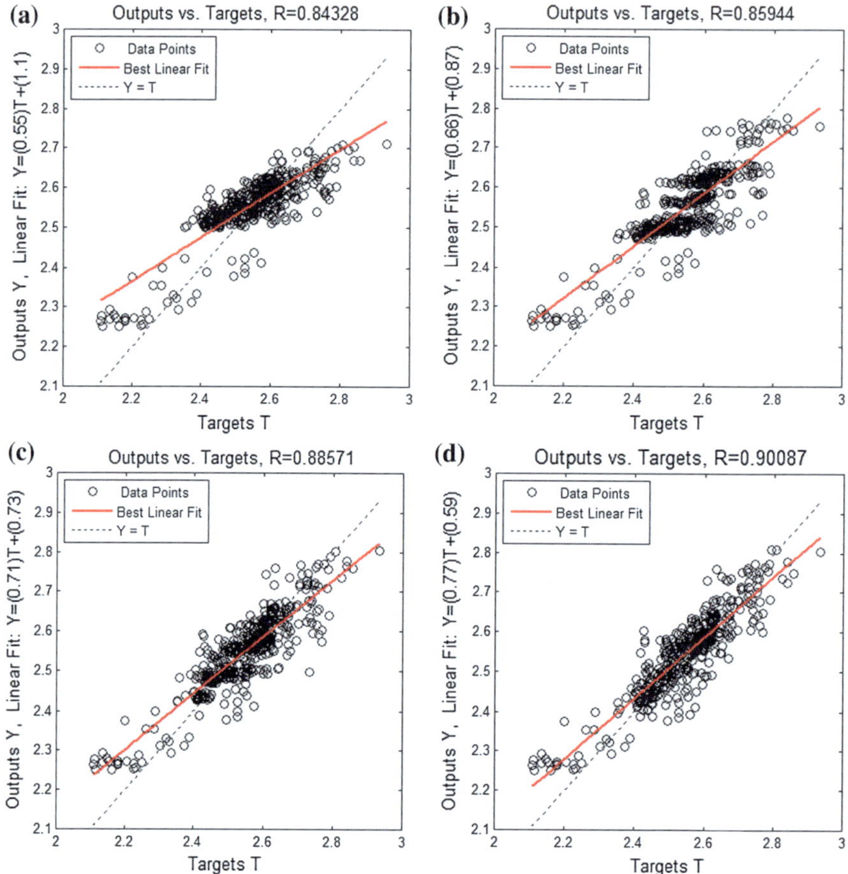

Fig. 9 *R* for measured and estimated outputs. **a** *R* at 3rd level of exploration; **b** *R* at 6th level of exploration; **c** *R* at 9th level of exploration; **d** *R* at 12th (final) level of exploration

also illustrates how ALM tries to converge nMSE to zero which is the most desired output of any types of estimating algorithm.

The most important part of ALM is the process of convergence to a stable state which is required for the validation of the algorithm. This goal is achieved as a result of the value of the given error. Results (Fig. 10) illustrate that ALM is converged to the stable state (no further exploration is observed in the process). The stable state is the final depth level of the ALM tree. The depth level indicates the maximum number of recursions that are allowed to hierarchically localize inside the parameter domain.

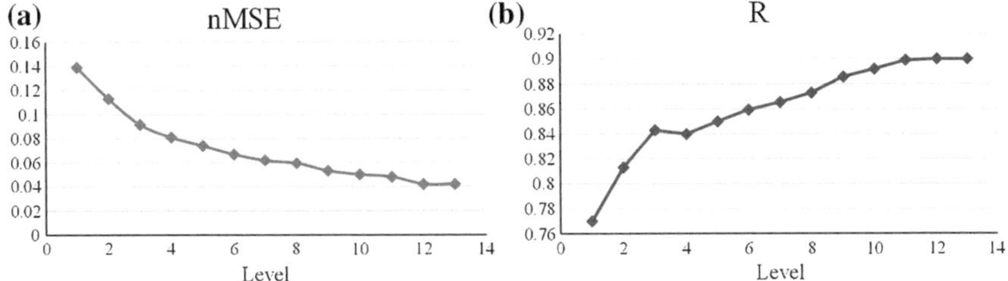

Fig. 10 The resulted R and nMSE gained by ALM at each depth level. **a** nMSE for each exploration level of ALM; **b** R of measured and estimated outputs for each exploration level of ALM

4 Discussion

Based on the nature of ALM, there are several advantages for ALM when it is compared with conventional estimation methods. In fact, the main characteristic of ALM is the way it deals with the complexities of estimation when it receives its raw data. Unlike other estimation methods, ALM does not tune weights or parameters for foreseeing the behavior it sees from the training data; instead, it extracts the general behavior of data locally and uses an aggregation mechanism to preserve the structure of the effectiveness of each input on the output. Then, speaking about ALM and its algorithm, there are some different points when compared with other methods. One is the overfitting, a major problem of conventional methods. Based on the learning methods and specifically, more related ones, like decision tree characteristics, ALM uses threshold as the early stopping factor to avoid overfitting (Jiang et al. 2009; Liu et al. 2008; Rynkiewicz 2012; Schittenkopf et al. 1997).

Figure 10 illustrates that the depth level of the ALM decision tree is equal to 12, and the algorithm has stopped at level 12. It means that the most localized sub-domain required 12 level divisions to gain the required accuracy.

The results of applying ALM and different modeling techniques are illustrated in Table 1 which includes a comparison between the performance of artificial neural network *ANN* (as a 3)20)1 trained by Levenberg–Marquardt learning algorithm), fuzzy logic *FL* (a Takagi–Sugeno-Kang Fuzzy Inference System TSK-FIS), radial basis function network *RBFN*, Neuro-fuzzy *NF* (fuzzy neural network), FFMM (Bahrpeyma et al. 2013), and the ALM; except for FFMM and ALM, the other methods are provided by the MATLAB 2008 toolbox.

Figure 11 illustrates a visualized comparison between the performance of popular modeling methods and ALM for modeling RHOB.

Table 1 The results of applying ALM and popular modeling techniques for estimating RHOB

	FFMM	ALM	ANN	FL	RBFNN	NF
nMSE	0.14	**0.042**	0.057	0.063	0.073	0.053
R	0.77	**0.90**	0.86	0.85	0.83	0.87

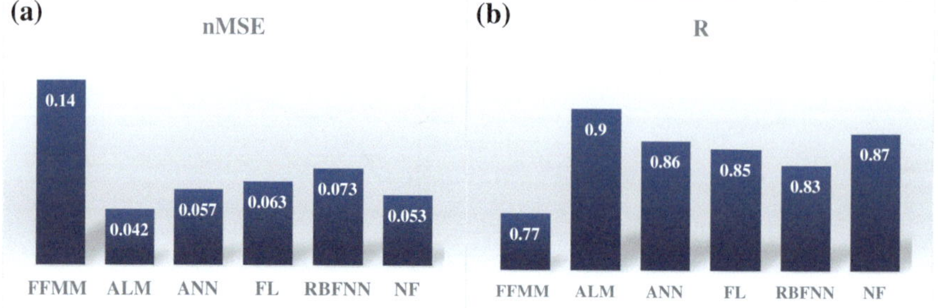

Fig. 11 Comparison between the performance of popular modeling methods (ANN, FL, RBFNN, and NF) and ALM for modeling RHOB. **a** nMSE, **b***R*

Comparison between the results gained by modeling RHOB via ALM and other modeling techniques (illustrated in Fig. 11) proves that ALM has acceptable performance and performs well in comparison with popular modeling techniques.

As a result of dealing with computational complexities naturally exist in the popular modeling methods through complex formulas, they generally suffer from lowness of the modeling speed. In this paper, ALM is presented as a modeling method that does not suffer from huge computational complexities. Therefore, ALM is expected to show an acceptable time performance. Figure 12 compares time performance between ALM and some popular modeling methods employed in this paper.

As shown in Fig. 12, ALM was successful to model RHOB without dealing with a huge computational complexity which results in a reduction in the time required for the modeling process. Although FFMM (Bahrpeyma et al. 2013) with the best time performance may be considered as the most successful method among the employed methods, the nMSE and *R* for FFMM are not so satisfying compared to other employed methods in this paper.

Fig. 12 Time performance in seconds(s) for modeling RHOB, comparison between ALM and some popular modeling methods addressed in this paper

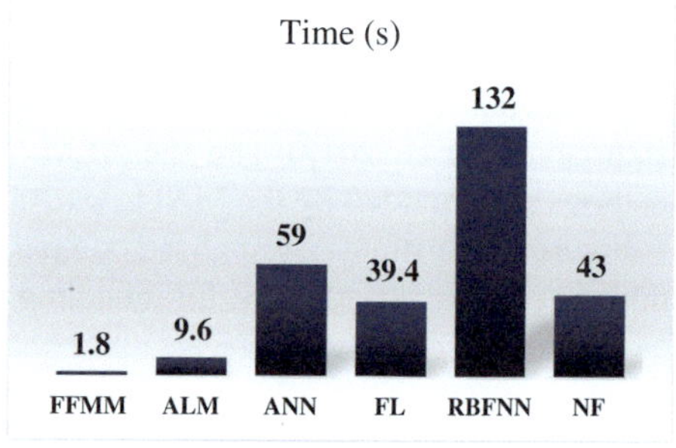

5   Conclusions

In the petroleum industry, the problems of incomplete data acquisition and loss of data in the process of measurement require supporting methods which are able to accurately estimate the unmeasured/lost data. This paper introduced a new soft computing tool for the application of estimating missing/unmeasured data called ALM. ALM is inspired by the ability of the human brain to model a multiple-input multiple-output system actively. Results (0.042 and 0.9 for nMSE and R, respectively) illustrate that ALM has excellent ability for modeling RHOB and can be used as a reliable approximator in the petroleum industry. Comparison between the results obtained by modeling RHOB via ALM and other modeling techniques proves that ALM has acceptable performance and performs well in comparison with other popular modeling techniques.

References

Afrakoti IEP, Ghaffari A, Shouraki SB (2013) Effective partitioning of input domains for ALM algorithm. In: First Iranian conference on pattern recognition and image analysis (PRIA), 2013. IEEE

Bahrpeyma F, Golchin B, Cranganu C (2013) Fast fuzzy modeling method to estimate missing logs in hydrocarbon reservoirs. J Petrol Sci Eng 112:310–321

Cranganu C, Bautu E (2010) Using gene expression programming to estimate sonic log distributions based on the natural gamma ray and deep resistivity logs: a case study from the Anadarko Basin Oklahoma. J Petrol Sci Eng 70:243–255

Cranganu C, Breaban M (2013) Using support vector regression to estimate sonic log distributions: a case study from the Anadarko Basin, Oklahoma. J Petrol Sci Eng 103:1–13

El-Sebakhy EA, Asparouhov O, Abdulraheem AA, Al-Majed AA, Wu D, Latinski K, Raharja I (2012) Functional networks as a new data mining predictive paradigm to predict permeability in a carbonate reservoir. Expert Syst Appl 39(12):10359–10375

Jack PCK (1983) Cross-validation using the t statistic. Eur J Oper Res 13(2):133–141

Jiang Y, Zur RM, Pesce LL, Drukker K (2009) A study of the effect of noise injection on the training of artificial neural networks. In: International joint conference on neural networks, 2009 (IJCNN 2009). IEEE, pp 1428–1432

Liu Y, Starzyk JA, Zhu Z (2008) Optimized approximation algorithm in neural networks without overfitting. IEEE Trans Neural Networks 19(6):983–995

Mori M, Stephen J (1991) A note on generalized cross-validation with replicates. Stoch Process Appl 38(1):157–165

Murakami M, Honda N (2004) Hardware for a new fuzzy-based modeling system and its redundancy. In: 23rd international conference of the North American fuzzy information processing society (NAFIPS'04). pp 599–604

Murakami M, Honda N (2005) A comparative study of the IDS method and feedforward neural networks. In: International joint conference on neural networks. Montreal, Canada

Murakami M, Honda N (2006) Model accuracy of the IDS method for three-input systems and a basic constructive algorithm for structural optimization. In: International joint conference on neural networks, 2006 (IJCNN'06). IEEE, pp 2900–2906

Rezaee MR, Kadkhodaie-Ilkhchi A, Alizadeh PM (2008) Intelligent approaches for the synthesis of petrophysical logs. J Geophys Eng 5:12–26

Rider M (2002) The geological interpretation of well logs. Whittles Publishing, Malta

Rynkiewicz J (2012) General bound of overfitting for MLP regression models. Neurocomputing 90:106–110

Saeed Bagheri S, Honda N (1997a) A new method for establishment and saving fuzzy membership functions. In: 13th fuzzy symposium. Toyama, Japan, pp 91–94

Saeed Bagheri S, Honda N (1997b) Outlines of a living structure based on biological neurons for simulating the active learning method. In: 7th intelligent systems symposium. pp 183–188

Saeed Bagheri S, Honda N (1999) Recursive fuzzy modeling based on fuzzy interpolation. J Adv Comput Intell 3:114–125

Sagha H, Shouraki SB, Beigy H, Khasteh H, Enayati E (2008) Genetic ink drop spread. In: Second international symposium on intelligent information technology application, 2008 (IITA'08). vol 2, pp 603–607

Saramet M, Gavrilescu G, Cranganu C (2008) Quantitative estimation of expelled fluids from Oligocene rocks, Histria Basin. Western Black Sea. Mar Pet Geol 25(6):544–552

Schittenkopf C, Deco G, Brauer W (1997) Two strategies to avoid overfitting in feedforward networks. Neural Netw 10:505–516

Taheri-Shahraiyni H et al (2007) Application of the active learning method for the estimation of geophysical variables in the Caspian Sea from satellite ocean colour observations. Int J Remote Sens 28(20):4677–4683

Taheri-Shahraiyni H, Mohammad RG, Bagheri-Shouraki S, Saghafian B (2011) A new fuzzy modeling method for the runoff simulation in the Karoon Basin. Int J Water Resour Arid Environ 1(6):440–449

Taheri-Shahriyani H et al (2006) Investigating the ability of active learning method for chlorophyll and pigment retrieval in case-I waters using seawifs wavelengths. Int J Remote Sens 28(20):4677–4683

Takagi T, Sugeno M (1985) Identification of systems and its application to modeling and control. IEEE Trans Syst Man Cybern 15(1):116–132

Improving the Accuracy of Active Learning Method via Noise Injection for Estimating Hydraulic Flow Units: An Example from a Heterogeneous Carbonate Reservoir

Fouad Bahrpeyma, Constantin Cranganu, and Bahman Golchin

Abstract Due to many reasons, in many occasions, reservoir engineers should analyze the reservoirs with small sets of measurements; this problem is known as the small sample size problem. Because of small sample size problem, modeling techniques commonly fail to accurately extract the true relationships between the inputs and the outputs used for reservoir properties prediction or modeling. In this paper, small sample size problem is addressed for modeling carbonate reservoirs by the active learning method (ALM). In this paper, noise injection technique, which is a popular solution to small sample size problem, is employed to recover the impact of separating the validation and test sets from the entire sample set in the process of ALM. The proposed method is used to model hydraulic flow units (HFUs). HFUs are defined as correlatable and mappable zones within a reservoir which control fluid flow. This study presents quantitative formulation between flow units and well logs data in one of the heterogeneous carbonate reservoir in the Persian Gulf. The results for R and nMSE are equal to 85% and 0.0042 which reflect the ability of the proposed method when facing with sample size problem.

Keywords Active learning method · Noise injection · Overfitting · Hydraulic flow unit · Ink drop spread · Carbonate reservoir

F. Bahrpeyma
Faculty of Electrical and Computer Engineering, Shahid Beheshti University G.C, Tehran, Iran

C. Cranganu (✉)
Department of Earth and Environmental Sciences, Brooklyn College of the City University of New York, Brooklyn, NY, USA
e-mail: cranganu@brooklyn.cuny.edu

B. Golchin
Department of Geology, Shahid Beheshti University, Tehran, Iran

   213

C. Cranganu (ed.), *Artificial Intelligent Approaches in Petroleum Geosciences*,
https://doi.org/10.1007/978-3-031-52715-9_8

1 Introduction

A fundamental approach to decrease uncertainty in the estimation of reservoir properties (which is due to heterogeneity in carbonates) is rock typing based on petrophysical characteristics. One of the popular methods to rock classification from geology framework and physics of flow is through pore network scale (pore throat geometric properties). Generally, the pore geometry is controlled by mineralogy (type, abundance, location) and texture (grain size, grain shape, sorting, packing). Different combinations of these properties can lead to identify distinct flow groups which have similar fluid transport properties. Each group is referred to as a hydraulic flow unit (HFU).

HFU is defined classically as the representative volume of total reservoir rock within which geological and petrophysical properties that control the fluid flow. They are internally consistent and predictably different from properties of other rocks (Ebanks 1987; Jude et al. 1993). Briefly, it can be identified as a zone in a reservoir where the similar flow of the oil or gas is continuous laterally and vertically. It relates to geological facies distributions, however, do not necessarily coincide with facies boundaries.

To identify HFU for formation evaluation, reservoir description, and characterization using flow zone indicator (FZI), permeability and porosity of the reservoir rock have always been considered as two of the most important parameters. FZI, which is measured based on core porosity and permeability, characterizes each flow unit, and it is popular to derive the FZI value and its corresponding flow unit for each sample. Conventional studies are executed on the core data to determine the FZI values, however, such works are expensive, time-consuming, and limited. To solve this problem, researchers use soft computing's modeling techniques such as artificial neural network (ANN), fuzzy logic (FL), support vector machine (SVM), and committee machines (CM) to estimate FZI and other reservoir parameters by well logs data. Kadkhodaei-Ilkhchi and Amini (2009) developed an FL model to determine FZI from well logs at un-core wells. Ghiasi-Freez et al. (2012) compared two types of CMs to estimate FZI. In this study, we developed a new soft computing tool called the active learning method (ALM), which is a fast and accurate modeling technique, for determining HFUs in heterogeneous carbonate reservoir in the Persian Gulf as one of the world's largest non-associated gas accumulations. ALM (Saeed Bagheri and Honda 1999) is a solution to model multi-dimensional systems without dealing with huge computational complexities that exist in traditional modeling techniques.

The data set in this paper contains a small set of samples that charges ALM with a small sample size problem. Due to many reasons, the small sample size problem is one of the common concerns in the petroleum industry when trying to estimate the properties in the carbonate reservoirs. On many occasions, this is not possible to provide an adequate sample size for the modeling techniques; it may degrade the performance of the estimation process. The problem is more crucial when, to provide a valid modeling procedure, the modeling techniques split the entire sample set into three subsets: 1-the training set, 2-the validation set, and 3-the test set. The training set is used for constructing the model. The validation set is used for modeling inter-level validation. And the test set is used to evaluate the performance of the constructed model. More importantly, for small data sets, splitting the data samples into training, validation, and test sets separates some of the useful information from the training set which meaningfully degrades the generalization ability of the modeling procedure.

One of the popular solutions to the small sample size problem is to improve the generalization ability of the model by reviving the separated information through noise injection (Rifai et al. 2011). In fact, noise injection can be regarded as a regularization method. The problem of regularization has been broadly studied in the context of learning methods specifically in the field of ANNs. Regularization is commonly used by learning methods to improve the generalization ability for small sample sets or when data samples are contaminated with noise, meaningfully. In the context of ANNs, weight decay and output smoothing that are used to avoid overfitting during the training of the considered model are the other popular methods (Yulei Jiang et al. 2009).

Noise injection has been broadly studied in the literature. In Raudys (2003, 2006), the impact of using noise injection technique to fight with sample size problem is studied. Rifai et al. (2011) used noise injection to the input with a regularized objective to improve generalization performance in the ANNs. Yulei Jiang et al. (2009) studied the effect of noise injection on the training of ANNs. Liu and Janusz (2008) proposed an optimized approximation algorithm which employed noise injection in ANNs to overcome overfitting. Skurichina et al. (2000) used the K-nearest neighbors algorithm to direct noise injection in multilayer perceptron training. Piotrowski and Napiorkowski (2013) compared popular methods, such as noise injection, for avoiding overfitting in ANNs training in the case of catchment runoff modeling.

In addition to the mentioned ability which noise injection provides, the smoothness of the curve which provides more generalization ability is another benefit of the noise injection technique; an example of smoothness provided by additional noise is illustrated in Fig. 1.

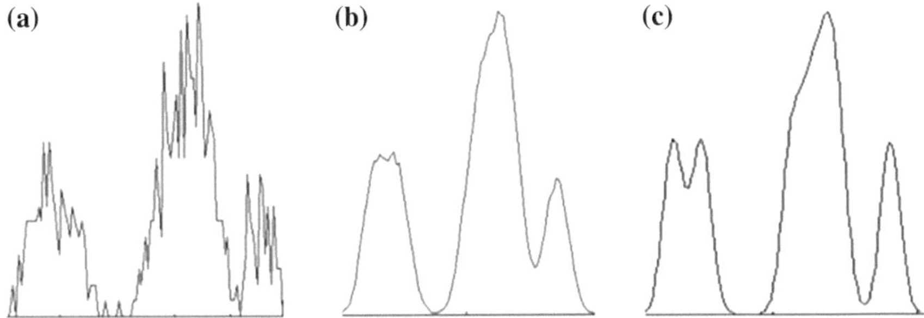

Fig. 1 Additive noise for improving the smoothness and penalization ability of a curve gained by training samples. **a** Actual arbitrary data curve, **b** adding 5% noisy samples to the data samples, **c** adding 10% noisy samples to the data samples

The rest of the paper is organized as follows: the next section is an overview on ALM. In Sect. 3, the proposed methodology for employing noise injection in ALM is described. The experimental results are illustrated and discussed in Sect. 4. Finally, Sect. 5 states conclusions derived from the work.

2 An Overview on Active Learning Method

ALM algorithmically models a system similar to the intelligent information management procedure of the human brain. ALM works fundamentally based on the hypothesis that the human brain observes an information system as an image in which the information is stored as patterns (Saeed Bagheri and Honda 1999). The main advantage of ALM is its ability to model multi-dimensional systems without dealing with computational complexities.

ALM uses a recursive approach in order to model multi-dimensional systems without dealing with computational complexities. This special approach is characterized by the usage of a fuzzification engine called ink drop spread (IDS) which is inspired by the way the drops of ink can create a continuous path that expresses the overall continuity instead of dealing with a discrete appearance (Saeed Bagheri and Honda 1997; Shouraki 2000). The main idea behind ALM is the projection of a multiple-input single-output (M.I.S.O.) system into some single-input single-output (S.I.S.O.) sub-systems and unifying the outcomes by an interpolation mechanism to generate the final output of the modeling system in a recursive manner.

2.1 *The ALM Algorithm*

A practical implementation of ALM can be performed through a recursive process which divides data hierarchically into small and smaller partitions, while proper partitioning is the goal of the entire process. Recursive partitioning enables to intuitively model a system based on the locality of the data distribution within the data domains. Hierarchical partitioning helps figuring out the locality imagination with the stopping parameter. The flowchart of ALM is illustrated in Fig. 2.

In the flowchart of ALM illustrated in Fig. 2, at each recursion, the IDS modeling method is responsible to model the corresponding data domain. Then, if the normalized mean-squared error (nMSE) of the model is greater than the threshold, the recursive process of ALM divides the corresponding data and partitioning does not continue for the corresponding data domain.

In Fig. 3, an example of applying ALM on a 2-input system is illustrated showing how the hierarchical division mechanism divides the domain of inputs in order to satisfy the predefined generalization threshold in a sub-domain.

In this example, starting from the entire domain, ALM divides training data into sub-domains and for each sub-domain, dividing is stopped when the terminating

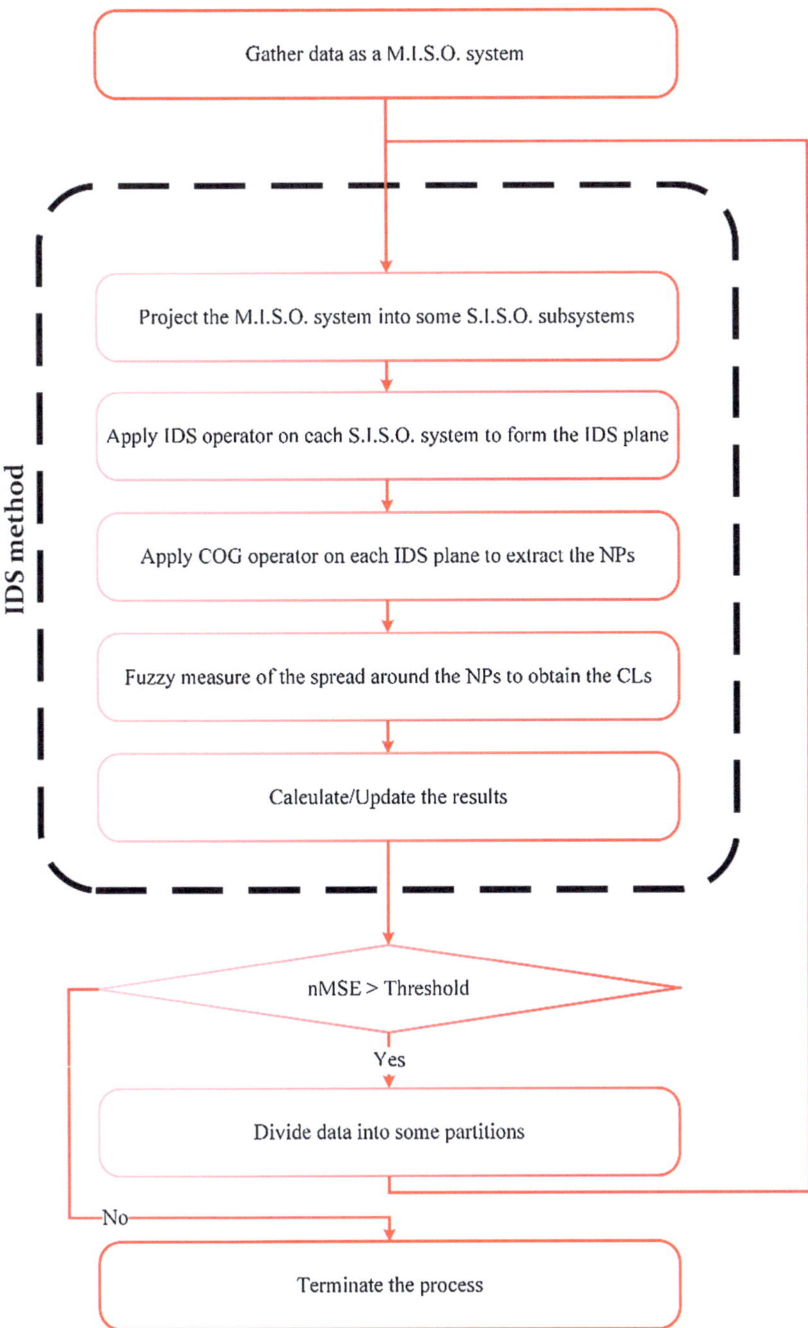

Fig. 2 The flowchart of ALM (Shouraki 2000)

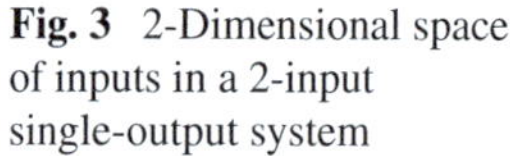

Fig. 3 2-Dimensional space of inputs in a 2-input single-output system

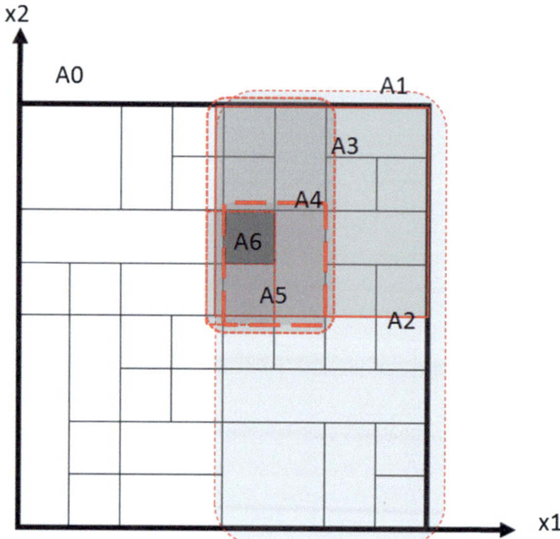

condition (nMSE's threshold) is satisfied. A particular tail of the hierarchy in Fig. 3 starts from the entire 2D domain (A0) to reach A6 area, which satisfied the terminating condition. Figure 4 illustrates that how modeling procedure (performed by IDS) is localized through the recursive process of ALM. In this paper, at each recursion, ALM divides data sub-domain into two partitions, i.e., the branching factor is 2.

As illustrated in Fig. 4, IDS separately models each localized sub-domain which is introduced by ALM.

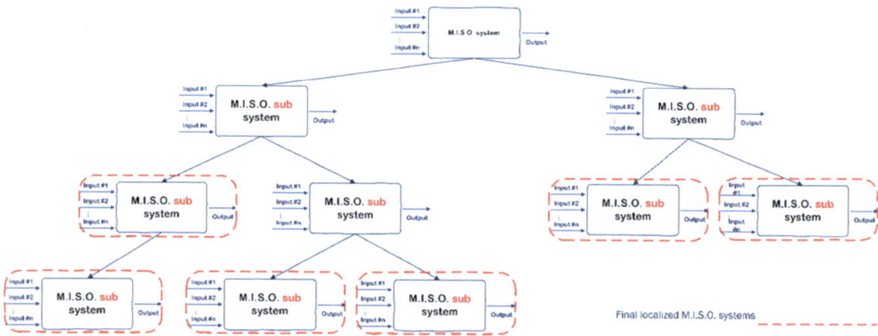

Fig. 4 Stages of recursively dividing data planes

2.2 IDS Modeling Method

As the engine of ALM, IDS modeling method models a multiple-input single-output (M.I.S.O.) system (or sub-system when the system, as a domain, is divided into some partitions) based on the projection of the system into single-input single-output (S.I.S.O.) sub-systems and then unifies the S.I.S.O. models into a single model through an aggregation mechanism. IDS models each S.I.S.O. sub-system separately through IDS and COG operators. Therefore, after projection, the process of the IDS method includes the application of IDS and COG operators, (consequently) on each S.I.S.O. sub-system and then unifying them into a single model.

- Applying the IDS operator on the projected data planes (fuzzification)

IDS operator roles as a fuzzification operator. The goal of the IDS operator is to provide continuous paths between data samples (in 2D spaces of the S.I.S.O. systems). IDS uses a different approach to devote membership values to the elements in a data set. IDS operator extracts the relationships between data points intuitively which are discrete in details, but the entire set is treated as a continuous pattern. Figure 5 illustrates the application of the IDS operator on a 2D data plane.

As shown in Fig. 5, while the midpoints between A, B, and C, which are the actual data samples, are not observed as the data samples, the application of the IDS operator on the 2D domain of the S.I.S.O. system creates the sense of continuity in an arbitrary unity. Therefore, the propagation of membership values to the neighbors from each actual data point which is attenuated by distance and reinforced by other actual neighbors creates the sense of continuity in the 2D input–output (I–O) domain. As Fig. 5 shows, while A, B, and C are the actual data samples and the domain formed by discrete units, quantization of the domain by the means of a board and propagation of the membership values by IDS extracts the continuity between actual data samples in each 2D I–O system.

- Applying center of gravity operator on the IDS planes (defuzzification)

Defuzzification in ALM is performed by COG operator (Saeed Bagheri and Honda 1997). COG defuzzifies the results gained by IDS operator so as to make a crisp world

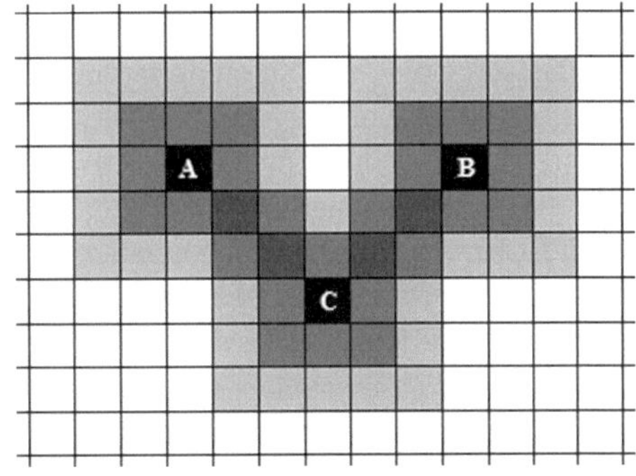

Fig. 5 Extracting the midpoints relationships between observed data samples in a S.I.S.O. system (Bahrpeyma et al. 2013)

induction. The outcomes of COG operator on the IDS planes are some curves called narrow paths (NPs) which express the overall behavior of the output (parameter) with respect to the input (parameter) in each S.I.S.O. system. This behavior ought to be unique for the input domain (acting like a function of input domain).

- Aggregating the S.I.S.O. candidate models

The NPs are the models which are extracted by the IDS method for each S.I.S.O. sub-system in the current M.I.S.O. sub-domain, separately. Each NP introduces a particular candidate S.I.S.O. model which has to be unified to a unique M.I.S.O. model. Therefore, at the final step of the modeling process in the IDS method, the NPs are aggregated through an interpolation mechanism to form a single M.I.S.O. model.

3 The Proposed Methodology

A very popular approach to deal with small sample size problem is the noise injection technique. Noise injection is known as a practical technique to improve generalization ability of many different regions such as classifiers, approximators, data-dependent analysis, and especially in NNs.

In this section, since the literature has not studied on the generalization ability and overfitting in ALM, we briefly explore the topics and then the proposed methodology is described.

3.1 Generalization in ALM

Inside ALM, generalization, which is the main goal of the entire modeling process, is evaluated after modeling each data sub-domain in order to recognize whether overfitting is occurred, or the desired generalization ability (measured by nMSE) is provided for the sub-domain. The goal of generalization is to provide a model so as to be used to approximate unseen data samples. Generalization in a modeling method is measured by the generalization error. Generalization error is a criterion that measures how good a modeling method generalizes to unseen data. Generally, the generalization error is measured as the distance between the error on the training set and the test set in the process of learning which can be regarded locally (for ALM) or in the entire training set.

In ALM, the recursive localization continues only if the generalization error (which is measured by nMSE) is greater than a predetermined threshold. If the generalization error is less than the threshold, ALM stops localization. In other words, ALM tries to reach a certain value of generalization performance to model the system and to avoid overfitting stops partitioning the data in the current level.

One of the most important methods for measuring the generalization error is the cross-validation technique. Using cross-validation for measuring the generalization error requires separating the entire training data set into two different subsets of data: *Training set* and *Validation set*. This separation reduces the number of actual training data which regarding that the entire data set is formerly separated into the training and test data obviously reduces the generalization ability of the trained model. This issue becomes very important when the size of data samples is insufficient of the generalization faces with lack of enough data.

3.2 *Overfitting in ALM*

One of the most challenging problems with modeling techniques, which has an important impact on the performance of the modeling process, is the problem of overfitting and underfitting. While modeling techniques construct models from a set of training samples, identifying the structure and the effective characteristics of data used for constructing the models are the matter of concern. Overfitting is one of the problems which is faced by the modeling techniques and is due to the data structure and characteristics. Overfitting happens when the learning phase of the modeling process converges to influence from a very specific part or characteristic of the samples belonged to the training data which meaningfully degrades the generalization ability. As a result, when overfitting happens, the output cannot be guaranteed to exhibit the true formulation of what is actually expected from the learning process through training samples and the output is unreliable.

One of the reasons for overfitting is due to overtraining the model from the training samples called overlearning. At the occurrence of overlearning, model trains until almost all the model parameters converge to express characteristics of the training samples (or specific parts).

Based on the modeling techniques, many different methods have been developed for preventing the modeling process from overfitting to the training samples such as early stopping, weight decay (in ANNs), and so on.

ALM uses the method called early stopping so as to prevent from overfitting. As the flowchart of Fig. 2 exhibits, there is a level in the process of ALM for stopping the localization (more localized learning) process. This level includes evaluating the performance of the model and comparing the accuracy (nMSE) against a predefined threshold of error, and stopping the learning/localization process for the corresponding data sub-domain if a predetermined accuracy is reached.

One of the most popular approaches for avoiding overfitting is to divide the entire data set into two separated sets: a set for training and another set for validation. This approach is called *early stopping*. Then, modeling and training are performed just by the training set, and evaluation of the model or the performance of the system is performed by the validation set. The validation data are independent of the training data. This is due to have a proper measure for generalization which is the goal of

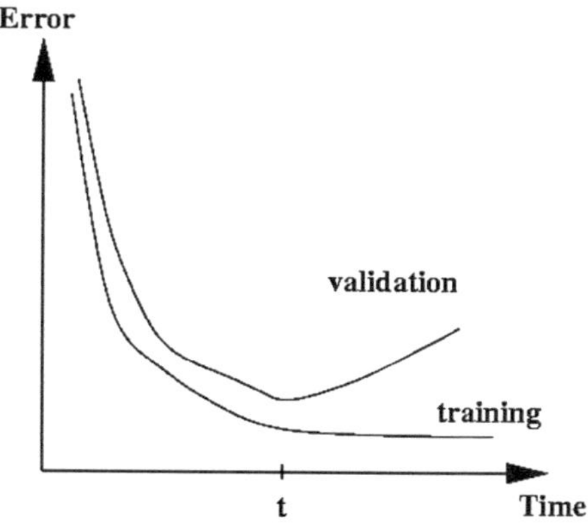

Fig. 6 Overfitting avoidance by stopping the learning process at time *t*

training. So, until the learning process continues, performance on the validation set improves with training.

A schematic view of the learning process illustrating error through the training and validation sets is shown in Fig. 6. To avoid overfitting, training stops at time *t* where performance on the validation set is optimal.

Choosing a threshold of error originates from the generalization disability of the learning method to identify well-sampled data and noise locations in the overall view of data and to prevent the learning process from minimization of overtraining from a particular part of data samples. This leads to choose from several methods so as not to lower the performance of the modeling process due to overlearning or overfitting.

One of the disadvantages of the early stopping approach is that commonly, most part of the data are not used for training which is broadly studied in the field of ANNs (Rynkiewicz 2012; Schittenkopf et al. 1997; Liu and Janusz 2008; Yulei Jiang et al. 2009). Another possibility for early stopping detection is to use the whole part of the data for training and perform validation on some other data samples which is not always accessible.

3.3 Noise Injection in ALM

This section describes where and how to use noise inside the ALM process to improve the generalization ability. Due to the need for independent training, validation, and test data sets, the early stopping used in ALM is only useful in a data-rich situation. The main idea behind using noise injection in this paper is to provide pseudo-training set for ALM in order to properly regularize the model when facing with small sample size problem.

In the process of ALM, noise injection should be addressed inside the IDS method which is responsible to construct the model inside the ALM process. The modified flowchart of the IDS process with the aid of the noise injection technique is illustrated in Fig. 7.

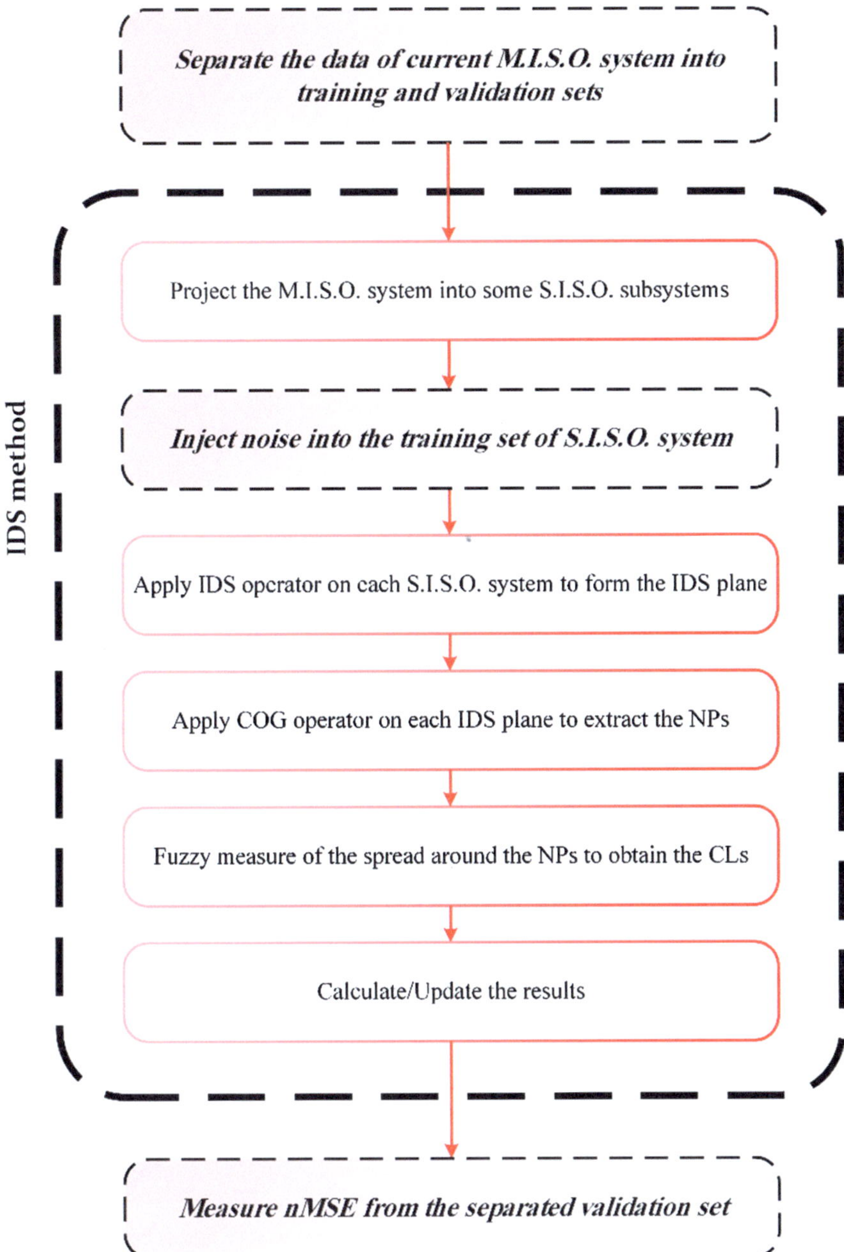

Fig. 7 Noise injection phase added to the algorithm

According to the flowchart of Fig. 7, first, the data of the current localized M.I.S.O. system (stated in the current recursion) is divided into two subsets: the training set and the validation set.

Based on the algorithm of the IDS method, the M.I.S.O. system is then projected into some S.I.S.O. systems with respect to the number of the input parameters.

For each S.I.S.O. system, noise is added to the actual data m times independently to increase the number of training samples and to smooth the 2D plane of the S.I.S.O. system through scattered noise around the actual samples. At each time the IDS method is addressed, noise is added separately in a random fashion to eliminate possible memory of noise so as not to degrade the training performance acquired by the pure behavior of the actual samples.

One of the most important characteristics which is required for preserving the pure behavior of the actual samples inside the training phase is the zero mean of the noise (Fig. 1).

This is important to note that the noisy data will not be participated to measure generalization ability (nMSE). The noisy data are just involved in the process of training, and the validation set is selected from the actual data in order to guarantee true measurement.

Noise is mainly characterized by three characteristics: 1-mean, 2-variance, 3-density, and 4-fashion. In this paper, the noise is characterized by zero mean value and scattered through normal distribution around each actual sample, independent from the other actual samples.

Noise injection is employed to produce pseudo-training or pseudo-validation data sets for enriching the sample sets. It can be done by adding random zero mean vectors to training data set (Yulei Jiang et al. 2009). According to (Yulei Jiang et al. 2009), to make pseudo sets, a zero mean and small covariance noise vector can be added to the training data set:

$$S_{NI} = S_P + N \tag{1}$$

where S_P is the actual training samples, N is the noisy data added to the actual training samples, and S_{NI} the increased training sample set polluted by noise.

From now on, the entire process before measuring nMSE is performed on the noisy data as the pseudo-training set. Implementation of the IDS operator requires regarding a $d\,\%$ margined operation board to provide continuity through fuzzy membership propagation. The operation board is a 2D board which enables the IDS operator to propagate fuzzy membership from actual data sample around to make a continuous fuzzy area.

The domain of ith S.I.S.O. sub-system for IDS operation board is defined as

$$\begin{cases} D_{X_i} : x \,|\, \min(X_i) < x < \max(X_i) \\ D_Y : y \,|\, \min(Y) < y < \max(Y) \end{cases}, \tag{2}$$

where D_{X_i} and D_Y are the domains of the input and the output of ith the S.I.S.O. system (X_i-Y).

Therefore, the domain of $d\%$ margined domain is regarded through:

$$\begin{cases} D_{\widehat{X_i}} : x \mid \min(X_i) - m \arg in_{X_i}^{d\%} < x < \max(X_i) + m \arg in_{X_i}^{d\%} \\ D_{\widehat{Y}} : y \mid \min(Y) - m \arg in_Y^{d\%} < y < \max(Y) + m \arg in_Y^{d\%} \end{cases}, \tag{3}$$

where $\mathrm{margin}_X^{d\%}$ is calculated by

$$\mathrm{margin}_X^{d\%} = [\max(X_i) - \min(X_i)] \times \frac{d}{100}. \tag{4}$$

Finally, for an $M \times M$ board, each unit $\left[U_{\widehat{X_i}}^M, U_{\widehat{Y}}^M \right]$ in ith S.I.S.O. sub-system is calculated by

$$\begin{cases} U_{\widehat{X_i}}^M = \dfrac{\max(\widehat{X_i}) - \min(\widehat{X_i})}{M} \\ U_{\widehat{Y}}^M = \dfrac{\max(\widehat{Y}) - \min(\widehat{Y})}{M} \end{cases}. \tag{5}$$

For applying the IDS operator on the 2D planes of each S.I.S.O. system (which is polluted by noise), the fuzzy membership μ which is propagated by each sample to the neighborhood is attenuated by distance. In this paper, attenuation is implemented through a linear function:

$$\mu = R - \sqrt{u^2 + v^2} + 1; \; -R \le u, v \le R,$$

$$\Delta d(x_s + u, y_s + v) \Rightarrow \begin{cases} \mu; & \text{if } \mu > 0 \\ 0; & \text{otherwise} \end{cases}, \tag{6}$$

where R is the radius of the IDS operator, (x_s, y_s) is the coordinates of the propagator, (u, v) is the distance between the receiver and

$$\Delta d(x_s + u, y_s + v) = \sqrt{u^2 + v^2}. \tag{7}$$

After forming the IDS planes, to extract the NP $\psi(x)$, the COG operator is applied on the IDS plane:

$$\psi(x) = \frac{\sum_{j \in Y(x)} \mu_j Y_j}{\sum_{j \in Y(x)} Y_j}, \tag{8}$$

where Y is the output axis of the operation board.

The NPs are in fact the candidate S.I.S.O. models for the current (localized) M.I.S.O. system which should be unified/aggregated to provide a true M.I.S.O. model. Therefore, a weighted averaging is employed as an interpolation mechanism to aggregate the candidate S.I.S.O. models:

$$y_{\text{final}} = \frac{\sum_{k=1}^{m} w_k y_k}{\sum_{k=1}^{m} w_k}, \tag{9}$$

In the weighted averaging of Eq. 9, the weight, as confidence level (CL), devoted to each S.I.S.O. candidate model corresponds to the reciprocal value of the spread:

$$w_k = \frac{1}{\text{Spread}_k}. \tag{10}$$

The Spread is calculated by

$$\text{Spread}_k = \frac{1}{n} \sum_{i=1}^{n} (\psi_k(x_i^k) - y_i)^2, \tag{11}$$

where x_i^k and y_i are the coordinates of ith data point and ψ_k is the projection of the ith point on kth NP.

At the end of IDS modeling process, if nMSE is greater than the predetermined threshold, the recursive algorithm of ALM divides data into two partitions from the midpoint of the input sub-domain in which its CL has the greatest value. This heuristic helps at least preserving the most reliable S.I.S.O. model for the two resulting subsets.

4   Experimental Results

This section describes and discusses the experimental results of employing noise injection for improving the generalization ability of ALM when is used for estimating FZI with the small sample size problem.

4.1   Data Preparation

Jude et al. (1993) presented a theoretical methodology to identify the flow units. He defined a concept called FZI, which is a unique and useful value to quantify the flow character of a reservoir and one that offers a relationship between petrophysical properties at small scale, such as core plugs, and large scale, such as well bore level. FZI is defined by the following equation:

$$FZI = \frac{1}{\left(\sqrt{F_s}\right)\tau S_{gv}} = \frac{RQI}{\phi_Z}, \tag{12}$$

where F_s is the pore throat shape factor, τ is the tortuosity, and S_{gv} is the effective surface area per unit grain volume. RQI is the reservoir quality index defined below.

$$RQI = 0.0314 \sqrt{\frac{K}{\phi_Z}}, \tag{13}$$

where K is permeability and ϕ_Z is the pore volume (ϕ) to grain volume ratio defined as

$$\phi_Z = \frac{\phi}{1 - \phi}, \tag{14}$$

The above parameters were derived from a modified form of the Kozeny–Carmen relation.

The present study used available data sets of two wells with the name of well A and well B from a carbonate reservoir in the Persian Gulf. These wells contain both log and core data. Core and log porosity values were plotted with respect to depth in order to check whether the data need any depth shifts. The three most commonly logs were selected as input to estimate FZI. These logs are neutron porosity (NPHI), sonic (DT), and density (RHOB). Data from well A were used for the training of the model (with 720 data samples) while data from well B were used for testing the reliability of the model (with 85 data points).

4.2 Numerical Results

In this paper, two statistical evaluation criteria were used to assess the model performance including correlation coefficients (R) and mean square error (MSE):

1. The correlation coefficient ranges between 0 and 1 which is defined as

$$R = \frac{\frac{1}{N}\sum_{i=1}^{N}\left(Y_i - \overline{T}\right)\left(T_i - \overline{Y}\right)}{\sqrt{\frac{1}{N}\sum_{i=1}^{N}\left(Y_i - \overline{T}\right)^2} \times \sqrt{\frac{1}{N}\sum_{i=1}^{N}\left(T_i - \overline{Y}\right)^2}} \tag{15}$$

where T_i and u_i represents the measured and estimated FZI for training data or testing data i, respectively, while T and u are the mean value of measured and estimated FZI, respectively, and n is the number of data in the training or testing data set.

Higher values of R indicate the better performance of the model. Legates and McCabe (David and Gregory 1999) argued that this indicator should not be applied as a fitness measure alone, and it is appropriate to quantify the error in the same unit as for the variables.

2. The mean square error (MSE) is one of the most commonly used measures of success for numeric estimation, computed by taking the average of the squared differences between each estimated FZI and its corresponding measured FZI. It is defined as

$$n\text{MSE} = \frac{\sum_{i=1}^{N} (Y_i - T_i)^2}{\sum_{i=1}^{N} (Y_i - \overline{Y})^2},\tag{16}$$

Model performance increases as MSE decreases.

For the implementation of the IDS operator, the operation board is considered as a 256×256 board with the radius of 8 for fuzzy membership propagation. For the experiments, noise injection has been characterized by zero mean value and variance of 8 units in the operation board for the I–O domain of the S.I.S.O. systems. The final result is the average of 5 experiments in order to reduce the possible impact of noise in the training process. At each stage of recursion, noise is added independently to each projected S.I.S.O. so as to eliminate the impact of "the history of noise" in the modeling process.

According to the algorithm, through the modeling process of the IDS method, for each recursion, FZI should be projected into some 2D planes with respect to the input parameters (DT, NPHI, and RHOB). Therefore, the three-input single-output system is converted into three S.I.S.O. sub-systems. The first application of the IDS operator (in the first recursion) on the projected noisy data planes is illustrated in Fig. 8.

ALM tries to reach a predetermined value of generalization error, and until the acquisition of this purpose, the process continues recursively and localized the IDS modeling procedure. In this experiment, ALM took 12 localization levels and noise-injected ALM (NIALM) took 13 localization levels to converge into a stable state. After convergence when no changes in the performance are observed, ALM stops the recursion and returns the simulated outputs.

Figure 9 illustrates the convergence process of ALM and NIALM convergence to the final state for R and nMSE when regarding a maximum of 20 levels for localization.

Figure 10 illustrates R for the final results of ALM and noise-injected ALM (NIALM).

Figure 11 illustrates a comparison between the performance of some popular methods such as ANN, RBFN, FL, and Neuro-Fuzzy (which are provided by the MATLAB toolboxes) with ALM and NIALM to demonstrate the effectiveness of noise injection into the ALM process.

Results illustrated in Fig. 11 show noise injection improves the performance of ALM while both ALM and NIALM express better modeling ability for FZI in comparison with conventional modeling techniques.

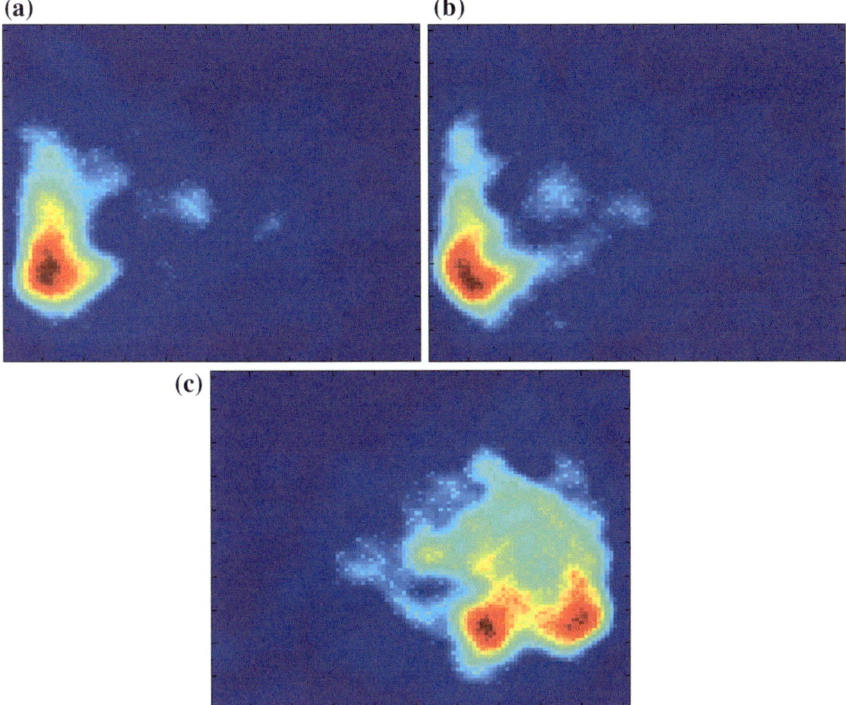

Fig. 8 The application of IDS operator on the projected X_i–Y data planes. **a** Applying IDS operator on DT-FZI plane **b** Applying IDS operator on NPHI-FZI plane **c** Applying IDS operator on RHOB-FZI plane

5 Conclusions

In this paper, the utilization of noise injection technique is addressed to improve the performance of ALM for estimating HFUs in a small sample set. While early stopping, which is used by ALM as the overfitting avoidance strategy, suffers from low performance when insufficiency in data samples is observed, the noise injection technique helps improving the performance of early stopping when separation of data samples set into training, validation, and test sets degrades the generalization ability of ALM modeling process as a learning technique. Results exhibit satisfactory performance in comparison with the use of original ALM and other conventional modeling techniques such as ANN, FL, NF, and RBFNN. Increasing nMSE and R by 20 and 10%, respectively, is a significant improvement which is the result of employing noise injection inside the modeling process of ALM.

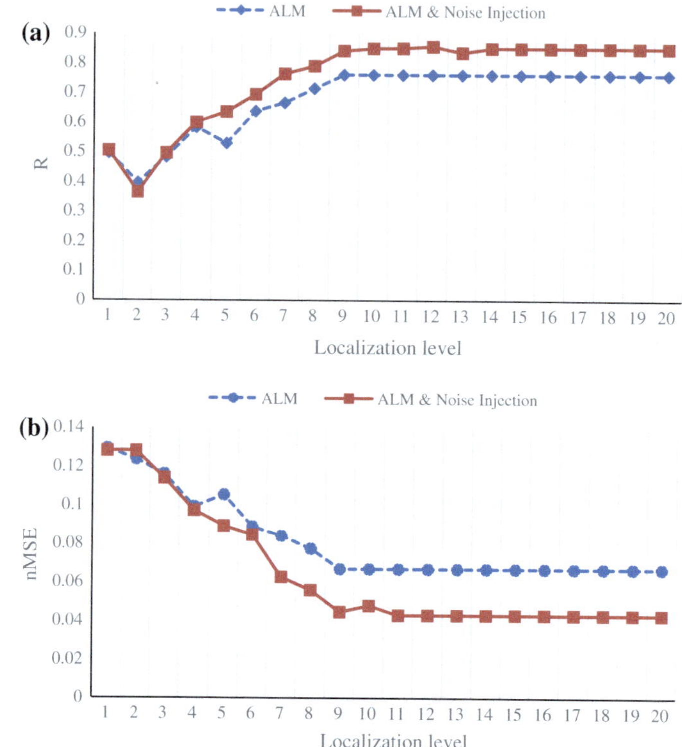

Fig. 9 The learning procedure of ALM and reaching the final state for R and nMSE

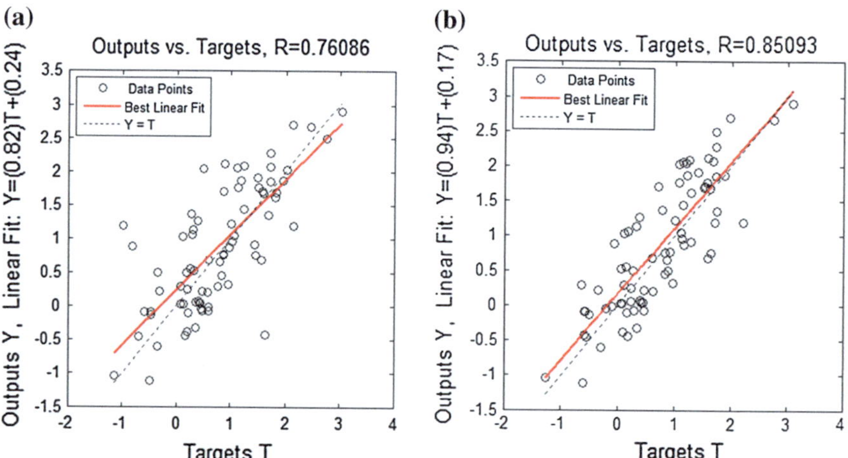

Fig. 10 R for ALM and NIALM. **a** R of ALM, **b** R of NIALLM

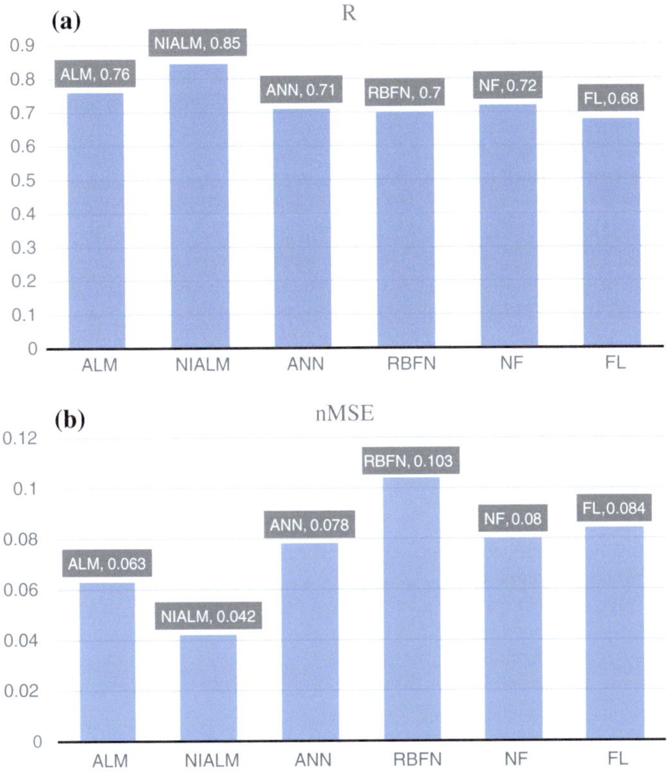

Fig. 11 Comparison between ALM, NIALM, and some popular methods. **a** R, **b** nMSE

References

Bahrpeyma F, Golchin B, Cranganu C (2013) Fast fuzzy modeling method to estimate missing logs in hydrocarbon reservoirs. J Pet Sci Eng 112:310–321

David RL, Gregory JM (1999) Evaluating the use of goodness-of-fit measures in hydrologic and hydroclimatic model validation. Water Resour Res 35(1):233–241

Ebanks W (1987) Flow unit concept-integrated approach to reservoir description for engineering projects. AAPG Meet Abstr 1(5):521–522

Ghiasi-Freez J, Kadkhodaie-Ilkhchi A, Ziaii M (2012) Improving the accuracy of flow units prediction through two committee machines, South Pars Gas Field, Persian Gulf Basin Iran. Comput Geosci 46:10–23

Jude OA, Altunbay M, Tiab D, Kersey DG, Keelan DK (1993) Enhanced reservoir description: using core and log data to identify hydraulic (flow) units and predict permeability in uncored intervals/wells. In: 68th annual technical conference and exhibition, Houston

Kadkhodaie-Ilkhchi A, Amini A (2009) A fuzzy logic approach to estimating hydraulic flow units from well log data: a case study from the Ahwazoil field. South Iran. J Pet Geol 32(1):67–78

Liu YS, Janusz AZ (2008) Optimized approximation algorithm in neural networks without overfitting. IEEE Trans Neural Netw 19:983–995

Piotrowski AP, Napiorkowski JJ (2013) A comparison of methods to avoid overfitting in neural networks training in the case of catchment runoff modelling. J Hydrol 476:97–111

Raudys S (2003) Experts' boasting in trainable fusion rules. IEEE Trans Pattern Anal Mach Intell 25:1178–1182

Raudys S (2006) Trainable fusion rules II. Small sample-size effects. Neural Netw 19:1517–1527

Rifai S, Glorot X, Bengio Y, Vincent P (2011) Adding noise to the input of a model trained with a regularized objective. Technical report dept. IRO, Universite de Montreal. Montreal (QC), H3C 3J7, Canada

Rynkiewicz J (2012) General bound of overfitting for MLP regression models. Neurocomputing 90:106–110

Saeed Bagheri S, Honda N (1997) A new method for establishment and saving fuzzy membership functions. In: 13th fuzzy symposium, Toyama, Japan 1997, pp 91–94

Saeed Bagheri S, Honda N (1999) Recursive fuzzy modeling based on fuzzy interpolation. J Adv Comput Intell 3:114–125

Schittenkopf C, Deco G, Brauer W (1997) Two strategies to avoid overfitting in feedforward networks. Neural Netw 10:505–516

Shouraki SB (2000) A novel fuzzy approach to modeling and control and its hardware implementation based on brain functionality and specification. The University of Electro-Communication, Chofu-Tokyo

Skurichina M, Raudys S, Duin RPW (2000) K-nearest neighbors directed noise injection in multilayer perceptron training. IEEE Trans Neural Netw 11:504–511

Yulei Jiang RMZ, Lorenzo LP, Karen D (2009) A study of the effect of noise injection on the training of artificial neural networks. In: Proceedings of international joint conference on neural networks, Atlanta, Georgia, USA

Well Log Analysis by Global Optimization-Based Interval Inversion Method

Mihály Dobróka and Norbert Péter Szabó

Abstract Artificial intelligence methods play an important role in solving an optimization problem in well log analysis. Global optimization procedures such as genetic algorithms and simulated annealing methods offer robust and highly accurate solution to several problems in petroleum geosciences. According to experience, these methods can be used effectively in the solution of well-logging inverse problems. Traditional inversion methods are used to process the borehole geophysical data collected at a given depth point. As having barely more types of probes than unknowns in a given depth, a set of marginally over-determined inverse problems has to be solved along a borehole. This single inversion scheme represents a relatively noise-sensitive interpretation procedure. For the reduction of noise, the degree of over-determination of the inverse problem must be increased. To fulfill this requirement, the so-called interval inversion method is developed, which inverts all data from a greater depth interval jointly to estimate petrophysical parameters of hydrocarbon reservoirs to the same interval. The chapter gives a detailed description of the interval inversion problem, which is solved by a series expansion-based discretization technique. Different types of basis functions can be used in series expansion depending on the geological structure to treat much more data against unknowns. The high degree of over-determination significantly increases the accuracy of parameter estimation. The quality improvement in the accuracy of estimated model parameters often leads to a more reliable calculation of hydrocarbon reserves. The knowledge of formation boundaries is also required for reserve calculation. Well logs do contain information about layer thicknesses, which cannot be extracted by the traditional local inversion approach. The interval inversion method is applicable to derive the layer boundary coordinates and certain zone parameters involved in the interpretation problem automatically. In this chapter, it is analyzed how to apply a fully automated procedure for the determination of rock interfaces and petrophysical parameters of hydrocarbon formations. Cluster analysis of well-logging data is performed as a preliminary data processing step before inversion. The analysis of cluster number log allows the separation of formations and gives an initial estimate for layer thicknesses. In the global

M. Dobróka (✉) · N. P. Szabó
Department of Geophysics, University of Miskolc, 3515 Miskolc-Egyetemváros, Miskolc, Hungary
e-mail: dobroka@uni-miskolc.hu

   233
C. Cranganu (ed.), *Artificial Intelligent Approaches in Petroleum Geosciences*,
https://doi.org/10.1007/978-3-031-52715-9_9

inversion phase, the model including petrophysical parameters and layer boundary coordinates is progressively refined to achieve an optimal solution. The very fast simulated re-annealing method ensures the best fit between the measured data and theoretical data calculated on the model. The inversion methodology is demonstrated by a hydrocarbon field example, which shows an application for shaly sand reservoirs. The theoretical part of the chapter gives a detailed mathematical formulation of the inverse problem, while the case study focuses on the practical details of its solution by using artificial intelligence tools.

Keywords Well-logging · Interval inversion · Global optimization · Simulated annealing · Cluster analysis · Calculation of hydrocarbon reserves · Hungary

1 Introduction

Geophysical surveying methods with their measuring and evaluation results support the exploration of the Earth and its outer environment. Borehole geophysics is abounding in observed information on the geological formations that are intersected by the drill hole. Well-logging data measured by different probes are recorded along depth in the form of well logs. The processing of open-hole logging data enables to determine some geometrical (e.g., thickness or dip of layers) and petrophysical properties such as porosity, water saturation, composition of rock matrix, and permeability that form an integral part of geological interpretation. Nowadays, there is an ever-increasing claim to the quality of well logs and interpretation results. This is especially important in oil field applications, where a precise calculation of hydrocarbon reserves should be made in complex geological environments.

The advent of inverse modeling (abbreviated as inversion)-based data processing methods was facilitated by the quick evolution of well-logging interpretation systems in the 1980s. In the early years, deterministic techniques for solving linear sets of equations or using cross-plot-based graphical methods were applied. These methods gave a solution in several consecutive steps at which the petrophysical parameters were extracted one by one in different procedures (Serra 1984). It was the increased storage capacity and processor speed of computers that promoted the use of simultaneous processing of well logs. The benefit of using the data and petrophysical parameters as statistical variables was unequivocal in the improvement of the quality of interpretation results. Nowadays, inversion methods are widely used in petrophysical practice as they give a quick, largely automatic, and reliable estimate to the vertical distributions of petrophysical parameters and their estimation errors. The biggest service companies offer inversion-based well-logging interpretation systems, e.g., Global by Schlumberger (Mayer and Sibbit 1980), Ultra by Gearhart (Alberty and Hashmy 1984), or Optima by Baker Hughes (Ball et al. 1987). The development of these methods is strongly focused on scientific research, too.

Local inversion is the most commonly used technique for the evaluation of borehole geophysical data. Several implementations used in the oil and gas industry are well-known. They have in common that a local value of any petrophysical property is estimated to one depth point using the data measured by different probes in the same depth. In the terminology of geophysical inversion, it is a narrow type of over-determined inverse problem, where the total number of data is barely more than that of the unknown model parameters. The data and the model are connected by probe response functions that are used to calculate theoretical logs in the forward modeling phase of the inversion procedure. By assuming a petrophysical model, one can calculate theoretical well logs, which are then compared with real measurements. The actual model is progressively refined until a proper fit is achieved between the predictions and observations. Local data processing comprises a set of separate inversion runs in adjacent measuring points for the logging interval. It is a general experience that in the inversion of the small number of observations, the inversion result is strongly influenced by the uncertainty of measured data. The noise of data highly affects the quality of parameter estimation; thus, the accuracy and reliability of local inversion results are relatively limited. The measurement accuracy of logging tools is prescribed that can be improved seldom with the use of any data processing method. It is a fundamental task to reduce the amount of estimation errors of inversion parameters. On one hand, one can develop more realistic probe response functions. This also means that one tends to set a model approximating the geological structure better. As a result, one can calculate such data by the response functions that are closer to the real observations. Petrophysical research deals with the development of these types of procedures that reduce model errors. By the above contexture, it is unequivocal that another alternative to improve the quality of parameter estimation can only be facilitated by the further development of the inversion procedures. The most important requirement of the development is the improvement of accuracy and reliability of parameter estimation. For this purpose, the most essential task is the increase of data used in one interpretation procedure. In the framework of local inversion, it leads to the expansion of log types, which is of course restricted and implies additional charges. There is a more effective technique to increase the amount of data without extra cost. In the so-called interval inversion procedure, all data of a longer logging interval are processed jointly to determine the characteristic values of petrophysical parameters of several rock units. As a result of the formulation of the interval inversion problem, at least one order of magnitude higher number of data than unknowns can be processed together compared with local inversion. This bears great influence on the accuracy and reliability of the extracted petrophysical parameters. The interval inversion method was introduced in Dobróka (1995), where depth-dependent probe response functions were used (instead of local ones) in the forward problem to give an estimate to the vertical distributions of petrophysical parameters for the entire logging interval. The interval inversion procedure allows to treat increasing number of inversion unknowns without a significant decrease of over-determination (data-to-unknowns) ratio. As a new feature, additional unknowns can be determined together with conventional petrophysical parameters in the same inversion procedure. In Dobróka and Szabó (2012), the possibilities of the determination of formation thicknesses were studied, where the starting model for layer boundaries was set by external procedure. In this chapter, we suggest a fully automatic inversion strategy using a series expansion-based parameter discretization scheme to estimate the formation boundary coordinates and petrophysical parameters in one inversion procedure for a more objective calculation of hydrocarbon reserves.

2 Inverse Problem of Borehole Geophysics

In well-logging inversion, the model parameters of the geological structure are determined in the knowledge of measurement data and approximate formulae of response functions. The aim of interpretation was the lithological separation of formations and the estimation of layer thicknesses and petrophysical properties of formations such as effective porosity, water and hydrocarbon saturation, shale content, mineral volumes, and permeability to infer the quantity and quality of mineral resources. Among them, only those parameters can be determined by inversion, which are contained explicitly in the set of probe response functions and to which almost all types of data are sufficiently sensitive.

The inverse problem of borehole geophysics is classically a joint inversion problem with the particular feature that the quantities included in probe response functions can be divided into two groups. The first group comprises the so-called zone parameters, which are either constants or varying slowly over a longer depth interval (e.g., pore-water resistivity and cementation exponent). The layer parameters form the second group that is nearly constant in a given layer (e.g., porosity and mineral volume). In the practice of well-logging inversion, the zone parameters are treated as external constants that are a priori given in the inversion procedure. This simplification is compulsory in local inversion because the total number of suitable well logs is no more than 10–12, which sets a limit to the number of designated unknown quantities. If the zone parameters were treated as unknowns, an underdetermined (ambiguous) inverse problem would be encountered. In the kth local response equation

$$\varphi^{(k)} = g^{(k)}(m_1, \ldots, m_P, M_1, \ldots, M_L) \quad (k = 1, 2, \ldots, S) \tag{1}$$

the layer parameters $(m_1, \ldots, m_P)$ are only determined by inversion, while the zone parameters $(M_1, \ldots, M_L)$ are fixed during the procedure. On the left side of Eq. (1), the calculated value of the kth logging data can be found (S is the number of applied probes). As $\varphi^{(k)}$ normally represents a nonlinear functional relationship, thus a nonlinear over-determined inverse problem is posed in the case of $S > P$.

2.1 Theory of Local Inversion

In formulating the local inverse problem, all data measured in a given depth point are collected in a column vector

$$\mathbf{d} = \{d_1, \ldots, d_N\}^{\mathrm{T}}, \tag{2}$$

where $(d_1, d_2, d_3 \ldots)$ represent different types of logs such as natural gamma-ray intensity, neutron porosity, and density (T is the symbol of matrix transpose). The theoretical values of the above data can be calculated by the response equations defined in Eq. (1). Let the computed data be represented by vector

$$\varpi = \{\varphi_1, \ldots, \varphi_N\}^{\mathrm{T}}, \tag{3}$$

where the kth response equation is as follows:

$$\varphi^{(k)} = g^{(k)}(m_1, \ldots, m_P). \tag{4}$$

The nonlinear functional relationship $g^{(k)}$ can be approximated by its Taylor series truncated at the first order

$$\varphi^{(k)} = \varphi^{(k)}(\mathbf{m}_\mathrm{o}) + \sum_{i=1}^{P} \left(\frac{\partial \varphi^{(k)}}{\partial m_i}\right)_{\mathbf{m}_\mathrm{o}} \delta\mathbf{m}, \tag{5}$$

where the series expansion is performed around point $\mathbf{m}_\mathrm{o}$, which denotes the vector of initial model parameters. Equation (5) is expressed in vector representation

$$\varpi = \varpi^{(o)} + \mathbf{G}\delta\mathbf{m}, \tag{6}$$

where $\varpi^{(o)} = \varpi(\mathbf{m}_\mathrm{o})$ and $G_{ki} = \left(\partial \varphi_k / \partial m_i\right)_{\mathbf{m}_\mathrm{o}}$ is Jacobi's (parameter sensitivity) matrix. The parameter correction vector $\delta\mathbf{m}$ is estimated by the damped least squares method, which minimizes the Euclidean norm of the following deviation vector

$$\mathbf{e} = \mathbf{d} - \varpi^{(o)} - \mathbf{G}\delta\mathbf{m} \tag{7}$$

with a side condition that $|\delta\,\mathbf{m}|^2$ is minimal. The objective function of the inverse problem is as follows:

$$E = \sum_{k=1}^{N} e_k^2 + \lambda \sum_{i=1}^{P} \delta m_i^2, \tag{8}$$

where λ is a positive damping factor. With the substitution of $\delta\mathbf{d} = \mathbf{d} - \varpi^{(o)}$, the following solution is derived

$$\delta\mathbf{m} = \left(\mathbf{G}^{\mathrm{T}}\mathbf{G} + \lambda\mathbf{I}\right)^{-1}\mathbf{G}^{\mathrm{T}}\delta\mathbf{d}. \tag{9}$$

By solving Eq. (9), the inversion procedure is continued in a given point of the model space

$$\mathbf{m} = \mathbf{m}_o + \delta\mathbf{m} \tag{10}$$

until a stopping criterion is met. The local inversion procedure differs from the Levenberg–Marquardt algorithm only in the consideration of a priori knowledge. Therefore some criteria for the lower and upper bounds of the unknowns as well as for the sum of the specific volumes of rock constituents (material balance equation) must be fulfilled. Besides applying these constrains, another program development question is that any parameter may be set fixed in the iteration procedure. The third group of unknowns of the well-logging interpretation problem is formed by the layer boundary coordinates or layer thicknesses. Their role in local inversion is unique, because they are not contained explicitly in the probe response equations. Thus, their estimation by local inversion is out of the question. The measurement data set does contain information on the boundaries that are of great interest in oil field applications, e.g., in the estimation of hydrocarbon reserves. The determination of layer-boundaries is realized commonly in well log analysis not within the inversion procedure.

2.2 Depth-Dependent Response Functions

For the calculation of layer thicknesses and zone parameters, a new inversion strategy called interval inversion was developed. Consider the petrophysical (layer) parameters $(m_1, \ldots, m_P)$ as the function of depth. Based on Eq. (4), the kth depth-dependent response function is as follows:

$$\varphi^{(k)}(z) = g^{(k)}(m_1(z), \ldots, m_P(z)). \tag{11}$$

In the general case, Eq. (11) contains also the functions of zone parameters $(M_1, \ldots, M_L)$, , which can be determined by the interval inversion method (Dobróka and Szabó 2011). The discretization of model parameters $m_1(z), \ldots, m_P(z)$ can be performed by several manners. In the case of a layerwise homogeneous model, a series expansion technique with proper basis functions including the coordinates of boundaries answers the purpose. Let $\left(B_1^{(i)}, \ldots, B_{Q_i}^{(i)}\right)$ be the series expansion coefficients of the ith model parameter $m_i(z)$. The response function in Eq. (11) takes the form as follows:

$$\varphi^{(k)}(z) = g^{(k)}\left(B_1^{(1)}, \ldots, B_{Q_1}^{(1)}, \ldots, B_1^{(P)}, \ldots, B_{Q_P}^{(P)}, Z_1, \ldots, Z_R, z\right), \tag{12}$$

where $(Z_1, \ldots, Z_R)$ denote the coordinates of layer-boundaries (Q_i is the requisite number of expansion coefficients describing the relevant model parameter). The above response function is valid in the entire interval, in which the series expansion coefficients must be chosen in such a way that the values of $\varphi^{(k)}(z)$ in each depth fit to measurement data $d^{(k)}(z)$ with the highest possible accuracy. The aim of the inversion procedure was the estimation of coefficients B, in which all data of the observed interval are inverted. This inverse problem is highly over-determined, because the number of data is several times higher than that of the unknown expansion coefficients. In local inversion, the over-determination ratio is at the best two. On the contrary, in interval inversion, the same ratio may reach 50–60. Under this circumstance, the boundary coordinates $(Z_1, \ldots, Z_R)$ can be treated also as inversion unknowns to determine them with the expansion coefficients without significant reduction of the over-determination ratio. The above procedure is called interval inversionincluding the depth interval where the series expansion is applied for the model parameters (Dobróka 1995). It is assumed that $d^{(k)}$ in a given depth represents a punctual data, i.e., the linear dimensions of the observed volume are smaller than the thickness of layers.

3 The Theory of Interval Inversion Method

In geophysical data processing, the term of joint inversion is used when different types of data sets are inverted together in one interpretation procedure. The data sets are measured either by different physical principles or by the same principle but in various measurement arrays. All data measured at different spread layouts carry information on the same geological structure. The theoretical values of data sets integrated into the joint inversion procedure are calculated in the knowledge of all model parameters by a proper forward modeling algorithm, that is, the data may depend on each model parameter. The more the parameters of the geological structure appear in the determination of different data sets, the more successful the solution to the inverse problem can be given. The use of such data sets that are depending only on separated groups of model parameters is unbeneficial compared with independent inversion. In the latter case, the solution will not be more accurate or reliable at all. The local inversion of well-logging data utilizes several data sets based on different physical principles (e.g., nuclear, acoustic, and electric methods), where each datum in the inversion procedure is acquired from the same depth. The observed datum does not depend on the parameters of outlying layers. In this case, therefore, the term of joint inversion can be used only in a restricted sense. It can readily be understood that in the interval inversion approach, it is easy to develop such procedures that allow to exploit all the advantages of joint inversion.

For approximating the depth variations of petrophysical parameters $m_1(z), \ldots, m_P(z)$, a series expansion technique is suggested as follows:

$$m_i(z) = \sum_{q=1}^{Q_i} B_q^{(i)} \psi_q(z, Z_1, \ldots, Z_R), \tag{13}$$

where $B_q^{(i)}$ are expansion coefficients, $\psi_q(z, Z_1, \ldots, Z_R)$ are properly chosen (known) depth-dependent basis functions including the layer-boundaries (Q_i is the requisite number of expansion coefficients describing the ith model parameter). Combining Eqs. (12) and (13), the total number of unknowns is $\sum Q_i$, while that of the data is $\sum N_k$. Let us define the data vector of the kth well log as follows:

$$d^{(k)} = \left\{ d_1^{(k)}, \ldots, d_{N_k}^{(k)} \right\}^{\mathrm{T}}. \tag{14}$$

The kth data in the jth depth is calculated by

$$\varphi_j^{(k)} = \varphi^{(k)}(z_j) = g^{(k)}\left(B_1^{(1)}, \ldots, B_{Q_1}^{(1)}, \ldots, B_1^{(P)}, \ldots, B_{Q_P}^{(P)}, z_j \right) \tag{15}$$

and the vector of calculated data is as follows:

$$\boldsymbol{\varpi}^{(k)} = \left\{ \varphi_1^{(k)}, \ldots, \varphi_j^{(k)}, \ldots, \varphi_{N_j}^{(k)} \right\}^{\mathrm{T}}. \tag{16}$$

The data vector of the joint inversion problem including S number of well logs is as follows:

$$\mathbf{d} = \left\{ d_1^{(1)}, \ldots, d_{N_1}^{(1)}, d_1^{(2)}, \ldots, d_{N_2}^{(2)}, \ldots, d_1^{(S)}, \ldots, d_{N_s}^{(S)} \right\}^{\mathrm{T}}, \tag{17}$$

and the vector of all calculated data analogously is as follows:

$$\boldsymbol{\varpi} = \left\{ \boldsymbol{\varphi}^{(1)}, \ldots, \boldsymbol{\varphi}^{(S)} \right\}^{\mathrm{T}}. \tag{18}$$

The series expansion coefficients in Eq. (13) represent the unknowns of the joint inversion problem, thus the combined parameter vector is as follows:

$$\mathbf{m} = \left\{ B_1^{(1)}, \ldots, B_{Q_1}^{(1)}, \ldots, B_1^{(P)}, \ldots, B_{Q_P}^{(P)} \right\}^{\mathrm{T}}, \tag{19}$$

with which the forward problem can be written as follows:

$$\boldsymbol{\varpi} = \mathbf{g}(\mathbf{m}) = \left\{ g_1^{(1)}(\mathbf{m}), \ldots, g_{N_1}^{(1)}(\mathbf{m}), \ldots, g_1^{(S)}(\mathbf{m}), \ldots, g_{N_s}^{(S)}(\mathbf{m}) \right\}^{\mathrm{T}}. \tag{20}$$

The measurement data are always contaminated with some amount of noise. On the other hand, the data calculated by Eq. (20) contain modeling errors resulting from discretization and other physical simplifications. The overall error between the two quantities is defined as follows:

$$\mathbf{e} = \mathbf{d} - \mathbf{g}(\mathbf{m}), \tag{21}$$

which is not a zero vector. Inversion methods find a solution at the minimum of some norm of deviation vector **e**. The response functions are usually nonlinear, thence the inverse problem can be solved by either global or linearized inversion methods.

In case of using linearized inversion, the starting point $\mathbf{m}_o$ in model space is considered not too remote from the solution **m**. The vector of parameter corrections in Eq. (10) is sought. The calculated data can be approximated by Eq. (5). The kth element of the deviation vector defined in Eq. (21) is as follows:

$$e_k = d_k - \varphi_k(o) - \sum_{i=1}^{P} \left(\frac{\partial g_k}{\partial m_i} \right)_{\mathbf{m}_o} \delta m_i, \tag{22}$$

where $\varphi_k(o) = g_k(\mathbf{m}_o)$. By introducing the notation $\delta d_k = d_k - \varphi_k(o)$ and $G_{ki} = \left(\partial g_k / \partial m_i \right)_{\mathbf{m}_o}$, the previous vector is as follows:

$$\mathbf{e} = \delta\mathbf{d} - \mathbf{G}\delta\mathbf{m}. \tag{23}$$

The determination of $\delta\boldsymbol{m}$ ensures to reach a closer point to the solution. The optimal estimate can be extracted by an iterative method. The correction of the actual model in the lth step is as follows:

$$\mathbf{m}^{(l)} = \mathbf{m}^{(l-1)} + \delta\mathbf{m}^{(l)}, \quad \delta d_k = d_k - \varphi_k(\mathbf{m}^{(l-1)}), \quad G_{ki} = \left(\frac{\partial g_k}{\partial m_i} \right)_{\mathbf{m}^{(l-1)}}. \tag{24}$$

If the solution is bound to the minimum of the L_p norm of the deviation vector given in Eq. (23)

$$E = \sum_{k=1}^{N} \left| \delta d_k - \sum_{i=1}^{P} G_{ki} \delta m_i \right|^p, \tag{25}$$

then the undermentioned set of conditions

$$\frac{\partial L_p}{\partial m_h} = 0 \quad (h = 1, 2, \ldots, P) \tag{26}$$

must be fulfilled. As a result of derivation, a nonlinear set of equations is obtained

$$\delta\mathbf{m} = \left(\mathbf{G}^{\mathrm{T}}\mathbf{W}\mathbf{G} \right)^{-1} \mathbf{G}^{\mathrm{T}}\mathbf{W}\delta\mathbf{d}, \tag{27}$$

which can be solved by the iteratively reweighted least squares (IRLS) method that re-calculates the diagonal elements $W_{ks} = |e_k|^{p-2}\delta_{ks}$ of the weighting matrix **W** in each iteration step.

3.1　Basis Functions Used in Interval Inversion

The selection of basis functions is not strictly limited, but the finding of suitable ones may greatly improve the accuracy and reliability of the inversion result. In case of proper basis functions, a relatively small number of additive terms are enough to be used, because the effect of truncation in Eq. (13) is negligibly small. One should tend to reduce the number of expansion coefficients to maintain the numerical stability of the inversion procedure.

In geophysical inversion, there are several applications of using simple models. In borehole geophysics, the layerwise homogeneous model is of high importance. This situation can be described easily by substituting a combination of Heaviside basis functions into Eq. (13)

$$m_i(z) = \sum_{q=1}^{Q_i} B_q^{(i)} \psi_q(z) = \sum_{q=1}^{Q_i} B_q^{(i)} \left[u(z - Z_{q-1}) - u(z - Z_q) \right], \qquad (28)$$

where Z_q is the depth coordinate of the qth layer and Q is the number of homogeneous layers. The basis function ψ_q introduced in Eq. (28) is always zero except in the qth layer, which is an element of an orthogonal sequence of functions

$$\int_0^{z_{\max}} \Psi_q \, \Psi_{q'} \, dz = \begin{cases} 0, & \text{if } q \neq q' \\ Z_q - Z_{q-1}, & \text{if } q = q' \end{cases} . \qquad (29)$$

It arises that the series expansion coefficient in the qth layer equals to the petrophysical parameter in the same layer, that is, $B_q^{(i)} = m_i(Z_{q-1} < z < Z_q)$. The series in Eq. (13) can be rewritten as follows:

$$m_i(z) = \sum_{q=1}^{Q_i} m_q^{(i)} \Psi_q(z), \qquad (30)$$

where $m_q^{(i)}$ is the value of the ith parameter in the qth layer. It is obvious that the inverse problem can be solved by the smallest possible number of unknowns. On the other hand, the layer-boundaries appear in the argument of the basis function ψ_q, which can be extracted by the interval inversion method.

The variation of petrophysical parameters within the layer can be approximated by polynomial series expansion

$$m_i(z) = \sum_{q=1}^{Q_i} B_q^{(i)} P_q(z), \qquad (31)$$

where $P_q(z)$ represents some polynomial, for instance Legendre polynomials. The meaning of expansion is not demonstrative than in Eq. (30). Similarly, the selection of parameter Q is less unequivocal. The number of unknowns can be much higher than in the homogeneous case. A trade-off must be taken between the vertical resolution of model parameters and the stability of the inversion procedure. To relieve the task, in some practical cases the combination of Eqs. (30) and (31) is used to get an adequate solution. If the layerwise homogeneous model contains an inhomogeneous layer, the following series expansion can be used

$$m_i(z) = \sum_{\substack{q=1 \\ q \neq q'}}^{Q_i} m_q^{(i)} \Psi_q(z) + \sum_{u=1}^{U} B_u P_u\big(z - Z_{q'-1}\big), \tag{32}$$

where U is the number of additive terms used in the approximation of variation in the q' th layer. The above problem was solved by Dobróka and Szabó (2005) using a combined inversion algorithm based on the subsequent use of global and linearized optimization methods.

3.2 Layer-Thickness Determination by Interval Inversion

The greatly over-determined interval inversion method allows to treat increasing number of inversion unknowns without a significant decrease of accuracy in parameter estimation. Some groups of the inversion unknowns are contained in local response functions (e.g., zone parameters); some are not included (layer thicknesses). The latter can be determined by the interval inversion of the combined data set defined in Eq. (17). Consider the model vector of the inverse problem including the layer boundary coordinates

$$\mathbf{m} = \left\{ B_1^{(1)}, \dots, B_{Q_1}^{(1)}, \dots, B_1^{(P)}, \dots, B_{Q_P}^{(P)}, Z_1, \dots, Z_R \right\}^{\mathrm{T}}. \tag{33}$$

The number of unknowns is $R + \sum Q_i$ that are substituted into Eq. (12) to calculate theoretical well logs in the forward problem. The number of data is specified in Eqs. (17) and (18). The connection between model and data $\varpi = \mathbf{g}(\mathbf{m})$ contains the layer boundary coordinates; therefore, Eq. (5) modifies as follows:

$$\varphi_k = g_k(\mathbf{m_o}) + \sum_{i=1}^{P} \left(\frac{\partial g_k}{\partial m_i} \right)_{m_o} \delta m_i + \sum_{r=1}^{R} \left(\frac{\partial g_k}{\partial Z_r} \right)_{m_o} \delta Z_r, \tag{34}$$

where $\delta Z_r = Z_r - Z_r^{(0)}$ is an element of the model correction vector $\delta \mathbf{m}$. In Eq. (23), the following Jacobi's matrix is used

$$G_{ki} = \begin{cases} \left(\dfrac{\partial g_k}{\partial m_i}\right)_{\mathbf{m}_\mathrm{o}}, & \text{if } i = 1, 2, \ldots, P \\[2ex] \left(\dfrac{\partial g_k}{\partial Z_i}\right)_{\mathbf{m}_\mathrm{o}}, & \text{if } i = P + 1, \ldots, P + R \end{cases} \tag{35}$$

The minimization of the L_p norm of the deviation vector leads to the solution of the inverse problem given by Eq. (27). The iterative method gives an estimate for the series expansion coefficients and layer boundary coordinates. The determination of layer-boundaries can be made easily by using a series expansion based on Eq. (28). However, if it is required, the method can be combined with the scheme of polynomial discretization. The estimation of layer-boundaries can be performed most efficiently by using a global optimization method.

4 Global Inversion by Simulated Annealing Method

The performance of inversion methods highly depends on how successfully the optimum of the objective function defined in Eq. (25) is found. Conventional interpretation systems offer linear optimization tools that give quick and satisfactory results in case of having a suitable initial model. The weakness of these gradient-based searching methods is that they tend to find a solution at a local optimum of the objective function. This problem can be avoided by using a global optimization method, which finds the absolute optimum of the same function. There is another typical problem of linear interval inversion. In case of linear optimization, the partial derivatives with respect to depth in Jacobi's matrix can only be determined in a rough approximation, because the difference quotient with a depth difference being equal to the distance between two measuring points. This may lead to a numerically instable inversion procedure. Global optimization does not require the computation of derivatives. For the solution of global inverse problems, artificial intelligence tools can be used effectively. Szabó and Dobróka (2013) published previously a float-encoded genetic algorithm-based interval inversion algorithm for oil field and hydrogeological applications. In this study, a fast simulated annealing (SA) method is suggested for the determination of layer parameters and formation thicknesses.

The conventional SA method was developed by Metropolis et al. (1953). In metallurgy, the removal of work-hardening is realized by a slow cooling manipulation from the temperature of liquid alloy state. This process reduces progressively the kinetic energy of a large number of atoms with high thermal mobility, which is followed by the starting of crystallization. Theoretically, the perfect crystal with minimal overall atomic energy can be produced by an infinitely slow cooling schedule. This is analogous with the stabilization of the inversion procedure at the global optimum of the objective function. A quick cooling process causes grating defects and the solid freezes in the imperfect grid at a relatively higher energy state. It is similar to the trapping of the inversion procedure in a local minimum. However, the atoms may escape from the high-energy state owing to a special process called annealing to achieve the optimal crystal grating by a slower cooling process. The SA algorithm employs this technology to search the global optimum of the objective (in the terminology energy) function such as E defined in Eq. (25). At first, the components of the model vector defined in Eq. (33) are modified properly. The modification of the ith model parameter in the lth iteration step is as follows:

$$m_i^{(l+1)} = m_i^{(l)} + b, \tag{36}$$

where $b < b_{max}$ is a perturbation term (b_{max} is decreased appropriately as the iteration procedure progresses). During the random seeking, the energy function $E(\boldsymbol{m})$ is calculated and compared with the previous one in every iteration step. The acceptance probability of the new model depends on the Metropolis criteria

$$P(\Delta E, T) = \begin{cases} 1, & \text{if } \Delta E \leq 0 \\ e^{-\Delta E/T}, & \text{otherwise} \end{cases}, \tag{37}$$

where the model is always accepted when the value of energy function is lower in the new state than that of the previous one. If the energy of the new model increased, there is also some probability of acceptance depending on the values of energy E and control temperature T. If $P(\Delta E) \geq \alpha$ fulfills, the new model is accepted, else it is rejected (α is a random number generated with uniform probability from the interval of 0 and 1). These criteria assure the escape from the local minima. Geman and Geman (1984) proved that the following cooling schedule is the necessary condition to find the global optimum

$$T(l) = \frac{T_0}{\ln(l)} \quad (l > 1), \tag{38}$$

where T_0 is a properly chosen initial temperature. The SA algorithm is very effective, but the logarithmic reduction of temperature in Eq. (38) is rather time-consuming. Several attempts were made to shorten the CPU time. Ingber (1989) proposed a modified SA algorithm called very fast simulated re-annealing (VFSR). Consider different ranges of variation for each model parameter

$$m_i^{(min)} \leq m_i^{(l)} \leq m_i^{(max)}. \tag{39}$$

The perturbation of the ith model parameter at iteration $(l + 1)$ is as follows:

$$m_i^{(l+1)} = m_i^{(l)} + y_i \left(m_i^{(max)} - m_i^{(min)} \right), \tag{40}$$

where y_i is a random number between -1 and 1 generated from a specified non-uniform probability distribution function. The global optimum is guaranteed when the decrease of the ith individual temperature follows

$$T_i^{(l)} = T_{0,i} \mathrm{e}^{(-c_i P \sqrt{l})} \tag{41}$$

Equation (41) specifies different temperatures to each model parameter, where $T_{0,i}$ is the initial temperature of the ith model parameter, c_i is the ith control parameter, and P is the number of model parameters. The acceptance rule of the VFSR algorithm is the same as that used in the Metropolis SA method, but the exponential cooling schedule assures much faster convergence to the global optimum than the logarithmic one suggested in Eq. (38).

5 Selection of Initial Model by Cluster Analysis

Multivariate statistical methods such as regression, factor, and cluster analyses help to find similarities between petrophysical properties of rocks, reduce problem dimensionality or explore non-measurable (latent) information from the observations, and arrange data into groups to reveal different lithological characteristics of the investigated formations. This a priori information can be useful in petrophysical modeling, facies, or trend analysis and in geophysical inversion to set an initial model as input for the inversion procedure. Clustering methods are applicable to sort data into groups in such a way that the S dimensional objects specified by well logs measured from given depths are more similar than other ones observed from different depths. From the point of the interval inversion method, it is of great importance that objects connected to the same cluster define approximately the same lithological character, while other clusters represent dissimilar ones.

Agglomerative cluster analysis builds a hierarchy from the observations by progressively merging clusters. At the beginning, we have as many clusters as individual elements. In the first step, the closest points are coupled together to form a new cluster. In each following step, the distances between objects are re-calculated and the procedure is continued until all elements are grouped into one cluster. Several distance definitions can be used as a measure of dissimilarity between the pairs of observed objects such as Euclidean (L_2 norm based), Manhattan (L_1 norm based), or Mahalanobis (sample covariance based). During the procedure, the distances between the elements of the same group are minimized, while they are maximized between the clusters simultaneously. For the reconnection of clusters, Ward's linkage criterion is followed that minimizes the deviances of $(x_i - C)$, where x_i is the ith object and C is the centroid (average of elements) of the given cluster (Ward 1963). The result of cluster analysis is a dendrogram that shows the hierarchy of clusters and the connections between them at different distances.

In this study, cluster analysis is used as a preliminary data processing step before inverse modeling. It is shown in Fig. 1 that clustering makes use of the complete wellbore data set originated from the entire logging interval. By finding the similarities between the well logs, the objects are grouped into clusters. The log of clusters correlates well with the lithology variation along a borehole. The change

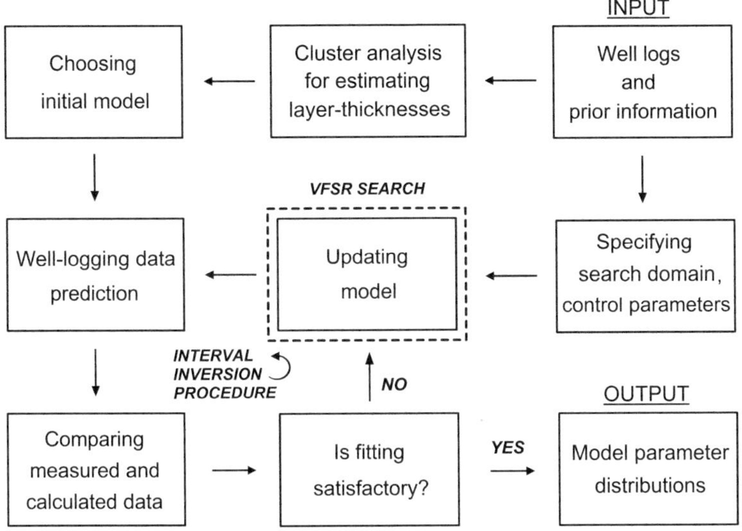

Fig. 1 Scheme of cluster analysis-assisted global inversion procedure

in the group number of clusters appearing on the log gives the positions of layer-boundaries, which can be read automatically by computer processing. The estimated layer boundary coordinates as important a priori information for constructing the initial model serve as input for the interval inversion procedure. In an earlier study, the layer-boundaries extracted by cluster analysis were fixed during the interval inversion procedure (Szabó et al. 2013). However, similar to the layer parameters, the layer boundary coordinates may be treated as unknowns in the interval inversion procedure.

6 Oil Field Application

6.1 Results of Cluster Analysis

In a Hungarian hydrocarbon borehole (Well No. 1), nine well logs were used for testing the interval inversion method. The following log types formed the input of clustering such as caliper (CAL), compensated neutron porosity (CN), gamma–gamma density (DEN), acoustic (primary wave) interval time (AT), natural gamma-ray intensity (GR), deep resistivity (RD), microlaterolog resistivity (RMLL), shallow resistivity (RS), and spontaneous potential (SP). In the first step of the procedure, hierarchical cluster analysis was applied to find a proper initial model for inversion processing (Fig. 1). In the processed interval, a sedimentary complex made up of seven unconsolidated shaly sandy beds were deposited. Three lithological categories were specified, namely, shale, shaly sand, and sand. At this stage of interpretation, this lithological resolution was enough for finding the layer-boundaries, because the relative volumes of rock matrix and shale could be estimated in the inversion phase. The standardized Euclidean distance evaluating each datum in the sum of squares inversely weighted by the sample variance was used for measuring the distance between data objects. A hierarchical cluster tree was created by using Ward's linkage algorithm (Fig. 2a). In Fig. 2b, the ordinal numbers of leaf nodes can be seen that were assigned to each object. Since some leaf nodes corresponded to multiple objects, the total number of nodes was 30. Three clusters can be separated if the tree is cut at centroid distance 1. The layer-boundaries can be traced out in the well log of cluster numbers. According to traditional interpretation, the inflection points of high amplitudes in the GR log indicate rock interfaces. The layer boundary coordinates can be well approximated by the steps between the clusters of sand and shale, i.e., cluster 2 and cluster 3. The depth coordinates indicated by black arrows in Fig. 2c were chosen as initial model parameters for the subsequent interval inversion procedure.

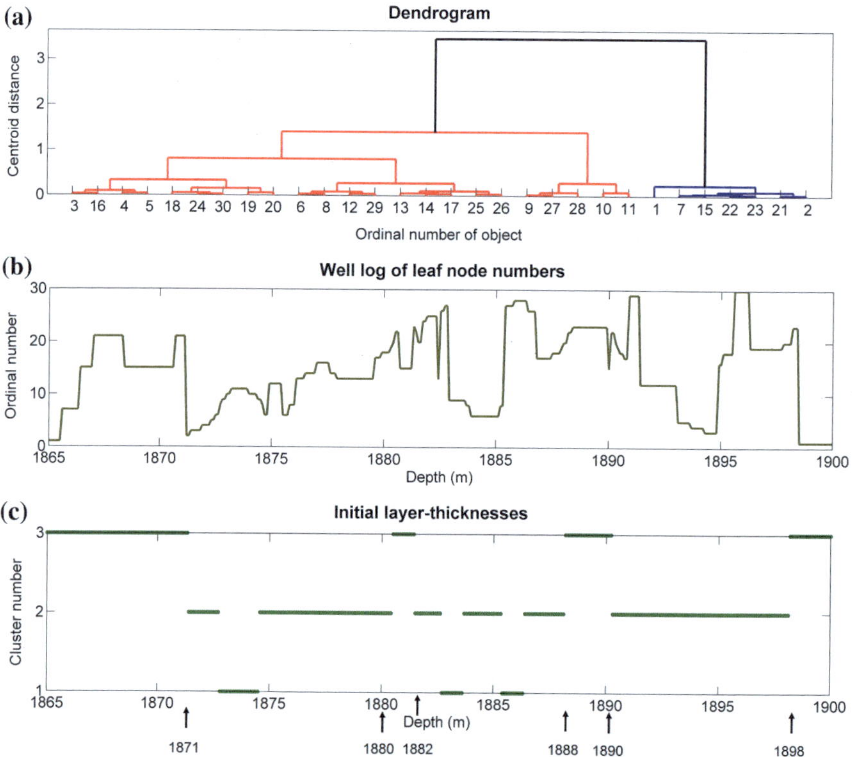

Fig. 2 Results of cluster analysis in Well No. 1. **a** Dendrogram. **b** Well log of leaf node numbers. **c** Initial layer thicknesses

The combination of well logs usually gives useful information on lithology, petrophysical, and zone parameters. In Fig. 3, the three-dimensional cross-plots of clustered well-logging data can be seen which specify several site-specific constants for calculating data in forward modeling. These constants can be used directly in the probe response functions. For instance, in Fig. 3a, the neutron porosity of sand (13%) and shale (25%) and the natural gamma-ray intensity of sand (45 API) and shale (140 API) can be chosen for the given hydrocarbon zone. Several observed data types show strong correlation with each other. In Fig. 3c, d, the nonlinear connection between natural gamma-ray intensity and resistivity is eye-catching. The detailed list of correlation relationships between the data is contained by the correlation matrix including Pearson's correlation coefficients (Table 1). The highest correlations are between lithology logs (GR and CAL, SP and GR) and saturation logs (RS and RD). Porosity-sensitive logs also show strong correlations with lithology logs (DEN and CAL, SP and CAL, SP and CN, GR and DEN, CN and GR). The negative elements of the correlation matrix show inverse proportionality between the data variables (GR and SP).

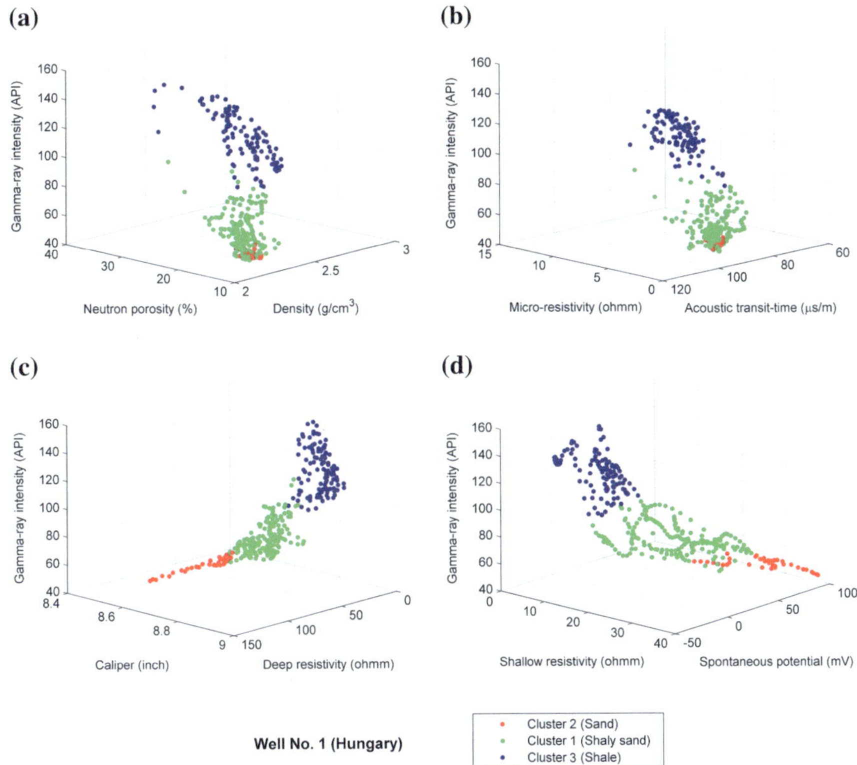

Fig. 3 Clustered data represented in the form of cross-plots in Well No. 1

Table 1 Correlation matrix of well logs measured in Well No. 1

	CAL	CN	DEN	AT	GR	RD	RMLL	RS	SP
CAL	1.0	0.67	0.89	−0.34	0.83	−0.53	0.82	−0.52	−0.74
CN	0.67	1.0	0.68	0.16	0.88	−0.59	0.47	−0.56	−0.76
DEN	0.89	0.68	1.0	−0.51	0.84	−0.58	0.82	−0.59	−0.73
AT	−0.34	0.16	−0.51	1.0	−0.01	0.16	−0.51	0.19	0.10
GR	0.83	0.88	0.84	−0.01	1.0	−0.59	0.68	−0.61	−0.81
RD	−0.53	−0.59	−0.58	0.16	−0.59	1.0	−0.41	0.92	0.68
RMLL	0.82	0.47	0.82	−0.51	0.68	−0.41	1.0	−0.41	−0.59
RS	− 0.52	− 0.56	− 0.59	0.19	−0.61	0.92	−0.41	1.0	0.71
SP	− 0.74	− 0.76	− 0.73	0.10	−0.81	0.68	−0.59	0.71	1.0

6.2 Forward Problem of Well-Logging

The mathematical relationships between the petrophysical properties and well-logging data are dominantly empirical. In the case study, the parameters of the initial model are effective porosity (POR), shale volume (VSH), water saturation of the invaded zone flooded by drilling mud (SX0), water saturation of the virgin zone occupied by the original pore fluid (SW), and sand volume (VSD = 1 − POR − VSH). These parameters and other derived ones underlie the calculation of hydrocarbon reserves. The following set of response functions was used to approximate observable data

$$
\begin{aligned}
\text{DENTH} = \text{POR}[(\text{SX0} \cdot \text{DEMF}) + (1 - \text{SX0})\text{DEHC}] \\
+ \text{VSH} \cdot \text{DESH} + \text{VSD} \cdot \text{DESD},
\end{aligned}
\tag{42}
$$

$$
\text{GRTH} = \text{GRSD} + \frac{1}{\text{DENTH}} \left(\begin{array}{l} \text{VSH} \cdot \text{GRSH} \cdot \text{DESH} \\ + \text{VSD} \cdot \text{GRSD} \cdot \text{DESD} \end{array} \right),
\tag{43}
$$

$$
\begin{aligned}
\text{CNTH} = \text{POR}[(\text{SX0} \cdot \text{CNMF}) + (1 - \text{SX0})\text{CNHC}] \\
+ \text{VSH} \cdot \text{CNSH} + \text{VSD} \cdot \text{CNSD},
\end{aligned}
\tag{44}
$$

$$
\begin{aligned}
\text{ATTH} = \text{POR}[(\text{SX0} \cdot \text{ATMF}) + (1 - \text{SX0})\text{ATHC}] \\
+ \text{VSH} \cdot \text{ATSH} + \text{VSD} \cdot \text{ATSD},
\end{aligned}
\tag{45}
$$

$$
\frac{1}{\sqrt{\text{RDTH}}} = \left[\frac{\text{VSH}^{(1-\text{VSH}/2)}}{\sqrt{\text{RSH}}} + \frac{\left(\sqrt{\text{POR}}\right)^{\text{BM}}}{\sqrt{\text{BA} \cdot \text{RW}}} \right] \left(\sqrt{\text{SW}}\right)^{\text{BN}},
\tag{46}
$$

$$
\frac{1}{\sqrt{\text{RSTH}}} = \left[\frac{\text{VSH}^{(1-\text{VSH}/2)}}{\sqrt{\text{RSH}}} + \frac{\left(\sqrt{\text{POR}}\right)^{\text{BM}}}{\sqrt{\text{BA} \cdot \text{RMF}}} \right] \left(\sqrt{\text{SX0}}\right)^{\text{BN}}.
\tag{47}
$$

On the left side of Eqs. (42)–(47), the theoretical (TH) values of the well-logging data stand, while on the right, the layer parameters (POR, VSH, SX0, SW, and VSD) and zone parameters can be found. The latter represent the physical properties of mud filtrate (MF), water (W), hydrocarbon (HC), shale (SH), and sand (SD). They are treated as unvarying quantities known from cluster analysis (Fig. 3) or other a priori information (laboratory and well-site reports). The textural constants, such as cementation exponent (BM), saturation exponent (BN), and tortuosity factor (BA), can be estimated from the literature, laboratory data, or the interval inversion method (Dobróka and Szabó 2011). In complex reservoirs, the rock matrix may be composed of several mineral components. Depending on the interpretation problem, the relative volumes of rock constituents can also be extracted within the interval inversion procedure (Dobróka et al. 2012). By using the response Eqs. (42)–(47), an estimate for the model (layer) parameters can be given by the inversion procedure.

6.3 Results of Interval Inversion

The interval inversion procedure was performed on suitable well logs (CN, DEN, AT, GR, RD, and RS) of Well No. 1. Based on the results of cluster analysis, the following depth coordinates were chosen as initial model parameters for interval inversion: 1871, 1880, 1882, 1888, 1890, and 1898 m. The last (seventh) coordinate was the depth at the bottom of the logging interval. As a result, seven shaly sandy layers were traced out. Within three permeable intervals, the separation between DEN and CN logs confirmed the presence of hydrocarbons. In the model approximation, constant layer parameters (POR, VSH, SX0, and SW) were assumed (VSD was calculated deterministically in every iteration steps), which represented a high over-determination ratio (62 with 2100 data, 28 layer parameters, and 6 boundary coordinates) and very stable inversion procedure.

The unknowns of the inverse problem were the series expansion coefficients given in Eq. (28) and the layer boundary coordinates. The VFSR algorithm was used to give a quick estimate to the global optimum of the L_2 norm based energy function. The maximal number of iteration steps was set to 10,000. The logarithmic cooling process based on Eq. (38) was applied with an initial temperature of 0.01. The lower and upper limits of layer parameters were 0 and 1, respectively. The layer-boundaries were allowed to vary within 0 and 10 m. The rate of convergence of the inversion procedure was smooth and progressive as it is seen in Fig. 4, where the root mean squared errors of the relative differences between the measured and calculated data were plotted. The data distance of the final result is influenced by the data noise and the model approximation. In Fig. 5, the change of layer thicknesses calculated from the boundary coordinates can be followed. Until the 7000th iteration step, all thicknesses had attained to their optima. Then, only the layer parameters showed considerable variation.

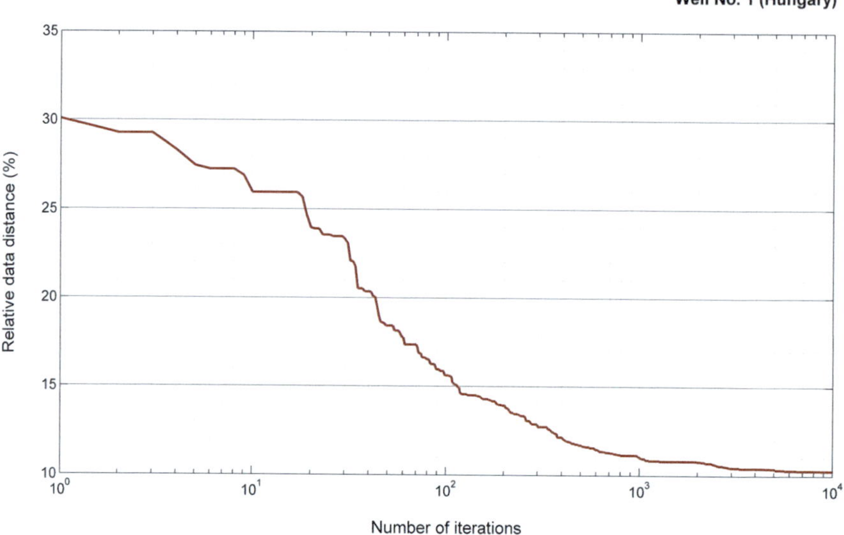

Fig. 4 The convergence of the interval inversion procedure

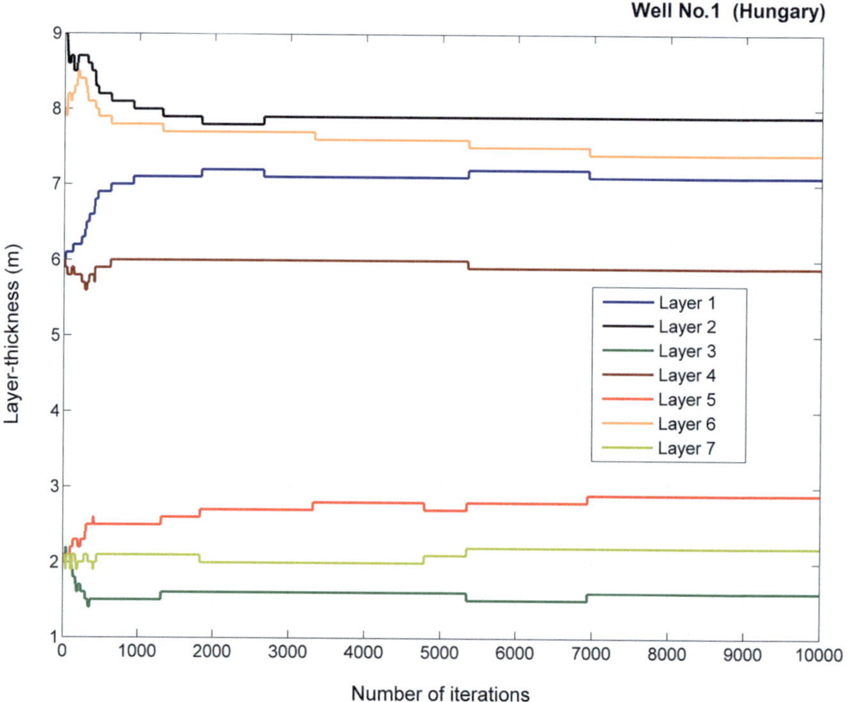

Fig. 5 The variation of layer thicknesses during the interval inversion procedure

In the depth scale of Fig. 6, the boundary coordinates estimated by the interval inversion procedure are marked. The method distinguished the permeable and non-permeable intervals within the hydrocarbon zone and gave a proper estimate to rock interfaces as they correlate well with the layer-boundaries inferred from the GR log. The well logs of estimated layer parameters are in tracks 6–8. The pore space was divided into two separate parts filled with salty water and hydrocarbons. The movable and irreducible hydrocarbon saturation were derived from the inversion results by basic equations (Movable HC = SX0 − SW, Irreducible HC = 1 − SX0). The hydrocarbon reserves are estimated from the movable hydrocarbon saturation, porosity, and the total volume of the reservoir rock. The latter can be estimated from ground geophysical surveys (e.g., seismic) and multi-borehole data. A model reliable estimate of oil and gas reserves is supported by the results of the interval inversion procedure. Szabó and Dobróka (2013) made a comparison between the local and interval inversion methods. It was shown that the estimation error of porosity and water saturation can be reduced significantly by the interval inversion method. This improvement bears influence on the calculation of hydrocarbon reserves. The unit volume of rock was composed of porosity, shale content, and sand volume in different proportions along the interval (track 7). The absolute permeability shown in track 8 for each formation was calculated by the knowledge of porosity and bound water saturation (chosen as 0.1) in the hydrocarbon reservoirs (Timur 1968).

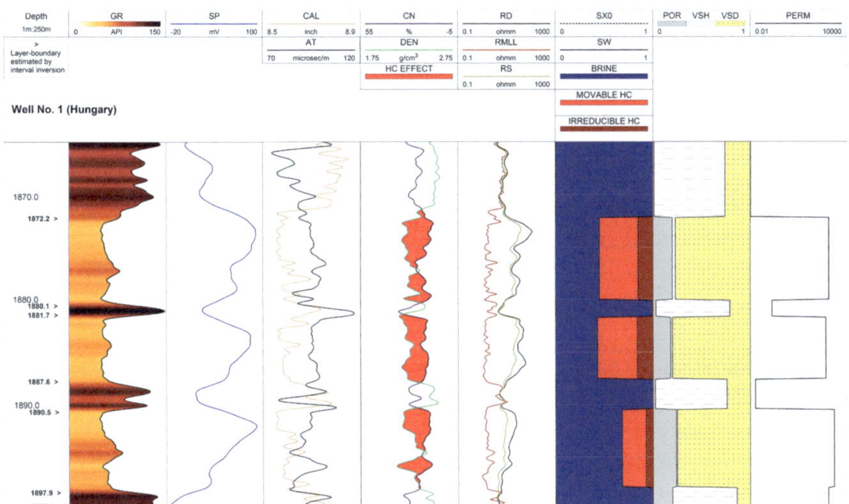

Fig. 6 The well logs of observed data and interval inversion results

7 Conclusions

Advanced data processing methods are essential to extract reliable petrophysical information from geophysical data sets. An intensive research of inversion methods is being made worldwide in all fields of geophysics. In oilfield applications, the proper interpretation of in situ borehole logging data is especially important, because these methods lay the foundations to hydrocarbon reserve calculations. In this chapter, a new inversion methodology was presented, which is now fully automatized by giving an estimate to petrophysical parameters and layer-boundaries in a joint inversion procedure. The cluster analysis-assisted interval inversion method assures a greatly over-determined inverse problem to give accurate and reliable solution along the entire borehole. The specialty of the method is that the basis functions of series expansion can be chosen arbitrarily. The optimal set of basis functions to be in use depends on the variation of lithology and pore fluid types along a borehole. In this study, a layerwise homogeneous model was chosen to reduce the number of inversion unknowns as far as possible. This approach keeps the numerical stability of the inversion procedure in view. However, there is nothing to prevent from the improvement of the vertical resolution of the interval inversion method, but it goes with the relative increase of the number of series expansion coefficients. As the problem is highly over-determined, it can be allowed to some extent. Practically, a trade-off must be taken between the number of unknowns (resolution) and stability of the inversion (unique solution) procedure as they are inversely proportional. An adequate solution is to choose orthogonal basis functions such as Legendre polynomials in the development of series, in which case the correlation between the estimated model parameters and the parameter estimation errors is relatively the lowest. It is suggested to apply preliminary cluster analysis to find lithological similarities in the data set, which separates such intervals where the polynomial discretization can be performed most effectively. This technical solution leads to the reduction of data and model distances. The inverse problem also affected by the quality of the forward problem solution. In the probe response equations, there are several zone parameters that could be chosen properly for the given well-site. An objective solution can be given by the interval inversion procedure as the zone parameters most sensitive to data can be treated as inversion unknowns. Another strength of the method is the possibility to extend the inverse modeling to multi-borehole applications by expanding the model parameters into series of bivariate basis functions. All of the above properties confirm the feasibility of the interval inversion methods and the use of global optimization techniques preferably very fast simulated re-annealing and float-encoded genetic algorithm in petroleum geoscience applications.

Acknowledgements The first author as the leading researcher of Project No. K 109441 thanks to the support of the Hungarian Scientific Research Fund. The second author as the leading researcher of Project No. PD 109408 thanks to the support of the Hungarian Scientific Research Fund. The second author also thanks to the support of the János Bolyai Research Fellowship of the Hungarian Academy of Sciences. The authors thank the Hungarian Oil and Gas Company's (MOL) contribution to the research work and the long-term cooperation. The authors also thank Hajnalka Szegedi for improving the manuscript formally.

References

Alberty M, Hashmy KH (1984) Application of ULTRA to log analysis. SPWLA 25th annual logging symposium, New Orleans, Paper Z, 10–13 June 1984

Ball SM, Chace DM, Fertl WH (1987) The well data system (WDS): an advanced formation evaluation concept in a microcomputer environment. In: Proceedings of SPE eastern regional meeting, Paper 17034, pp 61–85

Dobróka M (1995) The introduction of joint inversion algorithms into well log interpretation (in Hungarian). Scientific report, Geophysical Department, University of Miskolc

Dobróka M, Szabó NP (2005) Combined global/linear inversion of well-logging data in layer-wise homogeneous and inhomogeneous media. Acta Geod Geophys Hung 40:203–214

Dobróka M, Szabó NP (2011) Interval inversion of well-logging data for objective determination of textural parameters. Acta Geophys 59(5):907–934

Dobróka M, Szabó NP (2012) Interval inversion of well-logging data for automatic determination of formation boundaries by using a float-encoded genetic algorithm. J Petrol Sci Eng 86–87:144–152

Dobróka M, Szabó NP, Turai E (2012) Interval inversion of borehole data for petrophysical characterization of complex reservoirs. Acta Geod Geophys Hung 47(2):172–184

Geman S, Geman D (1984) Stochastic relaxation, Gibbs' distribution and Bayesian restoration of images. IEEE Trans PAMI 6:721–741

Ingber L (1989) Very fast simulated reannealing. Math Comput Model 12(8):967–993

Mayer C, Sibbit A (1980) GLOBAL, a new approach to computer-processed log interpretation. In: Proceedings of SPE annual fall technical conference and exhibition, Paper 9341, pp 1–14

Metropolis N, Rosenbluth A, Rosenbluth M, Teller A, Teller E (1953) Equation of state calculations by fast computing machines. J Chem Phys 21:1087–1092

Serra O (1984) Fundamentals of well-log interpretation. Elsevier, Amsterdam

Szabó NP, Dobróka M (2013) Float-encoded genetic algorithm used for the inversion processing of well-logging data. In: Michalski A (ed) Global optimization: theory, developments and applications. Mathematics research developments, computational mathematics and analysis series. Nova Science Publishers Inc., New York

Szabó NP, Dobróka M, Kavanda R (2013) Cluster analysis assisted float-encoded genetic algorithm for a more automated characterization of hydrocarbon reservoirs. Intell Control Autom 4(4):362–370

Timur A (1968) An investigation of permeability, porosity and residual water saturation relationships for sandstone reservoirs. Log Analyst 9:8–17

Ward JH (1963) Hierarchical grouping to optimize an objective function. J Am Stat Assoc 58(301):236–244

Permeability Estimation in Petroleum Reservoir by Meta-Heuristics: An Overview

Ali Mohebbi and Hossein Kaydani

Abstract Proper permeability distribution in reservoir models is very important in the determination of oil and gas reservoir quality. In fact, it is not possible to have accurate solutions in many petroleum engineering problems without having accurate values for this key parameter of hydrocarbon reservoir. Permeability estimation by individual techniques within the various porous media can vary with the state of in situ environment, fluid distribution, and the scale of the medium under investigation. Recently, attempts have been made to utilize artificial intelligent methods for the identification of the relationship which may exist between the well log data and core permeability. This study overviews the different artificial intelligent methods in permeability prediction with advantage of each method. Finally, some suggestions and comments to choose the best method are introduced.

Keywords Support vector machine (SVM) · General regression neural network · Imperialist competitive algorithm · Artificial intelligent method · Adaptive neural fuzzy inference system

1 Introduction

Permeability is defined as a measure of the ability of a porous material to allow fluids to pass through it (Ahmed 2001). The concept of permeability is vital in determining precise reservoir description and simulation, which are the most important means for reservoir management. Permeability is also essential in overall reservoir management and development for choosing the optimal drainage points and production rate, optimizing completion and perforation design, and devising EOR patterns and injection conditions. Therefore, before any modeling or calculation, this parameter must be determined (Biswas et al. 2003). The most exact method in permeability prediction is core analysis in a laboratory by application of Darcy's law, which is expensive

A. Mohebbi (✉) · H. Kaydani
Department of Chemical Engineering, Faculty of Engineering, Shahid Bahonar University of Kerman, Kerman, Iran
e-mail: amohebbi@uk.ac.ir

and time consuming (Timur 1968; Saemi et al. 2007). The well testing analysis is another expensive, time-consuming method in permeability prediction, which gives the average permeability of the porous media around the wellbore (Mohebbi et al. 2012).

To overcome these obstacles, different methods for permeability estimation have been proposed. The oldest method in permeability prediction was empirical correlations between permeability and other petrophysical properties (Timur 1968). These correlations have been used with some success in sandstone reservoirs (Weber and Van Geuns 1990); however, for heterogeneous formations, they cannot be applied (Molnar et al. 1994).

In any oil or gas field, all wells are logged using electrical tools to measure geophysical parameters. This availability leads to the attempts that have been applied to predict permeability from well log data (Mohaghegh et al. 1994, 1996; Huang et al. 1996). The complexity, vagueness, and uncertainty existence, in addition to the nonlinear behavior of most reservoir parameters, require a powerful tool to find relationship between permeability and petrophysical properties of reservoir rock. Therefore, intelligent techniques or meta-heuristics, such as artificial neural networks (ANNs), fuzzy logic (FL), support vector machine (SVM), and genetic algorithms (GAs), have played a noticeable part in permeability prediction from well log data. Previous investigations (Mohaghegh and Ameri 1995; Aminzadeh et al. 1999; Mohaghegh et al. 2001; Saemi et al. 2007; Saemi and Ahmadi 2008; Karimpouli et al. 2010; Al-Anazi and Gates 2010) indicated that intelligent techniques are superior to statistical methods in predicting permeability from well log data because of their excellent pattern recognition ability. In this chapter, an overview on different meta-heuristics in permeability prediction was done.

2 Permeability Estimation

Various techniques are available that provide the reservoir permeability estimation in porous media. There are two generally reliable ways of acquiring knowledge on rock permeability. These are (1) direct measurements of rock sample (cores) and analyzing well test data and (2) estimation models that relate permeability to other petrophysical rock properties. The first methods (e.g., coring and well testing) are very useful, but they are not sufficient to show the heterogeneity of reservoir, because, due to intensive time demanding and high cost, it is possible to drill only limited wells. Also, another restriction can be due to the unavailability of cores, missed cores in certain intervals, etc.

The well testing method for permeability determination is *pressure transient analysis*, which provides a volumetrically averaged permeability for the volume of the reservoir that has been investigated during the test. Tests should be designed so that they are long enough to achieve reliable and usable data. On the other hand, the longer the test time, the larger the volume represented by the calculated permeability. In any oil or gas field, all wells are logged using electrical tools to measure geophysical parameters. This availability leads to the attempts that have been applied to predict permeability from well log data.

The oldest method in permeability prediction was based on empirical correlations between permeability and other petrophysical properties. These correlations have been used with some success in sandstone reservoirs. The complexity, vagueness, and uncertainty existence, in addition to the nonlinear behavior of most heterogeneous reservoir parameters, require a powerful tool to overcome these challenges. In recent years, meta-heuristics such as ANNs, FL, and GAs have played a noticeable part in reservoir engineering applications. Each of these methods is explained briefly below.

2.1 Core Analysis

Coring is an essential part of the reservoir life cycle process, with cored wells selected to verify or provide maximum information for the geological, engineering, or production model of the reservoir (Levorsen 1996). Permeability is one of the parameters that is generally measured in the laboratory on the cored rocks taken from the reservoir. In this method, permeability is determined primarily by flowing nitrogen, air, or any nonreactive fluid through the sample. The core plug was inserted in a special holding device such as illustrated in Fig. 1. The permeability can be determined from obtained data at several flow rates by applying Darcy's law. Figure 2 shows typical plotting results with either liquid or gas.

It should be mentioned that a viscous flow is the best satisfied condition in permeability measurements. Although gas permeability measurements are significantly faster and less expensive, the industry is still debating the validity and utility of gas permeability data for liquid-producing reservoirs because of the effect of gas slippage (or Klinkenberg effect) on permeability measurement (Amyx et al. 1960; Norman 1984). The effect of overburden on the permeability measurement is also important. This effect is more pronounced in unconsolidated, low-permeability, and fractured

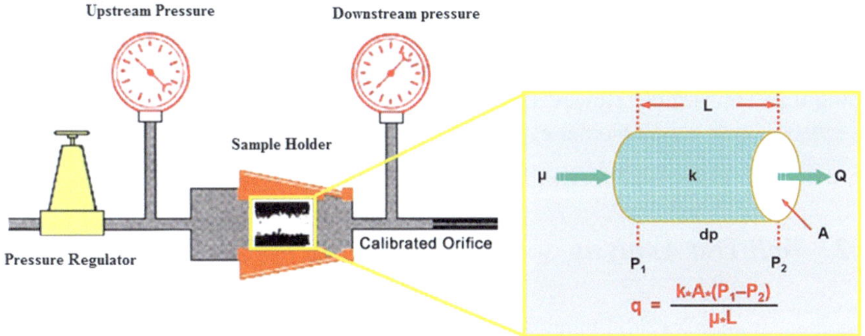

Fig. 1 A core plug in special holding device illustration for permeability measurement

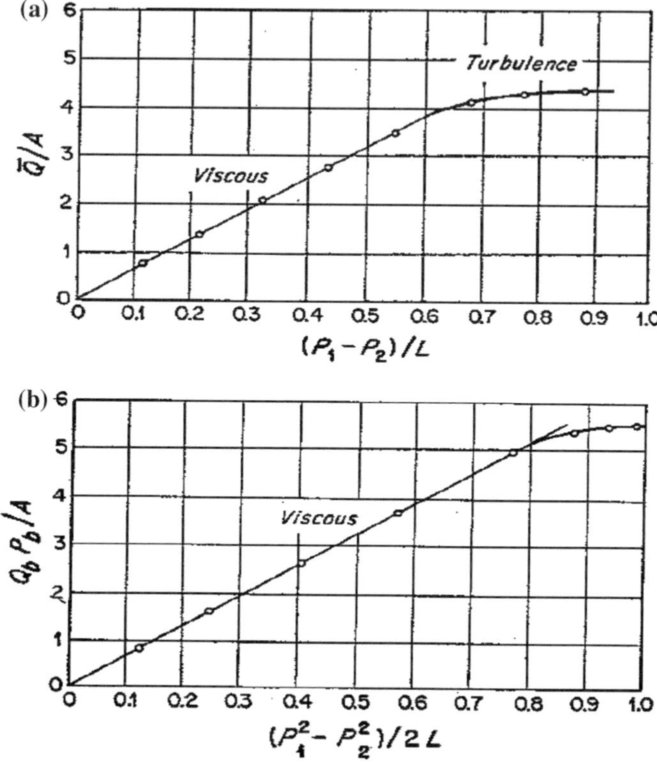

Fig. 2 A typical plotting results in permeability measurement with **a** liquid and **b** gas

samples. Consolidated-sample ambient-pressure permeability can be corrected for the effect of overburden by applying a correction factor determined by measuring a few samples at both overburden and ambient conditions (Tiab and Donaldson 2004; Amyx et al. 1960).

Among different techniques, the permeability obtained from core analysis in the laboratory is more valid than that obtained by the other methods. But, this technique cannot be widely used from an economic point of view, because of its high cost and being time consuming. However, the results from this reliable technique were used as target data to validate other estimation models (Saemi et al. 2007).

2.2 Well Test Analysis

Another popular way to obtaining reservoir permeability is *well test analysis*. The data that become available after a carefully designed well test help petroleum engineers to calculate a volumetric average of the formation permeability, among other

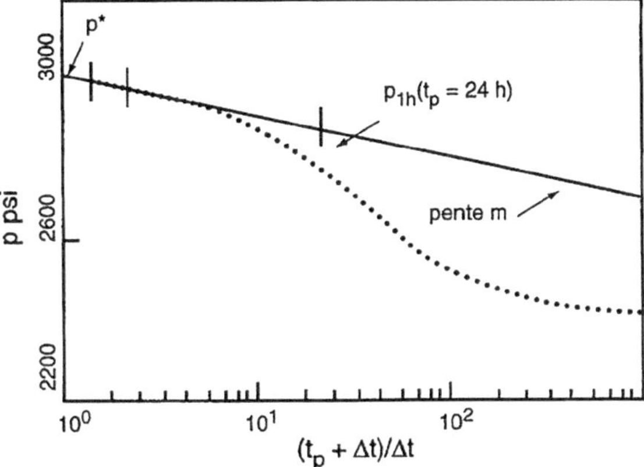

Fig. 3 A typical pressure curve response in buildup well testing

parameters such as skin factor and wellbore storage (Prasad et al. 1996; Jeirani and Mohebbi 2006). The basic of well test technique is to create a pressure drop in a bottom hole pressure, which causes reservoir fluid to flow in a certain rate from the reservoir rock to wellbore, followed by shut-in period. The production period is generally referred as pressure drawdown, whereas the shut-in period is called pressure buildup. By utilizing some analytical equations, which are solutions of diffusivity equations, from the response of the pressure versus time (pressure curve), some of the rock properties such as porosity and permeability can be obtained (Earlougher 1977; Horne 1995). A typical pressure curve response in the buildup test is showed in Fig. 3.

Although it is a valuable and necessary procedure, well testing is not a viable procedure for any developed reservoirs. Due to its cost of performing the test, in addition to the loss of production during the test and its limited radius of the formation investigated around the wellbore, this worthwhile method cannot be widely used. More information about well test analysis techniques can be found elsewhere (Matthewe and Russell 1967; Earlougher 1977; Horne 1995).

2.3 *Empirical Correlation*

The first equation relating petrophysical rock properties to permeability was proposed by Kozeny (1927). The lack of global applicability of his model has led researchers to modify it, and other parameters in this equation were considered. This equation was modified by Carman (1937) and expressed as follows:

$$k = \phi^3 / [5S_o^2 (1 - \phi)^2] \tag{1}$$

where S_o is the surface area of grains exposed to fluid per unit volume of solid rock and ϕ is porosity of the rock. The surface area parameter is the major drawback of this formula, because it can be determined only by core analysis and obtained only with special equipment. Tixier (1949) represented a mathematical correlation, which indicated the influence of resistivity gradients, water saturation, and capillary pressure on the permeability of rock. However, his model is physically limited in scope by the relative paucity of logs exhibiting valid oil–water contacts and the necessity for estimating the hydrocarbon density as it exists in the reservoir. Wyllie and Rose (1950) expanded the empirical relationship proposed by Tixier and investigated the effects of irreducible water saturation and tortuosity on rock permeability. Timur (1968) derived a similar expression that related permeability to porosity and water saturation in a generalized form as

$$k = A \frac{\phi^B}{S_{\mathrm{wi}}^c} \tag{2}$$

where A, B, and C are parameters that should be determined statistically. Also, a similar empirical correlation was proposed by some other investigators (Pirson 1963; Coates and Dumanoir 1974; Coates and Denoo 1981; Bloch 1991). Ahmed et al. (1991) presented some graphs used for permeability estimation from the various correlations (Fig. 4). Empirical equations have been used with some successes in some sandstone and fracture reservoirs.

Flow zone indicator (FIZ) is another way to estimate rock permeability (Prasad 1999; Perez et al. 2005; Uguru et al. 2005; Bagheripour and Shabaninejad 2011). FZI is a factor that includes the geological and petrophysical properties for permeability prediction. Low FZI can be a sign of fine-grained, poorly sorted sands, whereas high FZI can indicate clean, coarse-grained, and well-sorted sands. Different depositional environments and diagenetic processes control the geometry of the reservoir and consequently the flow zone index (Amaefule et al. 1993).

Among the available permeability correlations, the hydraulic unit concept is widely used in permeability prediction. This technique is based on the porosity–permeability empirical relationship, modified by Amaefule et al. (1993). The hydraulic unit permeability formula can be expressed as follows:

$$\log \mathrm{RQI} = \log \phi_{\mathrm{n}} + \log \ \mathrm{FZI} \tag{3}$$

$$\mathrm{RQI} = 0.314 \sqrt{\frac{k}{\phi_{\mathrm{e}}}} \tag{4}$$

$$\phi_{\mathrm{n}} = \frac{\phi_{\mathrm{e}}}{1 - \phi_{\mathrm{e}}} \tag{5}$$

where FZI is flow zone indicator (μm), ϕ_{n} is normalized porosity index, RQI is reservoir quality index (μm), ϕ_{e} is effective porosity, and k is permeability (md). This equation yields a straight line on a log–log plot of RQI versus ϕ_{n} with a unit slope. The intercept of this line is flow zone indicator. Samples with different FZI

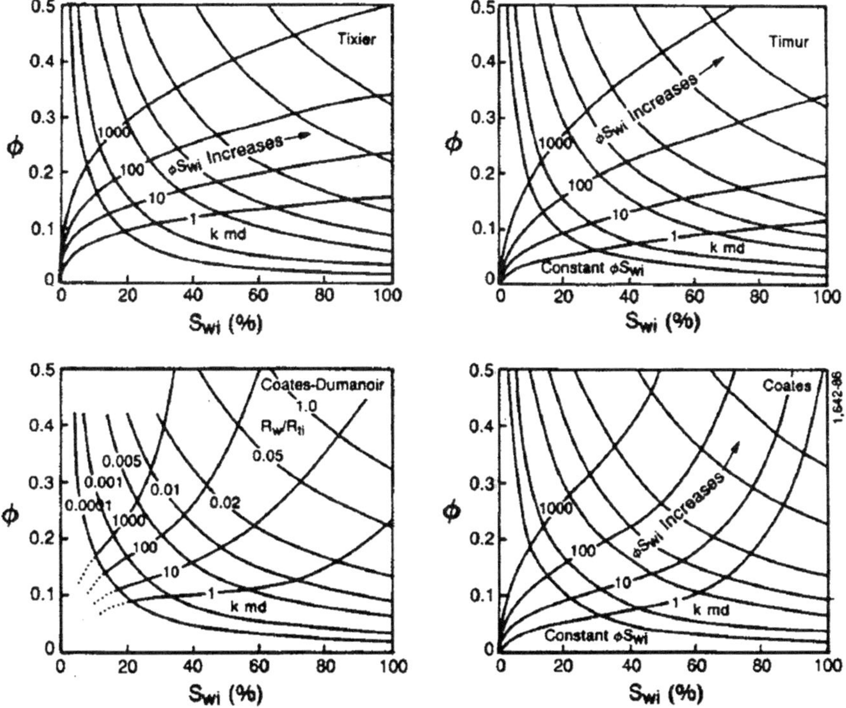

Fig. 4 Graphs used for permeability estimation from the various correlations presented by Ahmed et al. (1991)

values will lay on parallel lines. By the estimation of FZI, the permeability of the reservoir can be calculated according to the following equation:

$$k = 1014(\text{FZI})^2\left[\frac{\phi_e^3}{(1 - \phi_e)^2}\right] \tag{6}$$

Ghafoori et al. (2008) used FZI method for permeability estimation in one of the Iranian carbonate reservoirs and concluded that this method fails to predict permeability over the high permeable intervals.

Although this technique seems to be an ideal tool in permeability estimation, the vagueness and uncertainty existence, in addition to complex behavior of most reservoir parameters, caused this method to not be considered as powerful tool to overcome these challenges (Molnar et al. 1994; Saemi et al. 2007).

2.4 Meta-Heuristics Approaches

Recently, artificial intelligent methods (meta-heuristics) have become a notable part of petroleum engineering problems. The major reason for this rapid growth and application of meta-heuristics is their ability to approximate any function in a stable and efficient way by an inexpensive approach.

ANNs have been increasingly applied to predict reservoir properties using well log data. Moreover, previous investigations have indicated that ANNs can predict formation permeability even in highly heterogeneous reservoirs using geophysical well log data with good accuracy (Mohaghegh et al. 1994; Saemi et al. 2007). In spite of the wide range of applications, ANNs are still designed through a time-consuming iterative trial-and-error approach. This leads to a significant amount of time and effort being expended to find the optimum or near-optimum structure for a neural network for the desired task. In order to mitigate these deficiencies, design of neural networks using GAs has been proposed (Dehghani et al. 2008; Kaydani et al. 2011). Fuzzy modeling, as a powerful artificial intelligent method, can model highly complex nonlinear problems (Taghavi 2005; Ilkhchi et al. 2006). In the next section, permeability prediction by different artificial intelligent methods is discussed briefly.

3 Artificial Intelligent Approaches in Permeability Prediction

Availability of the well log and coring data for the wells in any oil and gas reservoir leads to attempts that have been applied to predict permeability from them (Mohaghegh et al. 1994; Malki et al. 1996; Wong et al. 1998; Kumar et al. 2000). The prediction of permeability in heterogeneous formations from well log data poses a difficult and complex problem (Saemi et al. 2007). A comprehensive approach for correlating permeability with geophysical well log data in heterogeneous formations was developed by Molnar et al. (1994). This approach combined gamma ray, deep induction, compensated bulk density well log responses, and detailed core analysis to subdivide the formation into several zones. Then, a reliable statistical correlation between permeability and bulk density was developed for each zone.

Alternatively, ANNs have been increasingly applied to predict reservoir properties using well log data (Mohaghegh et al. 1994; Mohaghegh and Ameri 1995; Wiener 1995; Boadu 1997; Arpat et al. 1998; Jamialahmadi and Javadpour 2000; Chang et al. 2000). ANNs are computing systems based on the interaction of large numbers of simple processing units, which are called nodes. A typical multilayer ANN consists of different layers of nodes as shown in Fig. 5. Mohaghegh et al. (1996) indicated that a neural network is a powerful tool for identifying the relationship among permeability and geophysical well log data. Moreover, Aminian et al. (2000, 2001) indicated that ANNs can be applied to predict formation permeability even in highly heterogeneous reservoirs with good accuracy.

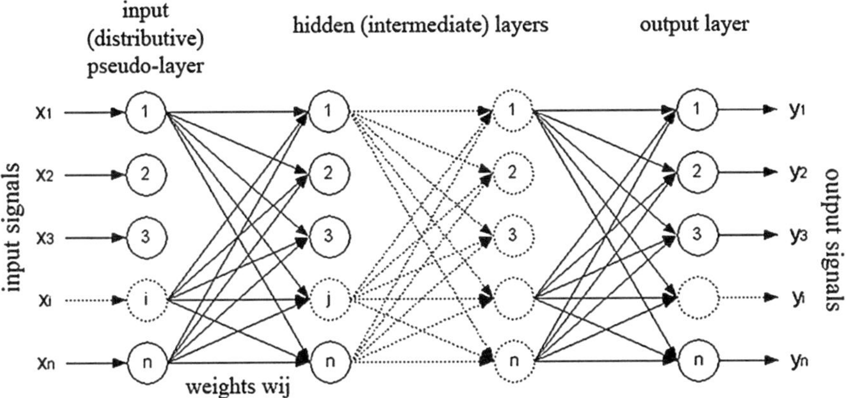

Fig. 5 A typical multilayer ANN with different layers of nodes

In spite of the wide range of applications, neural networks are still designed through a time-consuming approach. This leads to a significant amount of time and effort being expended to find the optimum or near-optimum structure for a neural network in a desired task (Niculescu 2003; Saxena and Saad 2006; Dehghani et al. 2008). In order to mitigate these deficiencies, automatic designs of neural networks have been proposed by Boozarjomehry and Svrcek (2001). However, these methods have been applied only for the design of neural networks used for simple tasks and not for more complex problems.

The researches by Van Rooij et al. (1996) and Vonk et al. (1997) have proposed the use of evolutionary computation techniques such as GAs in the field of ANNs to generate an optimal ANN architecture. Huange et al. (2001) and Chena and Lina (2006) applied this method in permeability estimation from log data and showed that it is highly effective to apply integrated GAs to ANNs in permeability prediction. However, these works did not cover the optimization of ANN parameters using GAs. Saemi et al. (2007) proposed a new method, whose design of topology and parameters of the neural networks as decision variables was done by using GAs in order to improve the effectiveness of forecasting when ANN is applied to a permeability predicting problem by a case study in South Pars gas field in Persian Gulf. Tables 1 and 2 show their results by two methods: trial-and-error approach and optimization with GA method, respectively. It can be found from these tables that GA was a good alternative over the trial-and-error approach to determine the optimal ANN architecture and internal parameters quickly and efficiently. Moreover, Kaydani et al. (2011) estimated permeability based on reservoir zonation by a hybrid neural GA in one of the Iranian heterogeneous oil reservoirs. They showed that permeability prediction based on designing separate networks for each zone is more accurately than designing single network design for all of zones. Tahmasebi and Hezarkhani (2012) proposed a method along four different neural network architectures to predict the permeability, and the obtained results were compared statistically. According to their results, they showed a modular neural network (MNN) as a new method, which

Table 1 The performance of Saemi et al. (2007) testing data set by trial-and-error approach

Parameter	Performance
MSE	0.3783
NMSE	0.2437
MAE	0.2272
Min absolute error	0.0011
Max absolute error	4.0766
r	0.8579

Table 2 The performance of Saemi et al. (2007) testing data set by ANNs optimization with GA

Parameter	Performance
MSE	0.1197
NMSE	0.0453
MAE	0.0681
Min absolute error	0.0001
Max absolute error	1.0864
r	0.989

had a very low computational time with high learning capacity and affordability for permeability prediction.

Kaydani and Mohebbi (2013) presented a comparison study of using optimization algorithms and ANNs for predicting permeability. They proposed a novel approach to estimate permeability by combining Cuckoo Optimization Algorithm (COA), particles swarm, and Imperialist Competitive Algorithms (ICA) with Levenberg–Marquardt (LM) neural network algorithm in one of the heterogeneous oil reservoirs in Iran. Figure 6 shows the proposed flowchart of the optimized LM neural network modeling with optimization algorithms for permeability prediction in their work. They concluded from a testing data set that the trained COA–LM neural model can efficiently accomplish permeability prediction. Also, the comparison of COA with particle swarm optimization and ICA showed the superiority of COA on fast convergence and the best optimum solution achievement (see Table 3). COA is a new evolutionary algorithm, proposed by Rajabioun (2011), and was inspired from special lifestyle of cuckoo birds. COA mimics the breeding behavior of cuckoos, where each individual searches for the most suitable nest to lay an egg in order to maximize the egg's survival rate, which is an efficient search pattern. Application of the COA in different optimization problems has proven its capability to deal with difficult optimization problems, especially in multi-dimensional problems. More information about this optimization algorithm is available in the literature (Rajabioun 2011).

Fuzzy set theory, a method to distribute linguistic fuzzy information by mathematics, distributes a set by using a membership function and extends the concepts of classical set theory (Zadeh 1965; Klir and Yuan 1995; Zeng and Singh 1996). Fuzzy modeling as a powerful meta-heuristics can model highly complex nonlinear

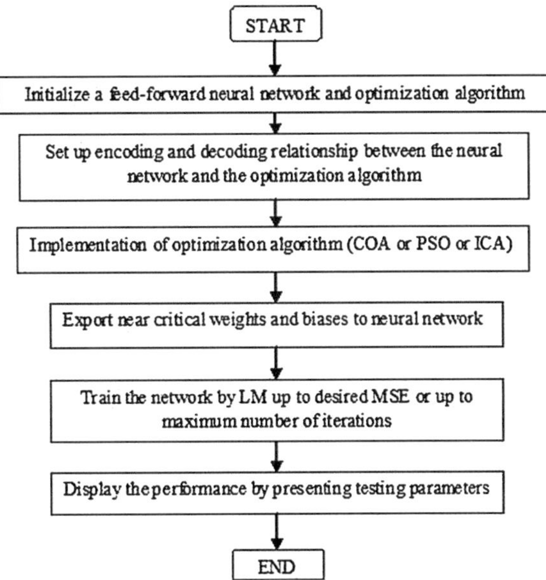

Fig. 6 Flowchart of the hybrid optimization—LM neural network proposed by Kaydani and Mohebbi (2013)

Table 3 Performance of neural network optimum model by different optimization methods

Parameter	COA-LM	ICA-LM	PSO-LM
AAD	0.099	0.104	0.119
NMSE	0.012	0.017	0.020
R	0.978	0.961	0.943
R-square	0.957	0.924	0.889

systems, such as multi-input and multi-output problems. It is an established fact that geosciences disciplines are not clear-cut and, most of the time, are associated with uncertainties. So, FL can be applied successfully in permeability prediction of porous media (Cuddy and Putnam 1998; Hambalek and Reinaldo 2003; Taghavi 2005; Lim 2005; Ali et al. 2006; Abdullraheem et al. 2007; Nashawi and Malallah 2010; Olatunji et al. 2011). Ilkhchi et al. (2006) used data from three wells of the Iran offshore gas field for the construction of FL models of the reservoir, and a fourth well was used as a test well to evaluate the reliability of the models. Their results showed that the FL approach was successful in the prediction of permeability in rocks of the gas field.

Combination of the explicit knowledge representation of FL and the learning power of neural nets yields an adaptive neural fuzzy inference system (ANFIS), which can be more useful in the prediction of the model. Nowadays, neural fuzzy systems have become more versatile approach to the problem in petroleum engineering (Nowroozi et al. 2009). One example of using this technique in permeability prediction was the work done by Gedeon et al. (1997). They incorporated fuzzy IF–THEN rules into neural networks to interpolate the reservoir properties. Kaydani et al. (2012) developed a neural fuzzy system for the prediction of permeability from wireline data based on fuzzy clustering in one of the Iranian carbonate reservoirs. They showed that by using a fuzzy c-means cluster technique in neuro-fuzzy model, the prediction of permeability in porous media can be improved.

Recent works on artificial intelligence techniques have led to introduce a robust machine learning methodology, called SVM. SVMs are supervised learning models with associated learning algorithms that analyze data and recognize patterns. SVMs, based on the structural risk minimization (SRM) principle (Stitson et al. 1999), seem to be a promising method for data mining and knowledge discovery. It was introduced in the early decade of 2000 as a nonlinear solution for classification and regression tasks (Burbidge et al. 2001; Jeng et al. 2003; Trontl et al. 2007). This technique aimed at predicting the permeability of the hydrocarbon reservoir (Al-Anazi and Gates 2010, 2012). As an example, Gholami et al. (2012) utilize the SVM for predicting the permeability of three gas wells in the Southern Pars field in Iran. Their results showed that the correlation coefficient between core and predicted permeability is 0.96 by using the SVM in the testing data set, which is illustrated in Fig. 7. Also, comparing the results of SVM with those of a general regression neural network (GRNN) revealed that the SVM approach is faster and more accurate than the GRNN in prediction of hydrocarbon reservoirs permeability.

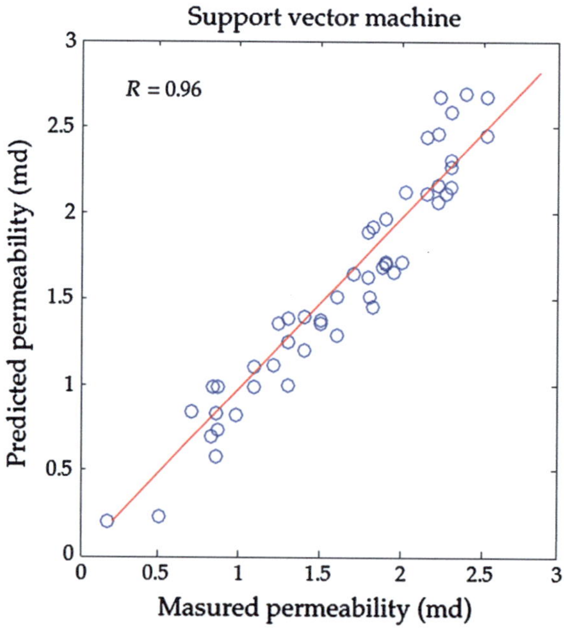

Fig. 7 SVM result for permeability estimation in three gas wells (Gholami et al. 2012)

4 Techniques Interrelationships and Comparisons

The ways for permeability determination can be categorized into two major groups: conventional and estimation methods. The conventional methods are core analysis and well test techniques. These methods are very expensive and time consuming. However, the essential information by conventional methods can be used in estimation methods for permeability determination. The oldest method for permeability determination is empirical correlations, which related permeability with other petrophysical properties of reservoir rock such as porosity and water saturation. Although this technique seems to be an ideal tool in permeability estimation in sandstone reservoirs, it fails to predict permeability over the high permeable intervals and heterogeneous formations.

The availability of the well logging data for the most wells in any oil reservoir motivates the researcher to predict permeability from them by using the multi-linear regression and artificial intelligent approaches. In these methods for estimation of permeability, logs, and cores data were used as input and target data, respectively. Intelligent systems, such as neural networks and FL, have much better solutions than multi-linear regression techniques in permeability prediction. Previous investigations indicated that artificial intelligence provides powerful tools for identifying the relationship among permeability and geophysical well log data even in highly heterogeneous reservoirs with good accuracy. In using intelligent techniques, the validations of coring data are essential and coring operation must be done more accurate in wells drilled in the reservoir.

In spite of the wide range of applications, a significant amount of time and effort is being expended to find the optimum or near-optimum structure for artificial systems such as ANNs or ANFIS for the desired task. To mitigate these deficiencies, design of them using optimization algorithms such as GA, PSO, and COA has been proposed. Application of the COA in different optimization problems has proven its capability to deal with difficult optimization problems, especially in multi-dimensional problems rather than other optimization algorithms (Rajabioun 2011). Moreover, zoning the reservoir according to geology characteristics and sorting the data in the same manner have been made to improve the proficiency of artificial intelligent results in permeability prediction especially in high heterogeneous reservoirs.

5 Conclusion

Permeability is one of the most important parameters in reservoir characterization, playing a major role in reservoir simulation, enhanced oil recovery, or well completion design. Therefore, before any field exploitation and development strategies design, this vital parameter must be determined. Core analysis provides the best fine scale permeability measurements. Nevertheless, the coring operations are very costly and time consuming and impractical to perform in all wells especially in horizontal wells. Also, permeability is commonly determined from transient well test analyses. However, well tests yield an average permeability value for the entire drainage area of the well.

Empirical correlations were the oldest estimation method for permeability prediction, which seem to be an ideal tool for homogeneous formations that have fairly constant porosity and grain size. But the complex behavior of most reservoir parameters caused this method to not be considered as a powerful tool in permeability prediction especially in carbonate formations.

Fortunately, well log responses are widely available in any oil and gas field. When these data are properly analyzed, wireline logs have the advantage of providing continuous permeability traces as opposed to scanty, discrete core data. Artificial intelligent approaches, such as ANNs and FL, are common methods that use wireline data in permeability prediction, even in high heterogeneous reservoirs, with good accuracy. The ability of soft computing in approximating virtually any function in a stable and efficient way caused a rapid growth in recent decade in permeability prediction. ANNs have been applied more than other soft computing techniques to predict reservoir performance. But optimum design structure of neural networks can be obtained with optimization algorithm especially COA, which has fast convergence in global optima achievement than other optimization algorithms.

Also, permeability prediction based on designing separate networks for each zone of the reservoir according to geology characteristics is more accurate than designing a single network design for all of the zones of the reservoir.

References

Abdullraheem A, Sabakhi E, Ahmed M (2007) Estimate of permeability from wireline logs in middle eastern carbonate reservoir using fuzzy logic. SPE paper 105350

Ahmed T (2001) Reservoir engineering handbook, 2nd edn. Gulf Professional Publishing, Houston

Ahmed U, Crary SF, Coates GR (1991) Permeability estimation: the various sources and their interrelationships. JPT 43(5):578–587

Al-Anazi AF, Gates ID (2010) A support vector machine algorithm to classify lithofacies and model permeability in heterogeneous reservoirs. Eng Geol 114(3–4):267–277

Al-Anazi AF, Gates ID (2012) Support vector regression to predict porosity and permeability: effect of sample size original research article. Comput Geosci 39:64–76

Ali KI, Mohammadreza R, Seyed AM (2006) A fuzzy logic approach for estimation of permeability and rock type from conventional well log data: an example from the Kangan reservoir in the Iran offshore gas field. J Geophys Eng 3:356–369

Amaefule JO, Altunbay M, Tiab D (1993) Enhanced reservoir description: using core and log data to identify hydraulic flow units and predict permeability in uncored intervals well. SPE paper 26436

Aminian K, Bilgesu HI, Ameri S et al (2000) Improving the simulation of water flood performance with the use of neural networks. SPE Paper 65630:105–110

Aminian K, Thomas B, Bilgesu HI et al (2001) Permeability distribution prediction. In: Proceedings of the SPE eastern regional conference, Oct 2001

Aminzadeh F, Barhen J, Toomarian NB (1999) Estimation of reservoir parameter using a hybrid neural network. J Pet Sci Eng 24(1):49–56

Amyx JW, Bass DM, Whiting RL (1960) Petroleum reservoir engineering: physical properties. McGraw-Hill Book Co., New York

Arpat GB, Gumrah F, Yeten B (1998) The neighborhood approach to prediction of permeability from wireline logs and limited core plug analysis data using back-propagation artificial neural networks. J Pet Sci Eng 20:1–8

Bagheripour HM, Shabaninejad M (2011) A permeability predictive model based on hydraulic flow unit for one of iranian carbonate tight gas reservoir. SPE paper 142183

Biswas D, Suryanarayana PV, Frink PJ et al (2003) An improved model to predict reservoir characteristics during underbalanced drilling. SPE paper 84176

Bloch S (1991) Empirical prediction of porosity and permeability in sandstones. Am Assoc Petrol Geol Bull 75(7):1145

Boadu FK (1997) Rock properties and seismic attenuation: neural network analysis. Pure Appl Geophys 149:507–524

Boozarjomehry RB, Svrcek WY (2001) Automatic design of neural network structures. J Comput Chem Eng 25:1075–1088

Burbidge R, Trotter M, Buxton B et al (2001) Drug design by machine learning: support vector machines for pharmaceutical data analysis. Comput Chem 26(1):5–14

Carman PC (1937) Fluid flow through granular beds. Trans Inst Chem Eng 15:150–166

Chang HC, Kopaska-Merkel DC, Chen HC et al (2000) Lithofacies identification using multiple adaptive resonance theory neural networks and group decision expert system. J Comput Geosci 26:591–601

Chena C, Lina L (2006) A committee machine with empirical formulas for permeability prediction. Comput Geosci 32:485–496

Coates G, Denoo S (1981) The producibility answer product. Tech Rev 29(2):55–63

Coates GR, Dumanoir JL (1974) A new approach to improved log-derived permeability. Log Anal 15(1):17

Cuddy SJ, Putnam TW (1998) Litho-facies and permeability prediction from electrical logs using fuzzy logic. SPE paper 49470

Dehghani SAM, Vafaie Sefti M, Ameri A (2008) Minimum miscibility pressure prediction based on a hybrid neural genetic algorithm. J Chem Eng Res 86:173–185

Earlougher RC (1977) Advances in well test analysis, 2nd edn. Society of Petroleum Engineers of AIME, New York

Gedeon TD, Wong PM, Huang Y et al (1997) Two dimensional neural-fuzzy interpolations for spatial data. In: Proceedings of GIS geo-informatics, vol 1, Taipei, Taiwan, pp 159–166

Ghafoori MR, Roostaeian M, Sajjadiain VA (2008) A state of the art permeability modeling using fuzzy logic in a heterogenous carbonate (an iranian carbonate reservoir case study). IPTC paper 12019, Kuala Lumpur, Malaysia

Gholami R, Shahraki AR, Jamali Paghaleh M (2012) Prediction of hydrocarbon reservoirs permeability using support vector machine. Math Probl Eng. https://doi.org/10.1155/2012/670723

Hambalek N, Reinaldo G (2003) Fuzzy logic applied to lithofacies and permeability forecasting. SPE paper 81078

Horne RN (1995) Modern well test analysis: a computer-aided approach, 2nd edn. Petroway Inc., Palo Alto

Huang Z, Shimeld J, Williamson M et al (1996) Permeability prediction with artificial neural network modeling in the Venture gas field, offshore eastern Canada. Geophysics 61(2):422–436

Huange Y, Gedeonb T, Wongc P (2001) An integrated neural-fuzzy genetic-algorithm using hyper-surface membership functions to predict permeability in petroleum reservoirs. J Pet Sci Eng 14:15–21

Ilkhchi AK, Rezaee M, Moallemi SA (2006) A fuzzy logic approach for estimate of permeability and rock type from conventional well log data: an example from the Kangan reservoir in the Iran offshore gas field. J Geophys Eng 3:356–369

Jamialahmadi M, Javadpour FG (2000) Relationship of permeability, porosity and depth using an artificial neural network. J Petrol Sci Eng 26:235–239

Jeirani Z, Mohebbi A (2006) Estimating the initial pressure, permeability and skin factor of oil reservoirs using artificial neural networks. J Petrol Sci Eng 50:11–20

Jeng JT, Chuang CC, Su SF (2003) Support vector interval regression networks for interval regression analysis. Fuzzy Sets Syst 138(2):283–300

Karimpouli S, Fathianpour N, Roohi J (2010) A new approach to improve neural networks algorithm in permeability prediction of petroleum reservoirs using supervised committee machine neural network (SCMNN). J Petrol Sci Eng 73:227–232

Kaydani H, Mohebbi A (2013) A comparison study of using optimization algorithms and artificial Neural networks for predicting permeability. J Pet Sci Eng 112:17–23

Kaydani H, Mohebbi A, Baghaie A (2012) Neural fuzzy system development for the prediction of permeability from wireline data based on fuzzy clustering. J Pet Sci Eng 30(19):2036–2045

Kaydani H, Mohebbi A, Baghaie A (2011) Permeability prediction based on reservoir zonation by a hybrid neural genetic algorithm in one of the Iranian heterogeneous oil reservoirs for permeability prediction. J Pet Sci Eng 86–87:118–126

Klir G, Yuan B (1995) Fuzzy sets and fuzzy logic: theory and applications. Prentice-Hall, Englewood Cliffs

Kozeny J (1927) Uber Kapillare Leitung des Wassers im Boot Sitzungsberichte, vol 136. Royal Academy of Science, Vienna, Paris. Class I, pp 271–306

Kumar N, Hughes N, Scott M (2000) Using well logs to infer permeability. Center for Applied Petrophysical Studies, Texas Tech University

Levorsen AI (1996) Geology of petroleum, 2nd edn. Freeman and Company Publishing, New York

Lim JS (2005) Reservoir properties determination using fuzzy logic and neural networks from well data in offshore Korea. J Pet Sci Eng 49:182–192

Malki HA, Baldwin JL, Kwari MA (1996) Estimating permeability by use of neural networks in thinly bedded shaly gas sands. SPE Comput Appl 8:58–62

Matthewe CS, Russell DG (1967) Pressure buildup and flow tests in wells. SPE, Dallas (Monograph series)

Mohaghegh S, Ameri S (1995) Artificial neural network as a valuable tool for petroleum engineers. SPE paper 29220

Mohaghegh S, Arefi R, Ameri S et al (1994) Design and development of an artificial neural network for estimation of formation permeability. SPE paper 28237

Mohaghegh S, Balan B, Ameri S (1996) State-of-the-art in permeability determination from well log data. SPE paper 30979

Mohaghegh S, Gaskari R, Popa A et al (2001) Identifying best practices in hydraulic fracturing using virtual intelligence techniques. In: Proceedings of 2001 SPE eastern regional conference and exhibition, SPE 72385, Oct 17–19, North Canton, Ohio

Mohebbi A, Kamalpour R, Keyvanloo K et al (2012) The prediction of permeability from well logging data based on reservoir zoning, using artificial neural networks in one of an Iranian heterogeneous oil reservoir. J Petrol Sci Tech 30(19):1998–2007

Molnar D, Aminian K, Ameri S (1994) The use of well log data for permeability estimation in a heterogeneous reservoir. In: Proceedings of SPE eastern regional conference, SPE 29175, pp 167–180

Nashawi IS, Malallah A (2010) Permeability prediction from wireline well logs using fuzzy logic and discriminated analysis. SPE paper 133209

Niculescu SP (2003) Artificial neural networks and genetic algorithms in QSAR. J Mol Struct Theochem 622:71–83

Norman JH (1984) Geology for petroleum drilling and production. McGraw-Hill Inc., New York

Nowroozi S, Ranjbar M, Hashemipour H et al (2009) Development of a neural fuzzy system for advanced prediction of dew point pressure in gas condensate reservoirs. J Fuel Process Technol 90:452–457

Olatunji SO, Selamat A, Abdulraheem A (2011) Modeling the permeability of carbonate reservoir using type-2 fuzzy logic systems. Comput Ind 62:147–163

Perez H, Gupta D, Misra S (2005) The role of electrofacies, lithofacies and hydraulic flow units in permeability predictions from well logs: a comparative analysis using classification trees. SPE Reservoir Eng Eval 8(2):143–155

Pirson SJ (1963) Handbook of well log analysis. Prentice-Hall Inc., Englewood Cliffs

Prasad M (1999) Correlating permeability with velocity using flow zone indicators. SEG conference Houston, Texas, paper ID 1999-0184

Prasad RS, Al-Attar EH, Al-Jasmi AK (1996) Reservoir permeability upscaling indicators from welltest analysis. SPE paper 36175

Rajabioun R (2011) Cuckoo optimization algorithm. Appl Soft Comput 11:5508–5518

Saemi M, Ahmadi M (2008) Integration of genetic algorithm and a coactive neuro-fuzzy inference system for permeability prediction from well logs data. Trans Porous Med 71:273–288

Saemi M, Ahmadi M, Yazdian A (2007) Design of neural networks using genetic algorithm for the permeability estimation of the reservoir. J Pet Sci Eng 59:97–105

Saxena A, Saad A (2006) Evolving an artificial neural network classifier for condition monitoring of rotating mechanical systems. J Appl Soft Comput 7:1568–4946

Stitson M, Gammerman A, Vapnik V et al (1999) Advances in kernel methods-support vector learning. MIT Press, Cambridge

Taghavi AA (2005) Improved permeability estimation through use of fuzzy logic in a carbonate reservoir from southwest Iran. SPE paper 93269

Tahmasebi P, Hezarkhani A (2012) A fast and independent architecture of artificial neural network for permeability prediction. J Pet Sci Eng 86–87:118–126

Tiab D, Donaldson EC (2004) Petrophysics, theory and practice of measuring reservoir rock and fluid transport properties, 2nd edn. Gulf Professional Publishing, Elsevier, USA, p 889

Timur A (1968) An investigation of permeability, porosity, and water saturation relationship for sandstone reservoirs. Log Analyst 9(4)

Tixier MP (1949) Evaluation of permeability from electric-log resistivity gradients. Oil Gas J 48:113

Trontl K, Smuc T, Pevec D (2007) Support vector regression model for the estimation of γ-ray buildup factors for multi-layer shields. Ann Nucl Energy 34(12):939–952

Uguru CI, Onyeagoro UO, Lin J et al (2005) Permeability prediction using genetic unit averages of flow zone indicators (FZIs) and neural networks. SPE paper 98828

Van Rooij AJF, Jain LC, Johnson RP (1996) Neural network training using genetic algorithms. World Scientific Publishing Co. Pvt. Ltd, Singapore

Vonk E, Jain LC, Johnson RP (1997) Automatic generation of neural network architecture using evolutionary computation. World Scientific Publishing Co. Pvt. Ltd, Singapore

Weber KJ, Van Geuns LC (1990) Framework for constructing clastic reservoir simulation model. JPT 42(10):1–248

Wiener J (1995) Predict permeability from wireline logs using neural networks. Pet Eng Int 68:18–24

Wong PM, Henderson DJ, Brooks LJ (1998) Reservoir permeability determination from well log data using artificial neural networks: an example from the Ravva field, offshore India. In: Proceedings of SPE Asia Pacific oil and gas conference, Kuala Lumpur, Malaysia, SPE paper 38034

Wyllie MRJ, Rose WD (1950) Some theoretical consideration related to quantitative evaluation of physical characteristics of reservoir rock from electrical log data. Trans AIME 189:105–118

Zadeh LA (1965) Fuzzy set. Inf Control 8:338–353

Zeng X, Singh MG (1996) Approximation accuracy analysis of fuzzy system as function approximators. IEEE Trans Fuzzy Syst 4:44–63

Index